Victorians and the Prehistoric

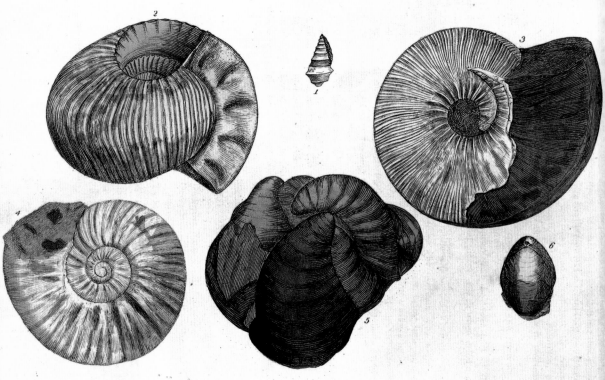

Fig: 1 Rostellaria   Fig: 4 Ammonites

   2 Ammonites sublævis Min:Con:t.54.   5 Gryphæa incurva Min:Con:t.112.

   3 A. Calloviensis Min:Con:t.104.   6 Terebratula ornithocephala Min:Con:t.101.

Victorians and the Prehistoric

Tracks to a Lost World

Michael Freeman

Yale University Press
New Haven and London

Designed by Sandy Chapman

Printed in China

Library of Congress Cataloging-in-Publication Data

Freeman, Michael.
  Victorians and the prehistoric : tracks to a lost world / Michael Freeman.--
1st ed.
      p. cm.
  Includes bibliographical references and index.
  ISBN 0-300-10334-4 (cl : alk. paper)
  1. Paleontology--Great Britain--History--19th century. 2. Paleontology--
Social aspects--Great Britain--History--19th century. I. Title.
  QE705.G7F74 2004
  560'.941'09034--dc22

                                                                    2004007555

A catalogue record for this book is available from the British Library.

Frontispiece: *Kelloways Stone* from William Smith's *Strata Identified by Organized Fossils*, 1816

Contents

Preface and Acknowledgements

Books can have strange origins. Some arise as an outcome of chance conversations, others as a result of chance encounters with hitherto unexplored documents. *Victorians and the Prehistoric* grew directly out of an earlier book, *Railways and the Victorian Imagination*, also published by Yale. The task of railway excavation brought surveyors, engineers and navvies face to face with a perspective on earth history that was as raw as it was vast. As contractors' gangs cut their way through successive bands of rock to try to make for a level permanent way, they exposed not only sedimentary formations in all their rainbow-like hues, untouched by the ravages of wind and water, but fossil beds by the score. Just as the speed of railway travel turned the Victorians' everyday time-world upside down (what once seemed quick became slow in comparison), so the view from the track-bed opened their eyes to a succession of long-lost time-worlds, hitherto the province of fable and fairy story.

This book would not have been possible without the remarkable resources of the Bodleian Library at Oxford. Helen Rogers and her staff in the Upper Reading Room have been unfailingly helpful. Clive Hurst, Julie-Anne Lambert and Sylvia Gardiner have provided valuable assistance in my searches of the John Johnson and Opie special collections. At the Oxford University Museum of Natural History, Stella Brecknell patiently guided me through the papers of William Smith and John Phillips. In the Oxford School of Geography and Environment Library, Sue Bird and Linda Atkinson have been constant sources of help and suggestions. Parts of the book were completed during a period spent as a Visiting Fellow at Yale University, at the Yale Center for British Art, and I must thank the Center's Director, along with all its staff, for their warm welcome. The Governing Body of Mansfield College, Oxford, made the task of writing the book easier still by granting me leave of absence as well as sabbatical leave during the academic year 2001–2.

Among individuals who must be singled out for their advice and assistance, I should first list my various colleagues in Mansfield

College, namely, Lynda Patterson, Philip Kennedy, John Muddiman, Walter Houston, Sarah Wood, Ros Ballaster, Lucinda Rumsey and Michael Freeden. Several of them kindly read and commented upon chapter drafts, although, as always, they cannot be held responsible for any errors that remain. At Yale, I must thank especially Susan Brady, Elizabeth Fairman, Malcolm Warner, Julia Marciari Alexander, Scott Wilcox, Martha Buck, Adrianna Bates and Christy Anderson. The book would not have materialized without the continuing interest and indulgence of John Nicoll, my publisher. He read and commented on the entire book, as, in turn, did Adam Freudenheim, my editor. Finally, I must thank the Press's anonymous referees for ideas for improving the text.

Many institutions and libraries have helped with picture research and also made generous concessions in the way of picture fees, Oxford's Bodleian Library and the Oxford University Museum of Natural History, especially. I must, however, offer special thanks to the staff of the Bodleian's own photographic studio for their skill and patience in responding to my almost continuous stream of orders.

Michael Freeman
Mansfield College, Oxford
March 2004

Photograph Credits

Prologue

Scene One

From the top of the high ground, forty feet in height, the visitor will see spread before him the extensive 'tidal lake', of thirty acres in area, in which are situated several small islands devoted to the reception and illustration of geological animals and strata. The first island, the Tertiary, is not so far advanced as that devoted to the Secondary Epoch; but it will be advisable, in our imaginary journey through the crust of the globe, to commence with the latest series. . . . The rocks of this Tertiary system are generally spoken of as 'basins', for they appear to have been formed in great hollows in the surface of the Chalk series. . . . The animals completed include a group of very fine figures of the Irish elk – an extinct creature . . . found in many parts of the bogs and marshes of Ireland. . . . The group of four graceful lama-like animals . . . are restorations of animals which once swam in the vast lake in which Paris now stands and which browsed amid the herbage and banks.

Scene Two

Leaving the latest formations, the visitor will now pass along the walk by the water's edge, and inspect the restoration of animals which existed in the vast and grand Secondary Epoch. These were ages when a luxuriant vegetation prevailed and huge creeping things and carnivorous monsters, roamed through tangled brake, or pursued their prey along the shallow banks of vast inland seas. Here creatures more terrible and appalling than poet's fancy ever dreamed of, lived and died. . . . Kent, Sussex and portions of Surrey [have] been called the metropolis of the 'dinosaurs' . . . covered with wild and rank vegetation . . . with vast freshwater lagoons and deposits. Several restored specimens of . . . fossil plants are placed in this stratum'.

 . . . The *Iguanodon* – there are two specimens of this animal; one is represented as standing upon its four legs, and the other is lying like a huge lizard upon the ground. . . . The Iguanodon is a

1 (*facing page*): 'The Primitive World' from C. G. W. Vollmer's *Wonder of the Primitive World*, 1855

2   The Crystal Palace and Park at Sydenham – Victorian precursor of a twenty-first-century theme park

colossal land lizard forty feet long. . . . Glistening with scales, and presenting the appearance of bright and dazzling shagreen. . . . A terrible scene it would have been to have witnessed on the sedgy banks of some old Thames or Medway the [*Megalosaurus*] hobbling down the margin of these muddy streams to slake his thirst . . . cast his saurian eyes over the livid expanse in search of some object he might drag down with him to his river den, and hear the ferocious howl and roar with which these two monsters would grapple with each other. Till the dark waters were reddened with their blood, and one or both of the combatants sank beneath their wounds.[1]

It seems that we are somewhere in or near London. There is an instant reminder of the BBC's famous television series *Walking with Dinosaurs*, or else the many recent movies and television documentaries that have featured dinosaurs, not least of them Stephen Spielberg's classic cinema film *Jurassic Park*. The setting could be a Disney theme park, or perhaps inside London's millennium dome.

However, it is nothing to do with any of these. The setting is Sydenham Park on the outskirts of London in 1853, a century and half before Tim Haines's fascinating computer animations of the world of dinosaurs. Here the early Victorians confronted the apocalyptic empires of the Mesozoic era, already dramatized in the artist John Martin's great 'Gothick' canvases and in the poet Tennyson's 'terrible Muses'.[2] Not far from the site used for the re-erection of the Crystal Palace, the Victorians created a 'mausoleum to the memory of ruined worlds'.[3] They built a whole series of life-sized dinosaurs in concrete:

> Scientific men who had devoted a long life to the accumulation and study of fossil remains, who had put together the skeletons of these gigantic monsters, and seen them in the imagination roaming over the pathless forests of our island, had never beheld the entire animal reproduced before them; – geologists were for the first time to gaze upon the fruits of their industry, and the results of their science.[4]

On New Year's Eve, 1853, the Directors of the Crystal Palace, in company with a group of invited scientists (among them, Richard Owen, who christened the dinosaurs[5]) held a terrific banquet inside the concrete carcass of the standing Iguanodon. Beneath an awning of pink and white drapery and banners bearing the names of distinguished geologists such as Conybeare, Buckland, Forbes and Mantell, they feasted and toasted in commemoration of the men whose labours had provided for the reconstruction within which they presided. At midnight, the company broke into such roaring song that observers outside might have been forgiven for thinking that a herd of Iguanodon was bellowing from the swamp and mud of Penge Park:

> For monsters wise our Saurians are,
> And wisely shall they reign,
> To speed sound knowledge near and far
> They've come to life again.[6]

The Megalosaurus had already imprinted itself on the Victorian imagination in the writing of Charles Dickens. *Bleak House*, which first appeared in serial form in March 1852, opens:

> London. Michaelmas Term lately over, and the Lord Chancellor sitting in Lincoln's Inn Hall. Implacable November weather. As much mud in the streets, as if the waters had but newly retired from the face of the earth, and it would not be wonderful to meet a Megalosaurus, forty feet long or so waddling like an elephantine lizard up Holborn-hill.[7]

The first Megalosaurus skeleton had been discovered in 1823, at Stonesfield in Oxfordshire.[8] By Victoria's reign, fossil reptiles were being despatched to London from 'every little England'.[9] Later still, they were turning up from all quarters of the Empire: dinosaurs apparently 'had cousins scattered over the face of the earth as widely as Queen Victoria's subjects'.[10] Eventually their vast carcasses were to feature in the great new gothic mueums that were rising in cities such as London and Oxford, their architectural decoration mimicking the sedimentary rock profiles from which the famous reptiles had been disinterred.

The recovery of these relics from lost worlds had already been elevated into spectacle. At Church Cliff in Lyme Bay in Dorset one sunny morning in July 1832, the country people flocked to witness Thomas Hawkins, in company with the fossil-hunter Mary Anning, direct the exposure of the skeleton of an Ichthyosaurus, a great marine reptile, or 'fish-lizard', on the tidal beach. The working men, with their spades and pickaxes, rejoiced alongside Hawkins at the first sight of the wonderful colossus. The hills and dells around echoed to their 'hurras' as each new bone was uncovered. By nightfall, the many pieces of the skeleton, within its matrix of lias, and weighing near a ton, had been carried away, packed and secured.[11]

Whilst Charles Dickens envisioned a dinosaur waddling along the streets of London, the poet Tennyson, in 1848, thought that the lost world of great lizards, marshes and giant ferns could best be recalled by standing alongside a railway line at night, the engine like some great Ichthyosaurus.[12] This was also how one of the earliest historians of the English railway, F. S. Williams, in 1852, described the experience of a train passing though a station at night: '. . . a great flaming eye . . . the thunder of its tread . . . the iron gullet of the monster vomit[ing] aloft red hot masses of burning coke . . .'[13] Lost worlds underground were being brought to imaginary life by the new world of the iron railroad on the earth's surface.

There are senses in which the dinosaur stood proxy for a whole range of currents within early and mid-Victorian thought. Just as the steam railroad, for all its terrible monstrosity and apocalyptic imagery, was a central emblem of the march of engineering science, so the unfolding underground world of extinct reptiles was central to the march of earth science. The fearsome world of 'great sea dragons', as Thomas Hawkins described the marine form of these reptiles,[14] came in due course to be an uncomfortable reminder of the desperate struggle for existence that was to emerge with the unfurling of Darwin's evolutionary theory. Henry De la Beche's classic pictorial reconstruction of untold primeval violence in which reptiles are displayed in an orgy of killing each other gave worrying counterpoint to contemporary theories like that of Thomas Malthus in which the human population was pitted in a continual struggle against subsistence.[15] For the scripturists, too, the pre-human world of the dinosaurs 'shrieked against a straight reading of Genesis where the implication was that pain and death had not entered the world until Adam's Fall'.[16] It was also decidedly out of step with the comfortable world of eighteenth-century natural theology, the viewpoint later most associated with William Paley in which the natural world in all its wondrous mechanisms was viewed as a direct function of divine wisdom and goodness.[17]

Victorians and the Prehistoric seeks to recover something of the new, extraordinary and often shocking accounts of earth history that Victorians saw unfolding before their eyes, especially over the early to middle decades of the nineteenth century. It is not intended as a history of geology or a history of palaeontology. It does not try to trace how geological discovery prepared the way for Darwin. Nor does it seek to address in any specific form how the accumulating knowledge of earth history contributed to the undermining of Christian belief. All of these objects would present worthy goals, but they could not be encompassed in a book of this length and they would be somewhat alien to the wider readership that it endeavours to engage.

What this book presents is a series of narratives, centred upon key themes in nineteenth-century earth history. Its sub-title, *Tracks to a Lost World*, can be taken both literally and figuratively. The tracks of the new railroads that were under construction all over the country in the 1840s proved to be a source of geological and palaeontological knowledge that was striking not just for the range of evidence that it revealed but the speed with which that evidence came to view. Tracks may also be related, of course, to the footprints left by long extinct animals as they roamed across muddy estuaries and sandy shores. For in stone quarries and on eroding sea beaches, Victorian field geologists and fossilists were constantly coming across these striking impressions left by creatures from former worlds. In another sense, tracks may be taken to allude to the voyages of intellectual discovery that an individual like Charles Darwin made during the course of his researches as a naturalist. This is revealed most plainly in his circumnavigation of the earth on HMS *Beagle* between 1831 and 1836, but it also comes to light from a detailed study of his notebooks and from his voluminous correspondence. Darwin's intellectual and scientific journey thus became itself a metaphor for the wider narrative of earth history of which it formed part. In museums, likewise, visitors were guided via showcase and gallery on tracks or journeys that took them back to lost worlds characterized by alien climates and strange collections of animals and plants. And in the 'open-air' museum that constituted part of the Crystal Palace park at Sydenham, visitors came face to face with the nineteenth-century equivalent of a modern 'virtual reality display', complete with mock geological strata and life-sized representations of reptiles as they basked amid man-made lagoons.

A book of this sort must inevitably gloss over the multiple histories of Victorian science that recent research has increasingly been uncovering.[18] Within Britain, the profound variation in the geography of Darwin's reception and impact, for instance, affords an especially telling illustration.[19] Alongside there are the variable relations of Darwinian theory within the separate spectrums of class and religion.[20] What has also distinguished much recent work on the history of science is the search for evidence that falls outside the traditional realms of investigation. Landscape painting and poetry, for instance, have acquired significance for the history of geology that few would once have imagined being able to accommodate.[21] In a wider sense, this forms part of a deliberate effort to connect science to ideas and beliefs of the times. There was arguably a seamlessness to the relations of science and society in the nineteenth century: science permeated society but was in turn confected by it. The variety of source materials that informs this book is in no way isolated from these wider concerns, as readers will soon register. However, their purpose is allusive rather then definitive, suggestive rather than argumentative. They help to colour and texture the broader narratives that the book's various chapters recount; they do not seek to structure them. Those familiar with the discourse of postmodernity will see nothing necessarily unusual in such a declaration. Narrative forms a part of postmodern exposition.[22] Familiar explanatory moulds in the humanities and the social sciences are firmly rejected by it. In this sense, the book will not appeal to those seeking definitive conclusions. There is no neatly delineated path towards 'closure' whereby a succession of developments and discoveries about earth history is demonstrated to lead inexorably towards some defined end-state. And certainly there was no linear succession of scientific insights that moved inexorably towards Darwin's theory of evolution by natural selection.[23]

The notion that Victorian science may have been a tensioned and fractured domain plainly raises the question as to whether one can really talk about the Victorians and the prehistoric in any unified manner. There exists, of course, a long tradition of studies that have marked out a kind of Victorian sensibility. A key early study was Walter Houghton's *The Victorian Frame of Mind* (1957). But even now the tradition remains vibrant, witness David Newsome's *Victorian World Picture* (1997) and A. N. Wilson's *The Victorians* (2002). According to much recent detailed scholarship, however, the overwhelming characteristic of Victorian society was its diversity.[24] Britain at the start of Victoria's reign was a 'social patchwork', reflecting a series of different and often conflicting opinions about nature, for example.[25] Thus it is quite logical to argue that any search for a core sensibility is fruitless. One response to this is to point out that a core sensibility does not necessarily have to be grounded in unity. It might just as easily be grounded in diversity. And even a hasty perusal of the chapter titles of Walter Houghton's study indicates that the Victorian frame of mind he describes was anything but unitary. One can discern not dissimilar features in David Newsome's study, with its multiple bases of 'triangulation': looking inward, outward, before, after, beyond and ahead. Ultimately, according to Newsome, the Victorian world picture was compounded of so many reflections from so many different vantage points that it was probably much more a collage than an ordered composition, and a perpetually changing one at that.[26] The student of postmodernism, of course, will recognize in the idea of 'collage' one of the key elements of a postmodern discourse.[27] What such broad studies plainly cannot do is look in any detail at the elements that comprise the collage. The differential reception of Darwin's ideas in different Victorian cities may thus be accommodated but not elucidated. And much the same applies in the case of issues of class, politics and religion.

Some readers of this book will quickly register that the label 'Victorian' has been invested with considerable licence. Some of the discussion, for instance, relates very plainly to a pre-Victorian era. However, one cannot begin to understand the Victorians' engagement with the prehistoric outside of its historical precursors. Moreover, there was much late eighteenth- and early nineteenth-century writing on earth history that did not obtain wider public view until the 1830s. By the same token, many of those who came face to face with the new narratives of the prehistoric after Victoria's accession were themselves members of a pre-Victorian generation. Indeed, it has been said that almost everything that came to maturity in the mid-Victorian decades had its roots in the first quarter of the nineteenth century.[28] The lesson is that historical continuums are not neatly divisible. Readers will simultaneously register that the book does not deal very much with the last decades of the Victorian era. The reason is that the challenges and spectres of the prehistoric had by then run their course. Darwin's *Descent of Man* (1871) brought little of the invective that his *Origin of Species* had done twelve years before. The Victorian mind had assimilated or, at the very least, accommodated the new narratives of earth history by the closing decades of the century. Subjects like geology, which had been at the cutting edge of science in the 1830s, had by then faded from central view. Unbelief, or agnosticism, as Huxley coined it, had in turn become an accepted feature of the cultural milieu by the 1870s. It was all a far cry from the Anglican state of Victoria's birth.

1

Tracks to a Lost World

... That splendid railway, the Great Western, by which geologists may be transported in five or six hours, from the Tertiary Strata of the metropolis, to the magnificent cliffs of mountain limestone at Clifton. . . .[1]

Prelude

To begin to understand how knowledge of earth history was extended from the start of the nineteenth century, one has first to grasp how the visual evidence upon which it depended was becoming progressively more extensive. There was initially no central or state agency for the accumulation and co-ordination of data on the country's rock strata.[2] Instead, an army of gentlemen scholars and, less often, surveyors or civil engineers, recorded and charted what occupied their interest or caught their eye. Traditionally, this had taken the form of natural rock exposures (in cliffs or in scarp slopes), quarry faces, the strata of coal or mineral mines and well sinkings. From the late eighteenth century, however, until the initial decades of Victoria's reign, new forms of rock exposure materialized for study and new impetus was attached to some older forms.

Canal-building was one catalyst, making necessary the excavation of level channels across often unlevel terrain. It was not just the incidental exposure of the geology that was centre-piece, but the necessity of grasping its particular order of layering so as to be able to produce a canal trench that retained water with only minimal leakage or minimal risk of collapse. One man, William Smith, a surveyor and civil engineer from Oxfordshire, came to dominate canal geology, and he later deployed his knowledge to produce the first geological map of England and Wales.

Where canals left off, railways began. By the late 1830s, they were beginning to criss-cross the landscape at a density neither canals, or river navigations before them, could have expected to attain. Railway

3 (*facing page*) Edwin T. Dolby's view of the cutting at the entrance to Abbotscliffe tunnel on the London and Dover Railway, the cleft in the rock face affording a fresh exposure of its geological form, *c.* 1850

engineers created vast hewn defiles through the English hills, contemporary observers casting such earth works as reminiscent of the genius of the ancient world. Geologists rushed in their scores to view and record the newly exposed rock series and to pick over the organic remains, or fossils, that came immediately to view. Representatives of the surveying and engineering professions also added to the profile of geological history by putting into effect schemes for agricultural improvement. Land drainage works, for example, could yield much the same kind of information about underlying rock strata. Indeed, William Smith arguably learnt almost as much about sedimentary rock sequences from such land improvement as he did from canals.

Finally, there was the coastal landscape, particularly its cliff faces. This was not by any means unfamiliar as a place for examining earth history, but it was one that acquired a greatly enhanced display as part of artistic preoccupations with the romantic, the picturesque and the sublime. Leisured society engaged on a project of 'discovery', making 'tours' in regions the length and breadth of Britain.[3] Topographers and antiquarians provided them with evocative written guides, whilst engravers illuminated their pages with scenes 'from nature' drawn by new breeds of artists who found the interweaving of what was romantic, picturesque and sublime irresistible. Thus a new spotlight was cast on coastal cliffs as fingerprints in the narrative of earth history.[4] And no more clearly was this displayed than in a small, then relatively unknown, island lying off the central coast of southern England.

An Island Jewel

> . . . Come to the Isle of Wight;
> Where, far from noise and smoke of town,
> I watch the twilight falling brown
> All around a careless-order'd garden
> Close to the ridge of a noble down.[5]

No spot in the world, perhaps, possesses in so small a compass such a diversity of interesting objects! – all that is pastoral, with all that is romantic; the richest landscapes, with views the most rude and rocky.[6]

If you love a green earth and a blue sky, and an ocean yet more deeply darkly beautifully blue, go to the Isle of Wight.[7]

Swinburne called the Isle of Wight a green jewel in a golden sea.[8] Sir Walter Scott remarked how the beauty of its coastal shores, once seen, was never forgotten.[9] For Jane Austen, it was that 'far-famed Isle'.[10] Tennyson lived there for nearly half a century – in a house on the

4   Thomas Webster's subliminal rendering of the chalk cliffs at the western extremity of the Isle of Wight, ending in the Needles, the three famous chalk stacks, 1815

downs near Freshwater Bay. Invalids and medical men claimed the bracing air of its south-eastern shores to be restorative of health.[11] By the early nineteenth century, Wight was regarded as among the most picturesque parts of southern England. Shaped like a lozenge, some thirty miles from east to west and twelve from north to south, it boasted the greatest variety of scenery of any equivalent area of the country's southern shores. The *Penny Magazine* could think of no other islets bordering the British coasts that could pretend to vie with such a 'gem of the ocean'.[12] Leading artists and engravers soon recorded it in a long succession of 'views'. By Victoria's reign, George Brannon's *Vectis Scenery* had become a bestseller. The young queen made a favourite home there, the twin campanile of her Italianate villa at Osborne forming an image that would forever lend to the island a flavour of the exotic. Wight also satisfied most contemporary senses of the sublime. Its cliffed and gullied southern coast, 'exposed to the impetuous tides of the

5   Queen Victoria's Italianate villa at Osborne, Isle of Wight, as drawn by Alfred Brannon in 1849 shortly after its completion

ocean',[13] ensured that it participated equally in the romantic and the sublime. Near Yarmouth, in the west, 'the hills rose with all the majesty of the Skiddaw mountains'.[14] At Alum Bay, the mountainous cliffs were 'terrific in the extreme'.[15] Meanwhile, at Blackgang Chine, the imagination could lead the inquisitive traveller 'to fancy that the earth had just opened her horrid jaws'.[16] Here, Wight displayed a truly awful grandeur.[17]

The *Penny Magazine*, in 1836, described Blackgang as 'wild, picturesque, and gloomily sublime'.[18] The cliffs, it claimed, were nearly five hundred feet high, the rocks almost black, with scarcely a trace of vegetation. The scene was reminiscent of a 'chasm in the Alps, or . . . of some of the lava recesses in the flanks of Mount Aetna'.[19] Equally startling was the 'Undercliff' west of Ventnor. Here, much like Blackgang, rock masses were perpetually slipping down to a lower level. The rents in the ground were 'frightful'.[20] The rock in some places had been 'ground to fragments'.[21] The old surface, including its vegetation, appeared to have been 'swallowed up'.[22] Trees could be seen with their roots in the air. It was altogether an awesome spectacle.

For all the superlatives used to describe its scenery, though, Wight had another very different appeal for some of its early nineteenth-century visitors. It was a geologists' paradise. In popular geological memory, and, indeed, in much modern geological instruction, it is

Arran, the picturesque Scottish isle near the entrance to the Clyde estuary, that comes most quickly to mind. In the annals of geology, Arran has an impeccable pedigree. James Hutton, one of the founders of modern geology, visited it in 1787.[23] The island soon became a mecca for geological study. Charles Lyell, whose best-selling *Principles of Geology* (1830–33) was read by Darwin whilst circumnavigating the globe, made his first visit there in 1836.[24] He found that Arran embodied an extraordinary range of the geology of Scotland in one place. Earlier, John MacCulloch had described it as offering to the geologist 'an epitome of the structure of the globe; forming a model of practical geology for the instruction of the beginner and for the study of all'.[25] In the first three or four decades of the nineteenth century, though, a stream of geology's founding practitioners discovered in Wight a whole raft of clues about the geological history of southern England. The island offered a keystone for configuring the almost immeasurably thick sequences of marine and freshwater sediments that had once overlain this portion of north-west Europe.[26] Its sometimes highly contorted rock formations also gave dramatic testament to the seismic forces that, over time, had affected those sediments – as one later geological text commented: 'raising the crust of the earth into giant billows, with crests and hollows, just as one may see any day, on a small scale, on the surface of the troubled sea'.[27] James Hutton had himself visited Wight in the 1750s, noting its ridge of chalk hills that could be traced west to the Dorset coast.[28] He seemed less aware of the rock strata adjacent to the chalk. Nor did he register how around some of Wight's southern coastal cliffs, fossils of every kind abounded. Vicious south-westerly winter storms were constantly disinterring lost inhabitants of former worlds, much to the delight of the island's geological visitors. Inland, meanwhile, quarrymen had spent lifetimes collecting fossils, selling many of their best specimens to eager and impatient observers.

If Wight today is less popular than Arran for mass geological study, the professional geological community remains clear as to its significance. For the Geologists' Association, 'no other area of comparable size in England has such a variety of formations in easily accessible exposures and containing such a diversity and abundance of fossils.[29] In all, the sedimentary rocks of Wight afford 120 million years of earth history, encompassing no less than ten separate formations in a manner unrivalled elsewhere in Europe.[30] The antiquaries and natural philosophers who visited Wight in increasing numbers from the 1780s and 1790s were necessarily much more hesitant in their understanding and appreciation of the significance of the island's geological structure; the subject was yet in its infancy. But what they saw was enough to excite interest and to bring a steady stream of investigators, Darwin among them,[31] across the water to survey the island's strata.

6 (*facing page*) Henry Englefield's careful delineation of the eroded and tumbled strata at Blackgang, Isle of Wight, 1815

Wight thus became one of the classic training grounds of Victorian geologists.[32]

In an age without railways and motorized road transport, the most convenient way for visitors to see Wight was from its seaward aspect. And this was inevitably how many of its serious geological observers first confronted it. Then, as now, there were three main boat crossings: from Lymington to Yarmouth, from Southampton to Cowes, and from Portsmouth to Ryde. The first was the shortest and gave the 'noblest view' of Wight, according to Hassell in 1790.[33] The cliffed coast, extending westward to the Needles, had rocks 'shelving over your head'.[34] In places, it rose nearly seven hundred feet above the surface of the sea at low water. Having set eyes on one coastal section, it was inevitable to wish to see more. And what better way than a 'circum-navigation', taking 'a view of the different shores of the island from the circumambient sea'.[35] At Alum, according to Hassell, 'the progressive operations of nature' in the formation of the near vertical strata were 'easily discernible', composed of a 'regular gradation of substances'.[36] Among them were the multi-coloured sands for which Alum Bay became famous. The tints of the cliffs were so bright and vivid that they had 'not the appearance of anything natural', according to one later writer. There was 'deep purplish red, dusky blue, light ochreous yellow, grey nearly approaching to white, and absolute black'. The succession of tints was 'as sharply defined as the stripes in silk'. And after rain, the sun shining on the cliffs gave an unrivalled brilliancy to some of these colours.[37]

Hassell's 'tour' by and large contained only incidental references to the island's geology, but it may have been enough to prompt Sir Henry Englefield, an antiquary and amateur scientist, to spend several summers there, beginning in 1799, exploring its 'geological phenomena' as well as its 'antiquities' and 'principal picturesque beauties'.[38] Sir Henry found the island 'peculiarly adapted' for the investigation of its geological structure.[39] The nature of its coasts afforded many 'rare and perhaps quite singular geological phenomena'.[40] He regarded it as 'quite unaccountable' that these should have been disregarded for so long.[41] 'It is no easy matter', he continued, 'to obtain correct ideas of the strata even of countries of mines[;] and in those extensive tracts where mines are not found, so few openings are made to any considerable depth, that it is impossible to obtain that sort of knowledge, which in the Isle of Wight . . . seems to force itself upon the eye'.[42] What struck Sir Henry, in particular, was the steep inclination of the beds of chalk that formed the backbone of the island. This inclination, in places near vertical, was highlighted by the presence of parallel lines of flints. But the flints were also observed to have been almost universally shattered – owing, so Sir Henry surmised, to the way in which

7   The multi-coloured vertical sands of Alum Bay, Isle of Wight, as painted by J. M. W. Turner in 1795

the various chalk beds had tended to slide one upon another during the 'convulsion' that altered them from a horizontal to a vertical position.[43] It was exactly this convulsion that had turned the wonderfully coloured sands of Alum Bay into their near vertical state.

Despite his several trips to Wight, Englefield considered that any published description or 'tour' merited a fuller account of its geology. To this end, he engaged Thomas Webster, a Scotsman, to survey the island.[44] Webster made visits in the summers of 1811 and 1812, hiring a boat on each occasion. With some training as an architect, Webster had been making much of his living by sketching for illustrated books. His architect's sense of structure and his artists's eye were put to good use in Wight, for as well as providing Englefield with a detailed geological account in a series of letters sent from the island, he produced a series of drawings, later to be engraved, many of which focused specifically on Wight's coastal geology.[45] Webster's letters, although he would not have been fully aware of it at the time, laid the foundations

8   Thomas Webster's depiction of the steeply inclined chalk beds at Culver on Wight's eastern coast, 1815

for the rapid extension and refinement of knowledge of sedimentary strata (stratigraphy) that characterized English geology in the second, third and fourth decades of the nineteenth century. Partly inspired by the work of the Frenchmen Cuvier and Brongniart, he registered, for instance, that the chalk hills of Wight, along with those of the South and North Downs on the mainland, were the remnant of a vast sheet of chalk that extended from Yorkshire down to Dorset and to continental Europe beyond.[46] Situated above the chalk, Webster identified an alternation of marine and freshwater sediments that could likewise be traced in the Hampshire (Wight), London and Paris basins. Below the chalk, meanwhile, were various sand beds that outcropped in Sussex and East Anglia. What Webster was describing was almost completely unknown at the time (in a geological sense) and subsequent writers on the subject of stratigraphy readily acknowledged his discoveries.[47]

If Webster's underlying subject matter had an element of the technical about it, the appeal of his sketches and the imagination of his prose style demonstrated that geology then had some of the elements of a 'poetic' discipline.[48] For in 'restoring the island's series of strata', 'now in ruins',[49] Webster felt he was putting together fragments of an ancient temple. He conceded that although the fabric being contemplated was of 'a nature too vast to be completely conceived as a whole', it was still true that 'great and little are so but by comparison; and that by an attentive observation of appearances and phenomena connected with great changes, the mode at least, may frequently be inferred by reasoning not unphilosophical'.[50] At Compton Bay, on Wight's south-western coast, Webster found 'fossil nuts of hazel', although no hazel then grew on the island. Local people called them 'Noah's nuts', associating them with the biblical flood.[51] Such was the 'awful grandeur' of the Bay, though, with its huge masses of rock 'just ready to fall' and the 'noisy fury of the waves', that he formed sensations 'little fitted for minute examination'.[52] He began reflecting, instead, on the 'strange revolutions which our planet has undergone, and the many different states it has been made to assume, before the present order of things became established'.[53] One of those revolutions, as far as the structure of Wight was concerned, involved an 'arch or vault' (what today is known as an anticlinal fold) that connected one end of the island with the other.[54] According to one recent commentator, this anticipated by some thirty years the first major description of folding of strata.[55] Wight's contorted rock structure was a relic of the great earth movements that created the Alps.

Several years before Thomas Webster's visits to the island, Charles Lyell, as a young boy, had gone to Wight with his family for a month's

9   John Phillips's three-dimensional drawing of Wight's geology, revealing the anticlinal fold that ran from one end of the island to the other

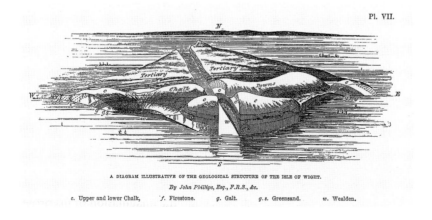

Pl. VII.

A DIAGRAM ILLUSTRATIVE OF THE GEOLOGICAL STRUCTURE OF THE ISLE OF WIGHT.

By John Phillips, Esq., F.R.S., &c.

c. Upper and lower Chalk,   f. Firestone.   g. Galt.   g. s. Greensand.   w. Wealden.

summer convalescence. From their base at Cowes in the north, they made frequent excursions in a sail-boat.[56] Charles's father had been teaching botany and their landfalls were much occupied in plant collecting. But passing a morning in Alum Bay, 'a spot little visited though more interesting as a picture than any other in the island', they marvelled at the surrounding cliffs, the westernmost rising to five or six hundred feet and terminating in the Needles, the easternmost (those of Alum) being striped in the 'most singular and brilliant manner with perpendicular veins of green, yellow, purple and black'.[57] The Lyells returned home that evening laden with rock as well as plant specimens, an outing that must have figured clearly in Charles's mind when, as a grown man, he returned to Wight in June 1822.

To any student of the history of nineteenth-century geology, the name of Charles Lyell requires little introduction. Although the nature and significance of his work has been much debated,[58] his *Principles of Geology* of 1830–33 was plainly the most important and influential work of the nineteenth-century discipline.[59] Charles Darwin, as remarked earlier, eagerly read its account of earth history while sailing round the world on HMS *Beagle* and whilst casting about for evidence to confirm his evolutionary theory. Back home, Lyell's work was reputedly snapped up and read as if it were a novel.[60] In order to support his thesis that the present held the key to the past, Lyell needed to be an assiduous field observer. Past processes could not be studied except by their 'frozen results'.[61] The Isle of Wight formed an important link within Lyell's accumulating configuration of geological pasts. It was not, though, the marvellously coloured vertical strata of Alum Bay he had seen as a boy that pre-occupied him, nor the chalk hills. Instead, it was the rock strata beneath the chalk that became Lyell's principal focus. It had become clear to Lyell and others that some of Webster's identifications of the strata exposed in Sandown Bay, on the island's eastern coast, were mistaken. Lyell's task was to confirm these errors and, in turn, to confirm the precise mainland beds with which they were properly correlated. The term 'Weald' typically describes the dome-shaped rock structure that lies south of the North Downs in Kent and in Sussex. Parts of it are exposed along the south coast, and Wight is included in this. Lyell visited the island in the summers of 1822 and 1823, on the second occasion in the company of William Buckland, the Oxford cleric-cum-geologist who had inspired his geological interests while a student.[62] Lyell's first visit was confined to studying the strata at Sandown Bay. On the second visit, he found in Compton Bay, on the island's south-western coast, a section that allowed one to view the whole geology of the Weald in one go, all within an hour's walk.[63] Buckland galloped over the ground without noticing it. But Lyell took him back to it and pointed out the characteristic

10   Thomas Webster's view of Compton Bay on Wight's south-western coast, where the Wealden rock beds are dramatically exposed, 1815

Wealden Clay, the pyritous coal like that at Bex Hill and the white sands of Winchelsea – 'magnificently exposed'.[64] What Lyell did not fully register at the time was that one section of this coastal strata, at Atherfield (the Lower Greensand), was of greater thickness than any other in England. It was also continuously exposed and accessible in all parts.[65]

A few years before the visits to Wight of Lyell and Buckland, two Cambridge men, Adam Sedgwick and John Stevens Henslow, had also made excursions there to study the island's geology. Sedgwick was Professor of Geology, Henslow later Professor of Botany. They first went to Wight in the Easter vacation of 1819, Sedgwick giving his travelling companion some basic lessons in practical geology, as he was much later to do for the young Charles Darwin.[66] Sedgwick paid a second very brief visit in the summer of 1820 and then made a longer stay in June 1821.[67] He subsequently wrote up his investigations in the *Annals of Philosophy*.[68] He was unequivocal in his praise of the 'diligence' and 'sagacity' of Thomas Webster's work of some ten years before.[69] His own attentions focused on the fossils contained in the iron sand and other beds beneath the chalk hills (i.e. the Wealden rocks).[70] He was also interested in the formations above the chalk, especially those at Alum Bay. The interpretative difficulty at Alum was the greatly displaced nature of the strata, rotated through almost ninety degrees. What Sedgwick did, therefore, was to search westward along part of mainland Hampshire's coast to find their continuation. Between Studland Bay and Hordwell Cliff, he was rewarded with a succession of good sections that offered 'the best possible commentary on the formation of Alum Bay'.[71] The paper for the *Annals* appeared in March 1822 and goes some way to explaining why, when Lyell went to Wight a few months later, his own attentions were largely centred on the correct identification of the Wealden rocks rather than re-visiting the Alum Bay he knew from his childhood. Sedgwick returned to Wight again in 1825, but this time approaching it by boat from the east, from Bognor in Sussex where the sand and shingle shores gave disappointingly little exercise for his hammer.[72] The wind and weather were so favourable that they had no need to shift sails during the whole journey. Making for Ryde on the north coast, he remarked how it was impossible to forget 'the glowing beauty of the shores' as the boat swept up channel. From Ryde he travelled to Freshwater in the west where his brother was living, remaining there a month.[73]

The geologist John Phillips, nephew of William Smith and later Professor of Geology at Oxford, arrived on Wight's shores in the 1830s. He already had a wealth of field experience behind him from his survey work on the Yorkshire coast.[74] For the Wight visit, he left an account of his investigations (one year in May) in the form of a small

11   Two sketches from John Phillips's field notebook of Wight: the bay at Freshwater, with its chalk stacks; and an unidentified chine along the south-eastern coast

field notebook replete with jottings and sketches on the rock strata and scenery.[75] He left London by the 'Rocket' stagecoach, which made the journey to Portsmouth in eight hours, and met the 4 p.m. 'packet' boat to the island. As the coach wound its way over the South Downs, he made sketches and notes of the geology around him. At Portsmouth, it was one of the new steamboats that took him across the water to Ryde with its amazing new pier stretching way out across the mud and sand of the bay, and, out of season, Phillips was able to find lodgings in Ryde for a guinea a week. Soon he was off 'geologizing' the island, exactly as Sedgwick had done some ten years before. He visited the eastern coastal cliffs around Sandown, the numerous gulleys or 'chines' of the southern coasts, and the great fold in the chalk at Freshwater and the Needles. It is not clear who accompanied Phillips on these various excursions, but among his island sketches are pictures of a young woman (perhaps his sister), leisured and elegantly attired, along with countryside scenes and renderings of some of Wight's antiquities. Here, once again, was that intermeshing of leisure and geological study that was such a hallmark of the period.

The singular nature of Wight's geology was not long in coming to the attention of Charles Darwin, following his return from the *Beagle* voyage, and with Lyell's *Principles* still sharp in his mind. Darwin wrote in 1836 to his second cousin and undergraduate contemporary, W. D. Fox, how there seemed to be 'few parts of the world more interesting to a Geologist'.[76] Fox had moved to the island on account of his poor health, and when Darwin made his visit, in November 1837, it was Fox who showed him the geological 'sights'.[77] On a much later visit, with his wife and children in the summer of 1858, Darwin was to recall vividly his first impressions of the solitary nature of the island, prior to its runaway growth as a Victorian resort.[78]

When Charles Lyell, in the mid-1820s, had unravelled the relations of Wight's Wealden rock beds to those from the Weald of Kent and Sussex itself, he communicated his observations to the Sussex-based surgeon-geologist Gideon Mantell, who was later to become a household name for his popular lectures and texts on geology.[79] In due course, Mantell, too, became a regular visitor to the island, later writing a guide to its geology.[80] The strata of which it was composed presented, in Mantell's view, 'phenomena of the highest interest', abounding in 'those "Medals of Creation" which elucidate some of the most important revolutions recorded in the early pages of the earth's physical history'.[81] Models were made of its fascinating rock sequences – 'coloured geologically', and available for purchase, priced from 5 to 42 shillings.[82] Examples were displayed in the Polytechnic Institution of London, along with organic remains collected from the various strata.[83]

12   Alfred Brannon's picture of a ship in a storm off Blackgang, Isle of Wight, in 1836

On one of his first visits, in July 1840, his small son Reginald accompanying him, Mantell travelled to Alum Bay, Shanklin Chine and Ryde in a hired carriage.[84] In July 1844, on another visit, he took the steamboat from Cowes to Yarmouth, affording him a seaward perspective of what Sir Walter Scott described as 'that beautiful island, which he who once see, never forgets'.[85] In the spring of 1845, Mantell was on the island again, this time exploring the Wealden rock beds exposed in Compton Bay.[86] Leaving his carriage on the clifftop, Mantell, accompanied by Reginald and the innkeeper from where they were staying in Yarmouth, clambered down Compton Chine where they found that the very low spring tide had exposed areas of petrified forest as well as reefs of Wealden 'sand-rock' stretching across the bay and far out to sea.[87] The petrified tree trunks were in places several feet long and a foot or more in diameter. Once hardened by exposure to the air, some of the wood proved so firm in texture as to be of domestic use.[88] This was the 'country of the Iguanodon', the land of colossal reptiles, many remains of which Mantell had already discovered in the Wealden rock beds at Tilgate Forest in his native Sussex.[89] That particular spring, a storm had wrecked a homeward-bound Indiaman just off the mouth of Compton Chine and the shore was still littered with silks and camphor which scores of men, women and children from local villages were busy scavenging.[90] Returning to his carriage at Brook Point, Mantell and his companions proceeded east to Atherfield where they descended 'by a perilous foot-track' to the seashore to collect specimens and make sketches of the cliffs.[91] By the time he got back to the inn at Yarmouth, Mantell was exhausted, not to mention chilled to the bone, for there was 'not a bud bursting on the trees, nor scarcely a single primrose in blossom', such was the severity of the north-east wind.[92]

The following September, Mantell was back on Wight, this time making his base at Ryde.[93] From there he went by coach to Ventnor on the island's south-east coast. This time, though, the weather was even more inclement: 'a hurricane and rain all night', 'rain and wind incessant by day'. At Ventnor, he almost slipped down the cliff into the sea. Offshore, an Indiaman was being driven towards the rocks, until a pilot managed to get aboard and steer the vessel clear. When Mantell returned to his London home a few days later, all he had to show for his visit were a few fossil bones procured from a man at Ventnor.[94]

By early 1846, Mantell had obtained his own geological model of Wight, taking it on 21 February to the Marquis of Northampton's soiree in London where it was examined by, among others, the Prince Consort, who proceeded to converse at length on the island's geological phenomena.[95] Soon Mantell was setting his mind to writing an account of Wight's geology and in August of that year he circumnavigated the island by steamer, from which, according to his own account, favourable views of the most interesting localities could be distinctly seen, as Thomas Webster some thirty-odd years before had discovered.[96] During the summer of 1847 Mantell lectured on the geology of Wight to Ryde's visitors, a practice he repeated in 1848, although on that occasion to a rather shrunken audience.[97] In the interim, his *Geological Excursions* of Wight and the adjacent Dorset coast had been published. In February 1847, he delivered a special copy of it to Buckingham Palace for the Prince Consort, to whom it was dedicated. Accompanying the book was a cased collection of nearly one hundred relevant fossils which survives today at Osborne House, kept in the Swiss Cottage.[98] In concluding the text, Mantell remarked how the geologist was 'often in the condition of the Antiquary': endeavouring to decipher an ancient manuscript where the original characters are obscured and partially obliterated by later superscriptions.[99] When the British Association met in Southampton in September 1846, the book (or an unpublished text of it) was used by Sir Roderick Murchison, in the company of a large party of other geologists and scientists, when

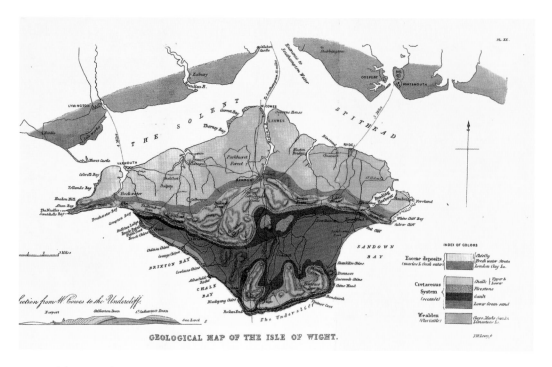

GEOLOGICAL MAP OF THE ISLE OF WIGHT.

13   Fold-out geological map of Wight from Gideon Mantell's *Geological Excursions* of the island, 1847

they went on an excursion by the steamboat Lady Saumarez around Wight one Saturday.[100] Mantell was ill and unable to go, but the trip was both a tribute to his popularization of Wight among the general scientific community as well as to the island's singular importance within the geological narrative. The excursion had been preceded by an evening lecture at the town's polytechnic institution in which a fellow of the Geological Society also illustrated the geology of Wight using models and drawings.[101]

Mantell's own particular fascination with Wight's geology was later matched by Joseph Prestwich. He seems first to have gone there in the late 1840s, writing to Lyell in August 1849 of the remarkable earth movements that had so distorted the island's rock strata.[102] According to his widow, he made repeated visits in later years, keen to unravel its *dynamic* geology.[103] Back there again in 1880, he filled eight pages of his notebook with new sections of Headon Hill, Totland and Colwell Bays on Wight's north-western shore.[104] The soft nature of the rock strata there meant that change was perpetual. Much the same was true of the south coast where narrow coastal gulleys of twenty-five years before had become wide open dells.[105]

Mantell's 1847 guide was primarily a popular one and ran to new editions in 1851 and in 1854, the latter posthumously. Sixty years later, however, a fellow of the Geological Society remarked how few of the island's thousands of visitors paused to view its geology, 'to turn over the pages of Nature's luminous manuscript'.[106] By then, geology had ceased to be at the leading edge of science and was also increasingly removed from popular gaze. The contrast was with almost a century before when geology had been firmly among the gentlemanly pastimes, part of natural philosophy, the pursuit of savants whose social backgrounds ranged from princes to tradesmen. Wight, moreover, was not only a geologist's paradise but a landscape that satisfied the contemporary fascination with picturesque and sublime scenery.

Beyond Wight

When Thomas Webster visited Wight to draw up an account of its geological structure in 1811, he ended his visit not by re-tracing his crossing to the mainland (from Cowes to Southampton) but by hiring a sailing cutter, probably from Yarmouth, to take him due west towards the Isle of Purbeck and the Dorset coast.[107] On a clear day the Purbeck hills are visible from West Wight and it may have been some similarity in their outline with the hills of Wight that prompted his voyage. When the boat reached Handfast Point, he found the chalk cliffs 'turned upwards into a curve, forming nearly the quarter of a circle'.[108] Upon returning to Wight a year later, in 1812, Webster again explored the Dorset coast, finding examples of near vertically disposed chalk as far west as Lulworth.[109] What he had discovered was the westward extension of the sharp fold that he had observed to bisect Wight. When Sir Henry Englefield published his descriptive account of the island in 1816, he incorporated the parts of the Dorset coast that Webster had

14   One of Thomas Webster's views of Lulworth, Dorset, 1815

viewed. Much like Wight itself, Lulworth and places nearby united 'great picturesque beauty with singularly instructive geological phenomena'.[110] The collection of plates that illustrated Englefield's account in fact included as extensive a coverage of Dorset as of Wight. Webster may have known of the picturesque nature of this coast from visiting Weymouth. Like Southampton in the late eighteenth century, it enjoyed a brief interlude as a 'resort' and 'watering place'. During the summer months 'water-parties' forayed from the town to view the adjacent shores, Lulworth among them.[111] By the time Mantell wrote his guide to the Isle of Wight in 1847, this coast had become as accessible from the western shores of Wight as from Weymouth, with regular steam packets from Yarmouth to Swanage and Poole in summer.[112] A geological excursion of Wight could thus be extended to Purbeck 'with but little inconvenience'.[113]

West beyond Purbeck the coastal geology was no less interesting, nor was it any less appealing to the scenic eye. For here, on the coast of west Dorset, was where the great Jurassic system, comprising largely limestones and clays, met the sea. These rock beds, 140 to 195 million years old, cut a highly distinctive swathe in a north-east direction across the centre of England, reaching the coast again in the sea cliffs of north Yorkshire. In the Cotswold hills, they provided the raw material for William Smith's early efforts at unravelling the geological puzzle in the 1790s. And Smith was, in any case, familiar with the west Dorset coastal cliffs, for one of his earliest section drawings was made at Charmouth.[114] Adam Sedgwick also visited this coast, in 1820, beginning at its western extremity at Sidmouth. He was enthralled by the magnificence of the cliffs and by the number of organic remains they contained. 'Almost the whole coast of Dorsetshire', he commented, 'presents a succession of rugged precipices of varied forms, arising from the peculiar disposition of the strata'. He stayed at Weymouth for three weeks to study its geology.[115] Around a year later, Sedgwick was viewing the equivalent rock series in Yorkshire. The chalk coastal range north of Bridlington left him to remark that the chalk cliffs of Wight had to 'yield to these in grandeur', even if the Wolds themselves had none of the peculiar beauty of Wight.[116]

What is so startling about Wight in the early nineteenth century is the extent to which it formed a comparative base for other districts of the country. It was not just that it embraced an unusually rich geology in a remarkably small area at a time when geological knowledge (especially of the sedimentary strata) was in a formative phase, but, in parallel, it became a perpetual 'marker' for geological investigation in similar strata elsewhere. Alongside, and shading into this, Wight also became an archetype for the sublime and the picturesque in coastal

15   Paul Munn, *Shanklin Chine, Isle of Wight*, around 1797

scenery, inviting comparisons that extended, as Sedgwick's comment
indicates, to districts hundreds of miles distant. Such significance was
compounded by the way Wight became a magnet for poets and artists
over the nineteenth century. Aside from Tennyson and his circle who
arrived in the decade after 1850, the island saw a stream of famous
personages crossing to its shores in the half-century or so preceding.
J. M. W. Turner's first oil was painted as an outcome of a visit there in
the late summer of 1795. He was just twenty years old at the time, the
picture forming a study of fishing boats in a squall off Freshwater Bay
at night.[117] While on the island, he made many sketches of its coastal
scenery, including a watercolour of Alum Bay, so fascinated was he by
the structure and colour of its cliffs.[118] Thirty years later he was to
return as the guest of John Nash at East Cowes Castle.[119] John Keats
made several visits to the island, composing his sonnet, 'On the Sea',
after walking down Luccombe Chine, near Shanklin. The 'Pot of Basil'
and 'The Eve of St Agnes', together with parts of 'Hyperion' and

'Lamia' were written whilst staying in a cottage at Shanklin in the summer of 1819.[120] The painter John Linnell visited in 1815, exhibiting a view of Shanklin at London's Spring Gardens in 1817. He walked his way over half the island and was struck by its solitariness. Near Niton, he caught a glimpse of the southward coast: 'nothing of the kind, before or since, was equal to that moment when I caught site of that high horizon of light blue sea'.[121] The painter George Morland did some of his best pictures on Wight whilst hiding there from a string of London creditors.[122] His footsteps were followed by the celebrated painter of epic scenes from the Bible, John Martin.[123] Later still, William Dyce, whilst engaged by Prince Albert to fresco the staircase of the royal couple's new villa at Osborne, did a number of watercolours on the island's east coast, including two views of the striking chalk cliffs at Culver.[124] Just as Wight became a classic training ground for Victorian geologists, so it was also a forcing house for aspiring young artists, their creative energies often tapping into the singularities of the island's geology. G. F. Robson, for instance, best known for his Highland pictures, sketched all the chines of Wight as a young artist.[125] The cartoonist Thomas Rowlandson toured Wight sometime in the 1780s, producing among his sketches a watercolour of the tumbled strata at the undercliff west of Ventnor.[126] Paul Sandby Munn was sketching on the island directly in the wake of Turner, his gouache of the entrance to the declivity of Shanklin Chine affording a fine illustration of the beauties of some of the island's southern shores.[127]

When leading geologists of the day visited Wight in the early nineteenth century, they were thus participating in much more than just a geological excursion. The island was inextricably part of contemporary literary and artistic inspiration, which was itself not in any sense isolated from geology's subject material. The island in turn had a 'social season' much like Weymouth, Southampton, Lyme and other Georgian resorts along the adjacent coastal mainland. When Adam Sedgwick was not on geological expeditions to every corner of the isle during his stay in the summer of 1825, he was at 'water-parties with the ladies'.[128] The apex that Wight provided for geological discovery was thus quite remarkable. And in an even broader sense, there was an uncanny irony in finding that who should be staying at Bonchurch in the summer of 1850, but Thomas Babington Macaulay, working on his *History of England*.[129]

Canal Geology

To sit forward in the chaise was a favour readily granted; my eager eyes were never idle a moment[130]

In old age, William Smith earned the accolade of the father of English geology. He was among the prototype field geologists and perhaps the last to come entirely from a non-academic background.[131] Born in 1770, the son of a blacksmith from deep in rural Oxfordshire, Smith had only a basic schooling. What he lacked in educational attainment, though, was more than compensated for by an exceptionally keen eye. He had quite astonishing sensitivities for observing the landscape or country around him. He was also fascinated by the mechanical arts. On his way to London, even at the age of twelve, he noticed the workmen cutting into the chalk hill at Henley and the way they had erected a rudimentary inclined plane, the downward carriages loaded with chalk spoil being made to bring up the empty ones from below. At eighteen, Smith became pupil to a land surveyor and so began a career that meant an entire working life spent largely in the open air, first in his native Oxfordshire and in its adjacent counties, later encompassing the length and breadth of England, together with parts of Scotland and Wales.[132]

Smith's training as a surveyor, involving measurement, levelling and orientation in the field and then translating this in to drawings and maps, does much to help us understand how a relatively uneducated man came to exercise such an influence over the development of English stratigraphy. It gave him a predilection to look for regularity and order in the way in which he viewed the landscape and rock surfaces around him. Moreover, when he was engaged on farm enclosure surveys, order was certainly the dominant ambience, for the restructuring ('enclosing') of the old open fields, including the fencing, hedging and road-making that accompanied it, was none other than an exercise in impressing order and regularity on what were largely disordered medieval farm structures.

The area of Oxfordshire and Gloucestershire in which Smith grew up came late to the enclosure movement and he appears to have spent a significant proportion of his early training in facilitating its implementation. At the age of twenty, he found himself required to attend the first meeting of the Commissioners for the Lower Swell Inclosure at the Unicorn Inn in Stow. This was late in April 1789. Much of the subsequent months were then spent outdoors taking dimensions and drawing up the plan. Work was still continuing in the autumn, in a year when, as Smith later recorded, there was rain on 116 of the 214 days from Whitsuntide.[133] As time passed, the range of projects in which Smith became involved extended to include land drainage, sea defences,

16 William Smith, a self-portrait sketch, 1808

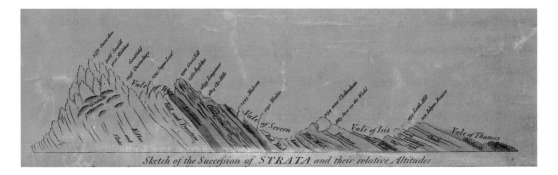

Sketch of the Succession of *STRATA* and their relative Altitudes.

17   Sketch of the succession of the English strata, from west to east, from William Smith's large geological map of 1815

prospecting for coal and, most important of all in terms of his developing geological expertise, prospecting the lines of canals. In the process, Smith was acquiring a steadily more extensive familiarity with the 'general aspect and character of the country'.[134] He began to register how vegetation, soil and rock type often coincided. He noticed how beds of rock in one county could be directly compared with those in another. But he also saw how some outwardly similar beds actually displayed subtle differences.

In 1791, Smith moved his working base from Oxfordshire to Somersetshire where he was engaged to survey the collieries around High Littleton, south-west of Bath.[135] Studying the coal seams, he once again registered how rock beds were often comparable and became convinced enough of this idea to make a model using materials collected from the pits themselves.[136] The key proof to the regular 'conformity' of rock strata, though, was found in a levelling survey that he undertook in 1793 for a projected canal, known later as the Somersetshire Coal Canal.[137] Conducting work in two parallel valleys, Smith established that the rock series of the area resembled, on a large scale, 'the ordinary appearance of superposed slices of bread and butter'.[138] The way in which levelling confirmed this was on account of the strata being laid not horizontally but inclined (dipping) in one direction – to the east. Thus what one traversed at the surface were the successive ends of the 'slices of bread and butter'. Along each of the levelled lines, the strata of 'red ground', 'lias' and 'freestone', for example, 'came down in an eastern direction and sunk below the level, and yielded place to the next in succession'.[139]

The Somersetshire Coal Canal received Parliamentary sanction in 1794 and the committee behind its projection called upon William Smith, in company with two committee members, to undertake a 'tour of enquiry and observation regarding the construction, management and trade of other navigations in England and Wales'.[140] News of the tour came to him as 'joyous intelligence': here was the chance for the 'truth and practicality' of his 'system' to be 'tested', to see whether his observed order of strata applied more widely than in the band of country extending from Warwickshire down into Somerset.[141] The slow pace of horse-drawn travelling, especially when making ascents, afforded Smith 'distinct views of the nature of rocks'.[142] He interpolated boundaries between rock types with variations in characteristic vegetation. He noted how, in Yorkshire, Tadcaster Moor was made of a limestone unknown in the south (the Magnesian limestone). Climbing to the top of York Minster, he took in the panorama of the Yorkshire Wolds, remarking how one could see they contained chalk 'by their contour'.[143] In Yorkshire, generally, he found the coalfield rocks 'everywhere so well developed' that he had ideas of being able to infer their details merely from looking at the surface.[144] The upshot of Smith's tour, which lasted around six weeks and covered 900 miles of ground, was that he found, as he had anticipated, the strata in the vicinity of Bath and Bristol to be 'prolonged into the north of England, in the same general order of succession with the same general eastward dip'.[145] In effect, he had identified the great Jurassic rock formations that described an arc across the centre of England.

When he returned to Somerset, Smith assumed the superintendance of the construction of the Somersetshire Coal Canal, a project that was to occupy him for some six years.[146] The line of the canal traversed a whole sequence of rock strata and Smith soon registered that he required a detailed knowledge of their dispositions in making a suitable bed for the waterway. As the ground of the canal was broken by the 'navigators', he discovered that the rock beds above the coal, the 'clift', abounded in the fossils of peculiar plants. He also found particular shells in the 'lias' or clay beds and in the 'freestone' or limestone.[147] From this he concluded that 'each stratum had been successively the bed of the sea, and contained in it the mineralized monuments of the races of organic beings then in existence'.[148] If one could systematize the identification of these fossil remains, it became quickly clear to Smith that they could themselves provide a means for identifying the different sorts of rock, sand and clay through which the navigation cut. What he had registered was that 'each stratum contained organized fossils peculiar to itself, and might, in cases otherwise doubtful, be recognized and discriminated from others like it, but in a different part of the series, by examination of them'.[149] Smith also made further observations concerning fossils that lay in gravel deposits as distinct from those in the regular rock strata. The former displayed a rounded form and were intermixed, whereas those in the regular strata were nearly always sharply preserved. The fossils in the gravel

18   The junction of the Somersetshire Coal Canal with the Kennet and Avon Canal at Limpley Stoke, 1864; the Coal Canal runs off into the right foreground, the first lock situated just right of the small bridge across the canal

series were later given the label of diluvial deposits, reflecting their random deposition by rivers and streams.[150] In continental Europe, Smith's ideas about fossils had been anticipated by Werner in Germany and Cuvier and Brongniart in France. However, most of this continental work related to the geology of later eras. Smith's work, by contrast, concentrated on older rocks. He qualified as the discoverer of the main natural divisions of this part of the stratigraphic profile, the so-called Secondary strata.[151]

   Smith's theory was, in more general terms, nothing new. Miners in England had long since identified their local sinkings of shafts with reference to the fossils found in adjacent rocks – and he recorded as much.[152] Masons and quarrymen could correlate beds of stone dug many miles apart in much the same manner.[153] Indeed, nowhere was this more conspicuous than in the lias quarries of Somersetshire. To begin with, Smith found the stratification of the stone 'something very uncommon'. He had no clear idea of 'what seemed so familiar to the colliers'.[154] However, after learning the technical terms of the strata and making 'a subterranean journey or two', he became fully accustomed to the working method and started to speculate upon its wider significance.[155]

When first identifying the fossils he discovered, Smith followed the fashion of the local colliers and quarrymen. He used dialect names such as 'pundits', 'snakestones', 'pundstones' and 'quoitstones'.[156] Eventually, however, with the aid of better educated friends, especially Richardson, he learned the appropriate taxonomy. Thus snakestones became ammonites and pundstones echinoderms.[157]

William Smith was far from the first person to be struck by the different fossils that the particular task of levelling a canal bed brought to light. Almost thirty years before the Somersetshire Coal Canal, the potter Josiah Wedgwood had remarked in a letter to a friend of 'the wonderful and surprising curiositys we find in our Navigation'.[158] This was part of the Trent and Mersey Canal in which Wedgwood played a key promotional role in association with his growing business interests in Stoke-on-Trent. Beneath a bed of clay on the south side of the

19   Impressions of plant remains in sandstone and other rocks, including examples of fossilized wood, from James Parkinson's *Organic Remains of a Former World*, 1804

Harecastle tunnel, the workmen had unearthed bones of a giant fish. Some of those who viewed these speculated that they belonged to Jonah's Whale. Others could not decide whether the animal was an inhabitant of sea or land.[159] North of the tunnel, in sandstone, the workmen were forced to blast a way for a suitable bed for the canal. Here, rather than animal remains, they found the impressions of ferns, vetches, crowfoot, hawthorn, yew and withey in the stone, not to mention the fossilized roots and trunks of trees, some of them two feet in diameter. The trees were all perpendicular to the strata (though not upright since the strata dipped). The engineer for Wedgwood's canal was the celebrated James Brindley. Whether he or any of the other men who worked on its construction made the kinds of connections between fossils and rock strata that William Smith was later to do in Somerset is only a matter for speculation. Josiah Wedgwood, though, having enumerated the 'curiousitys' in his navigation, appeared in the end to be dumbfounded: 'I have got beyond my depth – These wonderful works of Nature are too vast for my narrow microscopic comprehension'.[160]

William Smith circulated an account of his 'method for tracing the strata by organized fossils embedded therein', in 1799.[161] His 'card' of the English rock strata was soon being popularized by others, as well as plagiarized.[162] Among some of the gentlemen geologists of the day, he acquired the nickname 'Strata Smith'.[163] Within the surveying and civil engineering profession, particularly in the business of canal construction, his method came at a singularly appropriate moment, for in England the 1790s was a time of the 'canal mania'. This was a speculative boom involving canal promotion all across the land, the legacy of which lasted through the first and second decades of the nineteenth century. Where it came to their notice, Smith's work aided the endeavours of a whole generation of canal engineers and surveyors who, in turn, were able to add to the accumulating stratigraphic record up and down the country.

It had long been practice to 'improve' the course of natural rivers to render them continously navigable for some distance inland.[164] But the artificial 'cuts' and 'diversions' that this involved rarely extended to the excavation of much other than alluvium or gravel. The levels of river navigations were those of the river courses that they mimicked. The pattern was often altogether different with canals. Here surveyors and engineers defied the logic of natural river basins and evolved routes that crossed watersheds.[165] Among Smith's own notes on 'canals in relation to strata', he remarked how 'the choice of elevation admits of great variety according to the situation of the locks – and which variations of level will of course admit of great deviations in the line or course of the canal'.[166] The pound lock helped carry canals across the

Pennines. East and west coasts were linked by such canal projects as the Thames and Severn in the south and the Leeds–Liverpool in the north. In nearly all of these cases, surveyors and engineers found themselves traversing just the general sorts of rock sequences that Smith had faced in Somerset. To be able accurately to trace strata by their characteristic fossils thus came as a great advantage. There were some projects, though, notwithstanding improvements in puddling, where long summit sections stretched for too many miles over porous soil and 'hollow, jointy' substrata. Here, according to Smith, great difficulties had attended the engineering works as well as subsequent repairs. It was in such cases that Smith found his services in frequent demand.[167]

When Smith's work with the Somersetshire Coal Canal Company came to an abrupt end in the summer of 1799, he began casting around more widely to sustain his professional business. Over the following decade or so, this found him traversing almost the length and breadth of the country, taking in coal-mining districts in Lancashire, Yorkshire, North and South Wales, Staffordshire and Cheshire, where, essentially, he was performing the task of mineral surveyor. In North Norfolk he became involved in dealing with measures to protect the coast near Yarmouth from sea breaches. Other commissions took him to the North Yorkshire coast and to Northumberland.[168] In all of these extensive travels, Smith rarely missed an opportunity for recording the geology of the country across which he passed. He sketched profile sections of road cuttings. Natural cliffs were almost everywhere copied. He made records of borings and well sections. And all this was in addition to the numerous colliery sections from survey work done in the coal districts.[169] His draughtsmanship worked to a constant scale of eight yards to the inch and sections were differentiated using colours.[170]

The striking thing that Smith's diaries of this period reveal is the astonishing energy of the man. By his own account, he had been 'a tall and strong-grown boy'; and when a young man he was sometimes mistaken for a pugilist.[171] It was nothing for him to be up at the crack of dawn, off on horseback or by chaise to investigate some new commission. Staying near Gloucester in early April 1802, he was walking out before breakfast tracing the different strata up and down channel on the east banks of the River Severn. Then, later in the day, he went on to inspect some of the collieries of the Forest of Dean. The following morning he collected fossils on the road from Gloucester to Stow, not riding home to Churchill in Oxfordshire until after midnight.[172] A couple of weeks later, Smith travelled to Woburn, the Duke of Bedford's seat, to superintend the draining of some water meadows there. The levels and surveys made on the Duke's lands were already proving Smith's ideas about the regularity and dip of the strata in the neighbourhood. Days were spent, with an assistant, marking the lines of the strata around Woburn on sheets of John Cary's large map of England.[173] A month later Smith was back home in Oxfordshire 'long late studying the connections between the chalk of East Norfolk and those of Hants, Sussex and the South Downs – and recollecting the form of the country, depths of wells, pits and mineral springs'.[174]

In 1806, Smith spent much time in Norwich overseeing various projects in the vicinity. These included marsh drainage, inclosure, improvements to water meadows and dealing with coastal sea breaches. In July 1806, his diary records an attack of lumbago, but he was not so ill as to be unable to work and on various occasions he records walking the beaches and swimming and bathing.[175] In the four months of January, March, April and May 1809, Smith journeyed 2,250 miles on business. If one works on an average speed of travel of eight miles an hour, this was roughly twelve days worth of riding.[176] That year his commissions took him as far afield as Norfolk (sea breaches), Kidwelly Harbour in south-west Wales and Sussex (Ouse Navigation).[177] In February 1812, on one of his many journeys down to Sussex to deal with 'navigation business', the carriage he was travelling in overturned into a river swollen with heavy rains. One of the horses was drowned and Smith and his companion 'escaped with great danger to [their] lives'.[178] Even when he was close to fifty years of age, Smith's energy seemed undiminished. In 1819, for instance, his diary reveals him advising on a canal project in Yorkshire, on water supply for the resort of Scarborough, on colliery workings in the Forest of Dean, not to mention various 'geological journeys', necessary as part of the process of correcting the sheets for his geological atlas. On one occasion, deep in the winter of 1819, he records walking from Buckingham to Churchill over two days, a distance of more than thirty miles.[179]

The steady accumulation of geological data covering often widely separate tracts of country provided the basis for William Smith's geological map of England and Wales, the first such map ever produced and the most important mark of his subsequent geological fame. Smith's first geological maps had been of the area around Bath and for the county of Somerset.[180] By 1801, though, he had compiled a map that encompassed all of England and Wales.[181] And over the succeeding few years, as the extent of his travels widened, this map was progressively refined in conjunction with the additional stratigraphic knowledge that he acquired. The system for distinguishing rock strata was to colour the basal part of each rock formation and to fade that colour towards the outcrop of the next overlying formation.[182] This made corrections easier. Smith clearly had ideas of publishing his map at an early stage, but not until 1815 did the project come to fruition. It was at a scale of five miles to the inch, on fifteen separate sheets and

20   One of William Smith's first attempts at an outline map of the geology of England and Wales, 1801. The shaded line that runs from the coast just north of Whitby down to Somersetshire marks the westernmost outcropping of the famous Jurassic rock series. Comparison with a modern geological map reveals that Smith's outline was uncannily accurate

accompanied by fifty pages of explanatory notes.[183] It was published by John Cary, whose well-known general map of England and Wales (in its many separate sheets) had provided the base on which Smith's first working maps of the strata had been constructed during the years before. Among Smith's papers, there is a list of 122 subscribers to the map. The list includes business clients such as the Duke of Bedford, as well as geologists like Buckland, Conybeare and Greenough.[184] Two years after the map's publication, Smith's *Stratigraphical System of Organized Fossils* appeared, essentially a catalogue of some 3,000 specimens. There was a separate four-part work, published in 1816–19, under the title *Strata Identified by Organized Fossils*. In a prefatory comment, he asserted that the method of tracing strata by the organized fossils embedded in them was a 'science not difficult to learn'.[185] He observed to readers how, when he first made his discovery of an organized system of fossils, scientific acquaintances around Bath searched the quarries of different strata there 'with as much certainty of finding the characteristic fossils of the respective rocks, as if they were on the shelves of their cabinets'.[186] Smith went on to remark that the search for a fossil may be considered at least as rational as the pursuit of a hare – 'one the sport of infants, the other of adults; one squanders time and propriety, the other improves the mind and may afterwards extend such infant knowledge to the improvement of the estates he may enjoy'.[187] Those with mining interests increasingly had reason to avail themselves much more closely of the 'subterranean productions' of the land. The contents of quarries, pits and wells were now liable to be 'scrupulously examined'.[188] The way was now open for the 'virtuoso' to enter into such study. It would yield not only general economic utility but more general improvement through 'the early diffusion of science'.[189] Ever conscious of his own diminutive social station and the perpetual difficulties this had presented for the diffusion of his geological ideas, Smith wrote this preface partly with the object of trying to demonstrate that rock and fossil collecting was not necessarily just a gentlemanly pastime. In an age of capital, it had clear economic benefits for the landed classes. Herein, moreover, lay a measure of respectability for the practical man like Smith. It also marked the start of a coalescence of tradecraft (or technology) and scientific study, especially as the latter became increasingly separated from theology in the early decades of the nineteenth century.[190]

Smith's pioneering geological map was not a commercial success. It was eclipsed by the Geological Society's own map, published in 1820, which was claimed as the work of the Society's president, G. B. Greenough.[191] Although this production was a better piece of draughtsmanship and more accurate, it was largely a purloined version of Smith's map of 1815.[192] Much earlier in his career Smith had an idea

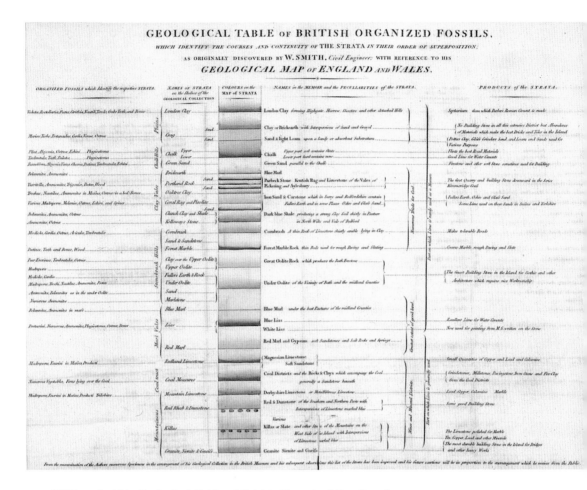

21   William Smith's Geological Table of British Organized Fossils, 1817

of writing a geological text that recorded not only the natural order of the strata but the relationship they bore to topography, vegetation and to economic utilization.[193] For many years, this project lapsed, but, following his publications of 1815–17, he returned to it. His notion was to offer 'a connected description of the face of the country'.[194] Every district had a peculiar character. The verdure of the fields, the colour of fresh-ploughed soil and the materials of the roads all provided interest for an eye accustomed to observation.[195] Where there was pen and paper to be had at an inn, Smith hastily transcribed the observations recorded in pencil while riding through the countryside. Sadly, though, the project was yet again cast aside almost as soon as it was taken up – probably owing to the financial straits in which Smith found himself by about 1818.[196] It was revived once more, shortly before his death in

1839, but he got no further than completing a section entitled 'Abstract Views of Geology'.[197] One of William Smith's underlying difficulties here may have been that writing did not come easily to him.[198] What cannot be disputed, though, is the formative impact that his writing had on some leading geologists. Adam Sedgwick traced Smith's footsteps through Wiltshire and the neighbouring counties with sheets of the 1815 map in hand. This was how Sedgwick learnt the subdivisions of the Jurassic beds.[199] Yet more interesting is that, annually from 1808, William Buckland, the Oxford geologist, independently made just the sort of 'geological tours' that Smith had regularly been making since the mid-1790s. And exactly like Smith, he coloured the results on Cary's large map of England.[200] Subsequently, these tours continued in the company of Greenough, but then with the object of providing a geological map of England that would eventually lever Smith's own map aside.[201] Had William Smith been a gentleman, his efforts would have placed him among the best geologists of the day. The questions he posed and the fieldwork he undertook were largely indistinguishable from the gentlemen scholars who appeared continually to steal his limelight.

In the last years of his life, however, William Smith did become a regular attender of the geological sections of the meetings of the British Association. It was at the Oxford meeting of the Association, in

22   Part of the geology of Wiltshire, from William Smith's *Geological Maps of English Counties*, 1818–1824

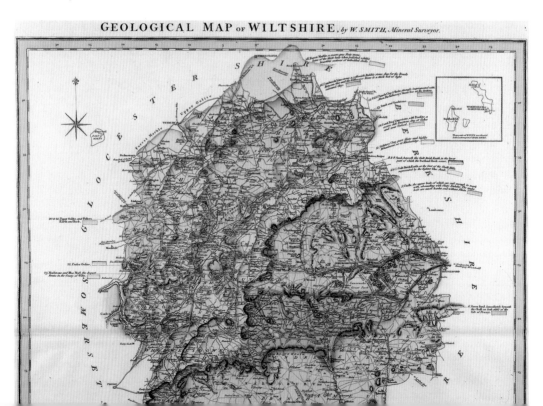

1832, that he was presented with the Wollaston Medal of the Geological Society.[202] This was the first formal recognition (first made public in 1831) of his contibution to geological study. Leading members of the geological circuit, conscious of Smith's precarious financial situation and increasing age, at the same time petitioned William IV for a pension in recognition of the services Smith had performed for the advance of the subject.[203] In the end, the King awarded him £100 a year to ease the poverty of his retirement at Scarborough in Yorkshire.[204]

Much is sometime made of William Smith's humble origins and how this limited the notice that his work gained in the corridors of power and among the great and the good. Even when accepted as part of the geological circuit in the 1830s, he was never a prominent participant.[205] However, one should not underestimate the significance of the contact that he had, as part of his profession, with some of the great landed proprietors. Not only did he deal with them professionally at their principal seats, but he also made professional calls at their various London houses. On 16 May 1804, for instance, he was at London's Grosvenor Square in conference with the Earl of Sefton. The same day he saw the Earl of Egremont in Grosvenor Place concerning a trial boring for coal on the Earl's lands. He also called on Mr Coke of Norfolk at Charles Street, Berkeley Square about some new water meadows. And as if this was not enough for one day, he yet had time to see the Duke of Bedford at his home in Stable Yard where, among other things, Smith raised the idea of showing his maps of the strata to the Board of Agriculture.[206] Many of the Board's newly published county surveys had contained significant geological description to which Smith was seeking to add. The day prior to this medley of calls, Smith visited St Paul's Cathedral 'where there was Anthems performing in the most sublime stile of music I ever heard for the benefit of sons of the clergy'.[207] Smith certainly lacked the classical education of his social peers, but he was plainly no cultural philistine.

The Iron Road

> Emerging from the abyss of the Linslade Tunnel into the ironsand excavation, the vivid red slopes of which, with the sudden transition from darkness to light, will produce the most powerful effect on the vision, and cause the traveller for a moment to close his eyes.[208]

Richard Fortey, a professional palaeontologist from London's Natural History Museum, begins his book, *The Hidden Landscape: A Journey*

23   J. C. Bourne's powerful rendering of the great limestone excavation at Blisworth Ridge, Northamptonshire, to make way for the line of the London and Birmingham Railway, *c.* 1838

into the Geological Past, with a train journey on the old Great Western main line out of London Paddington. The wonder of this journey is that, in geological terms you become a time-traveller. In the Thames valley you are in the age of mammals, with its soft sands and hard cobbles. Crossing the chalk hills, you are launched back to the age of the dinosaurs. Around Bath, you enter the great Jurassic belt, its limestones laden with corals and its shales with amazing fish lizards. Once under the River Severn, and into Wales, you are into the coal-swamps – a time when the Principality 'steamed and sweated' in the humid heat of the Carboniferous era and 'dragon-flies the size of hawks flitted in the mist'. At last, arriving at Haverfordwest, you came to the silts and muds of the Silurian sea, over four hundred million years old and astonishingly contorted with age.[209]

The westward line of railway out of Paddington was built when the young Queen Victoria was on the throne, well over a century before Fortey's imaginary journey into the geological past. His particular technique of geological instruction is also more than a century old, for

Gideon Mantell employed just such a method in introducing his geological guide to the Isle of Wight in 1847.[210] 'All the lines of railway, that proceed from London', so he wrote, 'traverse for the first ten or twenty miles beds of clay, loam, and loosely aggregated sand and gravel; hence the numerous slips that have taken place'. Following these so-called Tertiary sediments, the next geological unit was the chalk downland, which was 'invariably traversed by steep cuttings and tunnels'. Beyond the downland, a whole new series of older clays and sands appeared in rapid succession, to be followed by the Jurassic limestones on the Bristol and Birmingham lines, and by the Wealden beds on the Brighton.[211] For the Wight excursionist, Mantell was painting a picture of the entire geology of southern England for which (as already discussed) the island was a remarkable epitome.

Mantell was an early and enthusiastic rail traveller. He made his first trip on the Great Western from Paddington to Chippenham on 21 June 1841.[212] The following month found him on the Brighton railroad, on an experimental train prior to the line's formal opening in September 1841.[213] A few weeks later, he was 'railroading' from Paddington to Swindon, fossil-hunting and rambling over scenes of his boyhood where he had once attended a dissenting school nearby. By evening, he was so laden with finds that he had to engage a quarryman to carry the heavy load through the fields to the railway station before getting the train back to Paddington.[214] By 1845, Mantell was describing express trains going at fifty miles an hour as 'terrific to behold', raising 'clouds of dust from the whirlwinds induced by their rapid passage through the air'.[215] He also remarked on the frequency of train accidents, but this did little to deter his almost boundless investigative energy.[216] At Lewes, in Sussex, he climbed up Cliff Hill one sunny September morning in 1849. From his vantage-point, he commanded a view of all the principal roads into the town. What struck him was 'the entire absence of anything moving . . . not a horse and chaise, no vehicle whatever – not even a foot-paper boy!'.[217] The new Lewes railroad, he concluded, 'had engrossed everything'. Now, travellers had to 'move in shoals'.[218]

Much like the canals that went before them, railway construction involved excavation along levels. This is not to imply that the permanent way of the railroad was ungraded, but engineers, especially in the early days of construction, sought to make their lines as close to a level as possible for fear of difficulties in tractive adhesion. Unlike canals, though, railroads had no predisposition to look for routes that would guarantee a water supply. The tendency, therefore, was for railroads to follow much more direct lines of passage between centres of population. The outcome was that railway works involved a scale of excavation far greater than the canals that preceded them. This became

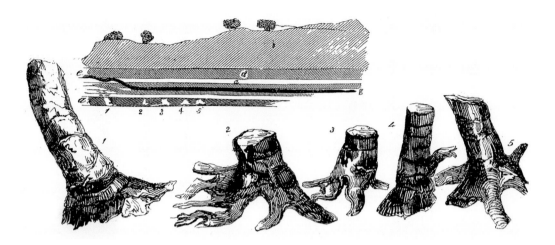

24   Drawings of the fossil trees found in the making of the Manchester and Bolton Railway, 1837–8 from Thomas Milner's *Gallery of Nature*, 1846

most starkly demonstrated on the route of the London and Birmingham Railway, the first major trunk line, constructed during the mid-1830s. Stretching from the London Clays of the London basin to the Coal Measures of the Midlands, it involved a 'geological journey' of 300 million years. The construction required a whole sequence of rock cuttings, some up to sixty or seventy feet deep in places and sometimes more than a mile long. At Blisworth Ridge in Northamptonshire, the engineers had to cut a chasm through the Jurassic limestone. And as the navvies toiled with their picks and shovels, their wheelbarrows and their horse-gins, they found fossils by the score. Soon geologists were flocking to examine these newly exposed sections.[219] While Mantell could describe the geology of the country as it appeared from the windows of a railway carriage, here was an opportunity to examine in detail the strata and organic remains of inland districts at sites easily as rewarding as the coastal cliffs of Wight or of Dorset. What is more, these were freshly cut sections, in places vividly coloured before prolonged exposure to the air.

It was not long before the professional geological community registered the importance that countrywide railway excavation held for the accumulating stratigraphic record. Whilst the Manchester and Bolton Railway was in the making in the late 1830s, its engineer, Sir John Hawkshaw, was president of the British Association. In the course of the line's excavation, the navvies had come upon a series of fossil trees standing upon the coal in perpendicular fashion.[220] Hawkshaw registered the trees' significance for geology, and straight away looked for measures for their preservation.[221] A shed was erected over the trees to protect them from decay, casts and models were made of them, the latter publicly displayed in the Owen's College Museum in Manchester.[222] Hawkshaw noticed in the clay under the coal immense quantities of fragments of a plant that was then known as Sigmaria.[223] Eventually, though, it was established that the clay was the soil in which the plant roots grew and that Sigmaria were actually the roots of the trees themselves.[224] The trees had not been carried there by some catastrophic flood event, but had been fossilized where they grew as part of a slow sequence of sedimentation. Hawkshaw reported his findings to the Geological Society in 1839 and again in 1840.[225]

The appearance of these papers in the geological press, along with some earlier notes on the geology of the proposed Birmingham and Gloucester Railway,[226] seems to have galvanized the geological community into pressing for efforts to record on a more systematic basis the valuable geological information being exposed in railway cuttings in different parts of England. Writing in 1840 about the cuttings of the Birmingham and Gloucester Railway, Hugh Strickland commenced by expressing regret at the irrecoverable loss the science had sustained in there being no proper system for recording such fresh exposures.[227] In the case of the Birmingham and Goucester line, its engineer, Captain Moorsom, had requested Strickland's help in undertaking a geological survey of it. Betweeen them, they drafted a series of elegant coloured sections.

Strickland's call was not long in being answered, for the British Association, at its Glasgow meeting in August 1840, decided to appoint a committee and make a grant of money 'to begin the important work of collecting and preserving information as to the structure and mineral riches of the country, which is now accessible in sections of the strata exposed in cuttings on the numerous railroads in various parts of the United Kingdom'.[228] The Institution of Civil Engineers had also expressed a zealous desire to co-operate in carrying the measure into effect. The committee had £200 placed at its disposal, the largest single sum of expenditure authorized under section C of its organization.[229] The Committee comprised the presidents of the Royal Society and the Geological Society; Roderick Murchison and Henry De la Beche, two leading geologists; and Charles Vignoles, a civil engineer. At the British Association meeting in Manchester in 1842, Vignoles reported that the entire sum had been spent and further liabilities incurred.[230] Sections had by then been completed for the lines from Bristol to Bath, Glasgow to Greenock, Greenock to Ayr, Rugby to Derby and Nottingham, Derby to Leeds, Manchester to Normanton, Manchester to Bolton and Hull to Selby.[231] Vignoles remarked how

25 William Mackenzie's view of navvies at work on the Glasgow and
Greenock Railway, 1841, one of the lines surveyed at the instigation of the
British Association on account of the geological information that the
excavations afforded

the committee's work had been 'aided in the most effective and
satisfactory manner by all the Railway Companies to whom they
ha[d] applied'.[232] The engineers had also taken great pains in assisting
the Association's task and special mention was made of the North
Midland Railways's engineer.[233] The committee's hope was that the
recording of future railway excavation would be taken up by the gov-

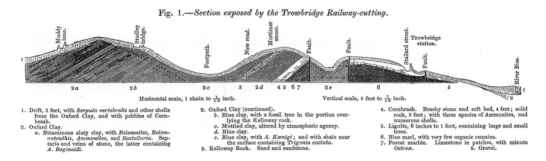

Fig. 1.—*Section exposed by the Trowbridge Railway-cutting.*

Horizontal scale, 1 chain to $\frac{1}{16}$ inch.   Vertical scale, 5 feet to $\frac{1}{16}$ inch.

1. Drift, 3 feet, with *Serpula vertebralis* and other shells from the Oxford Clay, and with pebbles of Cornbrash.
2. Oxford Clay.
  a. Bituminous slaty clay, with *Belemnites, Belemnoteuthis, Ammonites,* and *Rostellaria.* Septaria and veins of stone, the latter containing *A. Reginaldi.*

2. Oxford Clay (continued).
  b. Blue clay, with a fossil tree in the portion overlying the Kelloway rock.
  c. Mottled clay, altered by atmospheric agency.
  d. Blue clay.
  e. Blue clay, with *A. Kœnigi* ; and with shale near the surface containing *Trigonia costata.*
3. Kelloway Rock. Sand and sandstone.

4. Cornbrash. Brashy stone and soft bed, 4 feet ; solid rock, 3 feet ; with three species of Ammonites, and numerous shells.
5. Lignite, 6 inches to 1 foot, containing large and small trees.
6. Blue marl, with very few organic remains.
7. Forest marble. Limestone in patches, with minute *Ostreæ.*   8. Gravel.

26   Drawing of a geological section exposed in the making of the Trowbridge branch of the great Western Railway – from a paper by Reginald Mantell in the *Quarterly Journal of the Geological Society,* 1850

ernment and made to form part of the Geological Survey of Great Britain.[234] By the time of the British Association's next meeting, held at Cork in August 1843, further grants had been made available, but the firm recommendation of the then committee members, the Marquis of Northampton, William Buckland and John Taylor, was that all future work should be conducted as part of the Ordnance Geological Survey.[235]

Whatever the formal agency for recording the geology exposed in railway excavations, the subject's leading periodical, the *Quarterly Journal of the Geological Society (QJGS),*[236] rarely passed an issue over the course of the next twenty years in which there was not some kind of reference to sections uncovered from railway construction works. Aside from cuttings, they were drawn up from tunnel borings, and from temporary quarries made to extract stone for the structures of the permanent way and to dig gravel for ballast. Superintending engineers also sank bore-holes to ascertain the nature of the rock strata they were required to traverse. And not only were detailed records made of these holes but sometimes they were extended well beyond the immediate contingencies of a line's construction. When the South Eastern Railway was in the making near Hythe in Kent, F. W. Simms, in directing its works, not only sank shafts in connection with the principal tunnels at Saltwood, but afterwards extended his searches upwards in order better to illustrate the sequence of the strata. Later, he sank a shaft from the bottom of quarries at Hythe right down to the Wealden Clay. Simms reported his findings in an issue of the Geological Society's *Proceedings.*[237] The many railway projects that were being mounted in the south-east of England at this time came at an especially apposite moment in the unravelling of the area's geological history. William Smith had had little to say of it. In the 1820s, Lyell and Buckland were sorting out confusions in the identification of the Wealden beds of south-east England where they extended to the Isle of Wight.[238] It was arguably Gideon Mantell who did most to unlock the geology of Kent and Sussex, but there remained many loop-holes in knowledge of particular rock distributions and many broader questions about earth movements, about *dynamic* geology.[239] It was railway engineers who increasingly answered to these needs.[240]

The Weald of Kent and Sussex had by this time become legendary for the remains of giant extinct reptiles that were regularly being disinterred from within its sedimentary rock layers. Not surprisingly, many railway excavations added to these. At Bletchingley, for example, bones of an Iguanodon were discovered whilst excavating the tunnel and they were presented to the Geological Society, along with specimens of the fossil plant, Clathraria lyellii.[241] Roderick Murchison referred to cuttings of the 'Dover railroad' between Tunbridge and Folkestone where the teeth and tusk of a mammoth were discovered in the detritus. These organic remains were the more interesting because they were seventy feet above the local river levels, implying that they had been deposited as part of the silt and gravel of a river that extended over 'all the lower country between the hills of Lower Greensand on the north and those of the Hastings Sand on the south'.[242]

Outside of the Weald, the geological information revealed in railway excavation was no less important. Prestwich, when observing the long railway cutting at Sonning on the Great Western main line outside Reading, noticed how the uneven and wavy junction of the mottled clays with the brown sands almost exactly matched what he had seen in the Isle of Wight.[243] Prestwich, in particular, seems to have made this portion of southern England his very own geologizing territory, for he subsequently submitted to the *QJGS* a string of papers about railways sections here. One section, at Bushey in Hertfordshire, was viewed 'too late to see in good condition',[244] but elsewhere, as on the Newbury branch railway west of Reading, there were sections that were freshly dug, exposing fossil fish and other organic remains.[245] Edward Hull recalled in a paper in the *QJGS* of 1855 the discoveries made in the 'Cotteswold Hills' during construction of the Oxford, Worcester and Wolverhampton Railway in the late 1840s. A cutting in estuarine gravel near Ascott-under-Wychwood exposed the skeleton of a mammal some eighteen feet long. It turned out to be a primitive elephant and the navvies were continually digging up portions of it before they became aware of what they had found. Moreover, in the Lias Clay beneath they found the vertebrae of an Ichthyosaurus, and in the superficial gravel above the elephant's remains was a human skeleton. Thus, in a series of deposits only twenty feet thick, so Hull reminded readers, 'relics characteristic of three great epochs in the earth's history were entombed in the order of their relative ages'.[246]

Far away in Scotland, on the line of the Banff, Macduff and Turriff Extension Railway in Aberdeenshire, T. F. Jamieson, in 1859, gave an account of the navvies actually at work:

> At the time of my visit the excavation had reached a depth of from 10 to 15 feet and the 'navvies' were busy at work in the 'gullett', filling their waggons with the clay, which is a mass of a fine greenish-blue colour, very compact and tenacious . . . The most abundant fossil is the Ammonite, of all sizes, from individuals of a quarter of an inch in diameter to five inches . . . In some places every spadeful contained dozens of delicate thin-shelled Ammonites, much decayed, but still preserving their rainbow-like lustre.[247]

In southern Scotland, the construction of the Caledonian Railways's main line from Carlisle to Glasgow provided many excellent exposures where before geological observers had generally to make do with a few small quarries and brooks. Roderick Murchison remarked how it had 'laid open an excellent section of all the strata from Lockerby on the south by Beattock near Moffat'.[248]

In effect, there appeared to be no aspect of sedimentary geology (stratigraphy) that was not augmented by the work of railway excavation. The area of *dynamic* geology, of earth movements, was a much more difficult proposition, but as John Phillips remarked when describing the strata of railway cuttings at Stonesfield in Oxfordshire in 1860, 'the value of exact records of the pecularities of local sections is strongly felt by every geological reasoner who touches problems of the distribution of oceanic sediments, the boundaries of land and sea, the mixture or alternation of fresh and salt water, or the local origin and geographical diffusion of particular forms of life'.[249]

One terrible irony in the narrative that railway sections afforded the accumulation of geological knowledge was the death in September 1853 of Hugh Strickland whilst examining the cutting at the mouth of Clanborough tunnel near East Retford on the Manchester, Sheffield and Lincolnshire Railway. He was returning from attending the Hull meeting of the British Association. Notebook in hand, intent upon observation, he stepped across one line to avoid an approaching coal train, but was then killed instantly by an express train issuing from the tunnel on the opposite line.[250]

The British Association, from its foundation in 1831, proved a key in establishing geology at the leading edge of science in the middle decades of the nineteenth century. At its annual meetings, the geology section C easily drew the largest crowds.[251] And as the 1840s unfolded, it was by railway train that these crowds increasingly came. The Association's practice of selecting a different location for each of its annual meetings (the first was at York, the second at Oxford, the third at Cambridge) might otherwise have been a severe restriction on the number able to attend, but as the railway network grew, the very opposite prevailed. A correspondent of *Fraser's Magazine* in 1842 left an account of travelling to the Manchester meeting by overnight train from Euston. The train jolted, jumped, screamed and whistled as it rushed through the close and misty night. The writer's dreams became 'all smoke and fire'. It was 'an awful thing to be tied to the tail of a steam-engine'.[252] Once arrived in Manchester, he was startled and stupefied by the smoky, sulphurous atmosphere of the great cotton town. Long chimneys belched out dense, black volumes of smoke from all quarters.[253] Even so, the geology section of the meeting attracted 400 guests, many of whom went on to see the famous fossil trees on the Bolton Railway in special trains laid on by the railway's proprietors.[254] Five years later, in 1847, Charles Lyell expressed to the annual British Association (BAAS) meeting at Oxford at a mile a minute: 'it was a magical transfer from place to place'.[255] The year before, in the September, the meeting was at Southampton, where railway lines by then radiated from the town, north, east and west. Gideon Mantell travelled there by train,[256] as, almost certainly, did many of the other distinguished personages in attendance. The railway timetable was becoming as vital a part of the geologist's toolkit as the hammer and the lens.[257] Indeed, by the late 1840s, Mantell was living a life based around express timetables. On 16 May 1849, he was on the 10 o'clock from London to Brighton to deliver a lecture in the Town Hall there at 3 p.m., complete with drawings and specimens. He was back indoors at his London home by 9.30 p.m. the same day. Two days later, he was making the same lecture trip, leaving London Bridge station at noon and reaching Brighton at 2 p.m.[258] When he was not giving lectures, he was going by train to promising geological sites. Oxford's William Buckland, renowned for his fondness for field lectures, even used trains for the purpose, announcing at the end of one lecture: 'the class will meet at the GWR station at nine o'clock; when, in the train between Oxford and Bristol, I shall be able to point out and explain the several different formations we shall cross'.[259] The railway's startling and early facility in the annihilation of space by time was making it inseparable from the equally startling narratives that were unfolding about the history of the earth. Moreover, as geology itself became more and more detached from the Mosaic Cosmogony, so the railway, too, was evolving its very own signature in Nature. Railway tracks were carried across boggy fens once thought to have had their origin in the Biblical Deluge.[260] Vast mounds of earth were raised across water meadows as engineers sought a level line for the railway's permanent way. The railway became an effront to Paley's natural theology, that is to God's creation as expressed in the wonder of nature.[261] From the railway carriage window, one peered less at the wonders of Eden than at the abyss of time.

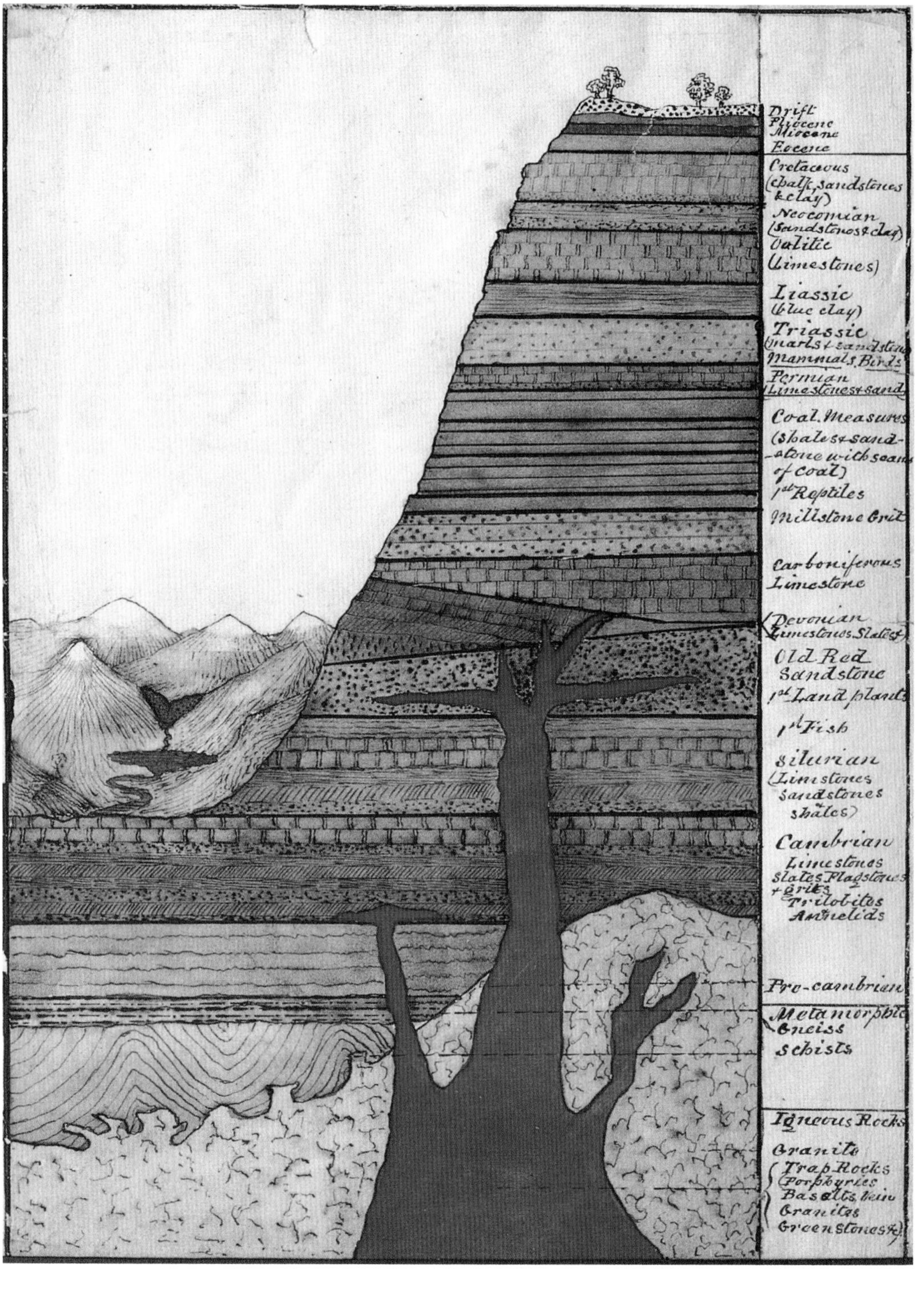

Drift
Pliocene
Miocene
Eocene

Cretaceous
(chalk, sandstones
& clay)
Neocomian
(sandstones & clay)
Oolite
(limestones)

Liassic
(blue clay)
Triassic
(marls & sandstones)
Mammals. Birds
Permian
Limestones & sand

Coal Measures
(Shales & sand-
stone with seams
of coal)
1st Reptiles
Millstone Grit

Carboniferous
Limestone

Devonian
(Limestones. Slates)

Old Red
Sandstone
1st Land plants

1st Fish

Silurian
(Limestones
Sandstones
shales)

Cambrian
Limestones
Slates Flagstones
& grits
Trilobites
Annelids

Pre-cambrian

Metamorphic
Gneiss
schists

Igneous Rocks
Granite
(Trap Rocks
Porphyries
Basalts vein
Granites
Greenstones &)

2

Time . . . that Unfathomable Abyss

Prelude

The sun was the earth's first watch-maker. The moving shadow cast by
its diurnal rising and setting was the basis for the sun-dial and, later,
the face of the clock. In turn, the earth's own satellite, the moon, provided in its four visible phases the basis of the month. Here was time
as cycle, a feature that found further echo in the seasonal routines of
farming and in the annual round of feasts and festivals.[1] But time could
also be viewed in its Biblical or Scriptural sense. That is the duration
of the earth according to Holy Writ. The earth had a beginning. At
some future date, it also had an end. Noah's Flood provided an inter-
mediate marker, as did the birth of Christ. For the Victorian generations, however, these seemingly immutable time-worlds were to be
turned upside-down. The appearance of the railway time table and,
then, railway time, was one central key. Time acquired a whole new
celerity.[2] Space melted in the face of the interlocking grid of the iron
road.[3] And so startling was contemporary time-space transformation
that Charles Dickens remarked that it was as if the sun had given in.[4]
Almost simultaneously, geology was revealing its very own time table.
Not one constructed around interactions in space, but one that peered
back in time, involving measures or quantities of time beyond all imagination. The successive geological epochs, the geological column, were
soon to 'trip off the tongue with the ready familiarity of a railway
timetable'.[5] It cast back hundreds of millions of years in a narrative
founded upon the fossilized remains of a weird and wonderful
sequence of earlier life-forms. The earth appeared to have its very own
chronology. But even those relics faded as the different geological
epochs piled one upon the other. Evidence of the most distant survives
rarely: 'only by a chain of minor miracles', according to one trusted
commentator.[6] And so Victorians were increasingly left to contemplate
a timeless abyss, unfathomable and terrifying, an extraordinary illustration of the contemporary sublime.

The Scottish natural philosopher James Hutton arguably let the
genie out of the bottle in the 1780s with his cleverly crafted model of

27 (*facing page*)
An idealized
geological column
from the mid-
Victorian period,
showing the
distinctive
succession of
sedimentary rock
types as well as
the trap (volcanic)
rocks and igneous
intrusions, the
latter prominently
identified here in
red to represent
the famous
Shropshire
dhustone, a hard
black dolerite
much used as
roadstone

an unending succession of former worlds that made biblical time appear redundant except as a footnote. His idea was re-worked first by a Scottish mathematician, John Playfair, and then, much later, by the geologist Charles Lyell, who was largely responsible for laying it open to wider public gaze. Biblical or Scriptural time, though, continued to inform more popular ideas of earth history for some decades to come. As Samuel Butler records in *The Way of All Flesh*, geology at the start of Victoria's reign remained 'just a little scare' for the average educated young person.[7] And for many among the uneducated, the biblical account was still treated literally. It was not until the dawning of glacial theory (the evidence of former ice ages) and the first full statement of the idea of organic evolution – in 1844, by Robert Chambers – that time began to be viewed seriously as endless, an extraordinary and awful abyss.

Worlds without End[8]

> Time, which measures everything in our idea, and is often deficient to our Schemes, is to nature endless and as nothing.[9]

This was the essence of a paper read to the Royal Society of Edinburgh in 1785. Its author was James Hutton, a member of the Scottish capital's intellectual elite, a man who had qualified as a doctor as a young man, had made an abortive study of the law, had had a lifetime interest in chemistry and had spent some of his middle years as a gentleman farmer, until returning to Edinburgh, his birthplace, in 1768. Hutton offered to his audience a theory of the earth. It was a theory that permitted no symptom of infancy or old age, no sign by which one might estimate either future or past duration.[10] He conceived of the earth as like a machine that was in perpetual motion. The solid earth was everywhere being wasted and degraded by the action of running water.[11] Continents were being sapped to their foundations, their material elements carried away and sunk at the bottom of seas. The process of demolition was constant. But what was destroyed could be renovated. The earth demonstrated reproductive and reforming powers.[12] The sediments that accumulated and consolidated at the bottom of oceans were uplifted to form new land masses. The cycle of degradation could thus begin all over again. A new world could be shaped from the ruins of the old.[13] And to use Hutton's own words: 'we are led to see a circulation in the matter of the globe and a system of beautiful economy in the works of nature'.[14] Here was the 'Huttonian Earth-Machine',[15] a wonderful piece of terrestrial economy.

28   James Hutton (left) in conversation with the famous chemist Joseph Black, from John Kay's *Series of Original Portraits and Caricatures*, 1842

Hutton's view of time was a cyclical one, an indefinite series of scenes that followed one upon the other.[16] Huttonian time had no historical narrative, no sequence, no progress. Just as blood cells in the human body undergo a perpetual cycle of decay and renewal, so, too, did the Huttonian Earth-Machine.[17] Just as the planetary motion of Newton formed a stately cycle of timeless repetition, so the earth described a similar motion in terms of the constant repetition of decay and renewal.[18] These outwardly parallel if differently derived perspectives had both specific and general contexts. Hutton's doctoral dissertation had been in medicine: he wrote on the circulation of the blood.[19] Much later, as a friend of James Watt, he was able to watch the slow motion of the reciprocating steam engine.[20] When Hutton described the earth as not merely a machine but an organized body,[21] he was recalling the ancient Greek concept of Gaia – the earth as a living being. And when invoking the idea of the earth as machine, he was echoing the contemporary fascination with mechanical contraptions. So it was not just the physically remote cosmology of Newton that informed Hutton's ideas but its earth-borne counterparts. The clockwork universe of Copernicus, Gallileo and Newton was being writ large in the technology of the industrial revolution.

Behind his earth machine, however, Hutton conceived of an omnipotent creator: 'the waters of land, sea and atmosphere functioned as servants of God performing an ordered plan of destruction and re-creation'.[22] Behind the inexorable cycles of decay and restoration was revealed a divine providence.[23] The Creation was not a one-off design but a world system characterized by a continuous and never-ending search for equilibrium. The earth machine enabled the fertility of the earth to be continuously maintained. Here was a specific illustration of the wonder of divine benevolence.[24] More widely, Hutton was offering a theology of nature.[25]

For the Scripturists, Hutton's theory nevertheless remained shocking. It rejected the linear vision of time within Christian thought and teaching. Hutton could trace 'no vestige of a beginning, – no prospect of an end'.[26] But the biblical story had both beginning and end. Most natural philosophers who interested themselves in the earth history took the account of the Creation in Genesis as their basic text.[27] It was a text that was constantly interrogated. It served as an almost unending repository of truth, such was the belief in Scripture's infallibility.[28] If discoveries of science cast doubt on the literal truth of the biblical account, there were nevertheless ways in which such discoveries could be harmonized. In the field of geology, for example, the Scottish churchman Thomas Chalmers argued that there had been a vast gap in time between initial creation and day one of Genesis. William Buckland subscribed to this viewpoint, whilst other scriptural geologists evolved their own characteristic perspectives.[29] For ordinary folk, however, the biblical account needed no such accommodations. They heard the story of the Creation from the pulpit on Sundays. They read it for themselves in the treasured and much-thumbed bibles that, for some households, provided the only book they knew. Genesis gave the primeval world its vital contours. And if such ordinary doubters had any further cause, there was the terrible spectre of the Day of Judgement to concentrate minds. Man's earthly paradise had a fixed duration. In the early seventeenth century, some thought that 1657 was the year of fate. Others thought in terms of a few more millennia, but few doubted the eventual imminence of conflagration.[30] Some students of the Scriptures had even managed to compute, with some apparent accuracy, the actual date of the Creation. The most well-known was James Ussher, Archbishop of Armagh, who calculated it to have been October 4004 BC, even specifying the day and the hour.[31]

29   A page from a mid-nineteenth-century English Bible, showing the date of Creation, 4004 BC

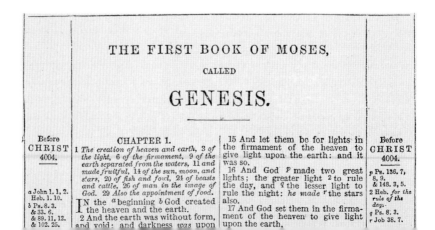

Noah's Flood was likewise dated – as December 2349 BC.[32] Both dates eventually found their way on to the margins of the Authorised Version of the Bible.[33] The idea that the earth was some five and a half thousand years old was thus widespread.

For Hutton, whilst no great antiquity could be attributed to the human race at the time he wrote, the monuments of natural history had proved that inferior species had long existed, so prompting him to 'procure a measure for the computation of a period of time extremely remote',[34] and certainly one far distant from scriptural chronologies like that of Ussher. He desired to know 'the time which had elapsed since the foundation of the present continent had been laid out at the bottom of the ocean'.[35] Fossils could become the 'clocks' to achieve this purpose.[36] He admitted that the data were deficient but found confidence in his method: '. . . we proceed in science, and shall conclude in reason . . . we seek not to know the precise meaning of anything; we only understand the limit of a thing; in knowing what it is not, either on one side or the other'.[37] He did not see himself as a 'mere empiricist', but someone whose faith was pinned to 'rational deductive methods'.[38] Hutton saw no need to invoke catastrophic causes in understanding the earth. Its former history could be adduced from a study of processes observable in the present. As Lyell was later to record, 'all past changes . . . had been brought about by the slow agency of existing causes'.[39] The mountain stream eroding its course towards the sea was a microcosm for the entire Earth-Machine. But it was a machine upon which no time limit could be placed. In Hutton's scheme, time was indefinite and unfathomable – in his own words, 'endless and as nothing'.[40] Or, as Lyell again later remarked, the imagination was at first fatigued and then overpowered in endeavouring to conceive the immensity of time required for the annihilation of worlds beyond worlds by 'so insensible a process'.[41]

The city in which James Hutton lived and studied was described by contemporaries as a modern Athens.[42] It was claimed that the spectacle of Edinburgh's spires and castles had no match in any other British city. From high on Arthur's Seat, an Athenian monarch could gaze out across its clustered towers and terraces and under the arches of its spacious bridges to the distant Pentland Hills, 'forming an aerial path from the grandeur of one place to the grandeur of another'.[43] Even in rain and mist, the city did not fail to enchant observers. When Dorothy Wordsworth first visited in September 1803, she found the Castle rock looming 'exceedingly large through the misty air', while the city itself was overhung by a cloud of black smoke that 'combined with the rain and mist to conceal the shapes of houses' and lent 'an obscurity which added much to the grandeur of the sound that proceeded from it'.[44] The spectacle reminded her of one of the vision-

30   Edinburgh Castle from
the Grassmarket, from a mid-
nineteenth-century lithograph

ary cities of the Arabian Nights.[45] Its scale was the more magnified for
being imperfectly seen.[46]

Whatever the atmospheric conditions, the city's thoroughfares
teemed with life: young dandies lounging in Princes Street, plaid and
plumed highlanders swaggering in self-importance, dour ministers of
the kirk in their parson's gray, and lowland farmers in uniform garb of
home-made blue. There were aged crown lawyers puffing and blowing
under the weight of iron-hilted swords, bonneted and feathered barris-
ters hurrying to their briefs, and attenuated striplings of the quill close
behind,[47] for the law was Edinburgh's Alpha and Omega.[48] The city
was also a renowned seat of learning, its university awash with new
ideas, a seedbed for challenging orthodoxies. In the second half of the
eighteenth century, there occurred a great intellectual upsurge, some-

times described as the Scottish Enlightenment.[49] The Edinburgh medical school was soon famous, turning out graduates like Erasmus
Darwin and his father before him. More widely, the university
embraced the sciences in a way that English universities did not.
Dissenting families found their sons favoured in its midst. And from
Edinburgh, the brightest pupils went on to attend universities on the
continent, in due course bringing back the latest in European thought
and ideas to its classrooms and lecture halls.[50]

In no other city, it seems, was there more licence for conversation,
more general freedom from all manner of restraint. Edinburgh's
Athenians were also 'the most religiously irreligious people' that could
be imagined.[51] Subscription libraries mushroomed in the wake of the
French Revolution. Apron-clad artisans read literature and philosophy
in their evenings. Gentlemen formed 'little disputing societies' where
they discussed the gravest and most profound questions.[52] Among
the younger men and apprentice boys, there were nightly drinking
bouts and much carousing. Overindulgence was, it seems, habitual
and deep.[53]

This was the vibrant city to which James Hutton had returned in
1768, to a house on St John's Hill, in the company of his three sisters
and supported by considerable independent means.[54] He was soon part
of the lively intellectual round, with the Royal Society of Edinburgh,
founded in 1783, eventually coming to form its apex. But Hutton met
some of the city's more famous intellectual sons in the clubs that had
long flourished in the city's inns and hostelries: men like the political
economist Adam Smith, the chemist Joseph Black, and the inventor of
the condensing steam engine, James Watt.[55] Hutton belonged at one
point to the Oyster Club which, for a winter season, met in what
turned out to be a house of notorious ill-repute at South Bridge.[56] It
was Hutton himself who helped organize the venue and who discovered the error when, arriving late for a meeting one evening, he accidentally came face to face with 'a whole bevy of well-dressed but
brazen-faced young ladies' making for an adjoining room.[57] The club
was much the resort of strangers who visited Edinburgh and it was by
this means, whether the objects were concerned with arts or science,
that it derived an extraordinary variety of interest.[58] In this sense, it
was powerfully emblematic of the synthetic philosophical traditions of
the Scottish Enlightenment.[59]

Hutton's theory of the earth seems to have been formulated in its
basic form perhaps as many as twenty years before its publication, in
other words at roughly the time he returned to Edinburgh to live.[60] The
City's striking topography, the combined creation of fire, water and ice,
could not have failed to influence Hutton's evolving geological ideas.
The Castle and Arthur's Seat were both sites of volcanic plugs. Calton

31   Edinburgh's striking topographical site, *c.* 1781, castle and town looking down on a shrunken freshwater lake

Hill was made up of lava flows. The Castle and the Royal Palace of Holyrood were linked by a glacial ridge. All looked down towards a shrunken freshwater loch, laden with sediments.[61] Hutton's close acquaintance with Adam Smith also had its own bearing on the theory. The famous political economist's *Wealth of Nations* (1776) presented a 'view of moral and political economy as the self-sustaining equilibrium of opposed human forces'.[62] Free Trade opened up the possibility of endless economic expansion. Hutton offered direct parallels for the 'economy' of the natural world, the opposing forces in this case provided by the cycles of erosion and deposition, the earth in reproduction and decay, set within the vastness of space and time.[63] Indeed, the force of this analogy was reinforced by Hutton's direct participation in the economic system that Adam Smith so famously delineated. For Hutton was a capitalist entrepreneur and rentier. He had established a sal ammoniac plant in the city and he received rents from houses and shops he owned.[64]

Hutton's theory of the earth seems to have been circulated in manuscript among fellow club members in the early 1780s, if not before.[65] It was then formally presented to the scientific world at two meetings of Edinburgh's Royal Society in March and April 1785, with a printed abstract made available shortly after.[66] Its reception at the Royal Society meetings was, apparently, cool. It bore too many similarities to the 'many fantastic and fabulous speculations on the earth's origin then in print'.[67] Although Joseph Black formally read out the March instalment of the paper (Hutton was ill), the prose style and its author's diffidence in championing his own cause apparently generated little excitement.[68] Much the same fate appears to have fallen the full printed version of the theory which made up part of the first volume of the Edinburgh Royal Society's *Transactions*, published in 1788.[69] Among learned reviews, one described Hutton's theory as a product of dreaming.[70] Another regarded as quite shocking the part of the paper that suggested a continuous succession of earths from all eternity.[71] Some found the theory intriguing but disagreed with much of it. Erasmus Darwin, in his much-read poem *The Botanic Garden*, footnoted Hutton's paper extensively, though he contested its geology.[72] The third edition of the *Encyclopaedia Britannica* (1797) devoted twelve pages to it, but the author of the entry roundly refuted the claims it contained.[73] One of the most influential rejections of Hutton's theory, though, came from the Swiss geologist De Luc, who had come to live in England in 1773 and published in 1778 a geology text that conceived each of the six days of the Creation as referring to former geological epochs.[74] He could find no way of contemplating that soil, vegetation and parent rock could be washed away eventually to form the strata of new land masses: 'stratum is as distinct from the original soil, as oil paint from cloth, wood from metal, over which it is laid'.[75] Nor could De Luc accept the erosive force that Hutton ascribed to river action: no gravel went into the sea from continents; rivers did not transport it.[76] More widely, he could not countenance the high antiquity attributed by Hutton to the existing continents.[77] Another high profile objector to Hutton's theory was Richard Kirwan, an Irish lawyer-turned-scientist who lived in London for a time, becoming a member of the Royal Society there. Later he settled in Dublin to become president of the Royal Irish Academy, from which position he produced a wholesale refutation of Hutton.[78] He protested, like De Luc and others, that the whole apparatus of the Huttonian Earth-Machine was flawed to its core. Worst of all was its implied atheism and infidelity.[79]

As a centre of radicalism, social, political, religious as well as intellectual, Edinburgh would have been hard to beat as a stage-set for displaying the kind of novel theoretical scheme of earth history that Hutton was advancing. But this fertile seedbed was unfortunately also the basis of Hutton's undoing. Radical talk in the classroom and taverns concentrated the minds of the political, social and religious establishments. Some among the higher orders feared deeply about the rise of Godlessness in society. And for all the deistical notions behind Hutton's Earth-Machine, its outward face spoke of heresy,[80] undermining, in a quite unequivocal way, basic Christian orthodoxies. All science deriving from speculative natural philosophy became quickly suspect.[81] Edinburgh afforded a macrocosm of such tensions in the famous King's Birthday Riot of June 1792 which embodied a complex and heady mix of popular radicalism, social theatre and deep class tension. Earlier, the building of the New Town, in all its classical architectural splendour, had allowed emerging social segregation to be expressed in spatial segregation, adding to class consciousness. The French Revolution had a profound impact on Scottish politics and society. 'Committees for the Revolution' were formed, one eyewitness recalling how 'everything, not this or that thing, but literally everything, was soaked in this one event'.[82] The King's Birthday Riot had been preceded by a flood of handbills and anonymous letters. Some of these announced how effigies of the French king and queen and other members of the French aristocracy would be paraded around the street and then burnt. Edinburgh's fiery political dissenters were soon raising their game and, in the process, progressive (Whiggish) Edinburgh gathered a powerful following.[83] However, the Tory establishment, in England as well as in Scotland, were quick to observe cracks appearing in its hierarchies of social control and thus reacted accordingly. It was once said that Edinburgh's Tories could hit a man whose politics they did not like through the medium of his banker. Whether true or not, the city's renowned legal machine went into overdrive.[84] For the country at large, treason and sedition became new watchwords. Habeas Corpus was suspended. Combination among the working classes was soon proscribed.[85] In Edinburgh itself, the clergy preached religious order as the key to social order.[86]

Against such a background, Hutton's theory quickly faded. But stung by the wave of misrepresentation and emotional revulsion and with the encouragement of friends,[87] he revised and expanded the 1788 paper to make the two-volume *Theory of the Earth, with Proofs and Illustrations*, published in 1795. A third volume remained in manuscript form until it was published posthumously, over a hundred years later, under the auspices of the Geological Society of London.[88] Unfortunately, the two-volume work was prolix in the extreme, replete with lengthy quotations from French geologists and footnotes that turned into commentaries

32   John Kay's sketch of James Hutton as field geologist, from *A Series of Original Portraits and Caricatures*, 1842

on the text itself.[89] Even so, a remarkable four or five hundred sets of the book were printed.[90] The third volume, which lay undiscovered for many years, contained material that might well have answered some of Hutton's critics in his lifetime. It afforded much of the field evidence to support his ideas and, under the eulogising of Sir Archibald Geikie, it became (in the twentieth century) the basis of a view of Hutton as the consummate field geologist, formulating his theory of the earth out of a medley of research on the ground.[91] This certainly did not directly reflect Hutton's research method, nor the manner of its formal presentation in 1785, 1788 or 1795. However, he did gain great skill as a geological observer over the years and acquired a considerable mineral collection. This came not just from the well-known field excursions made in Scotland in the company of friends in the 1780s, but from earlier tours in Cheshire, Wales, Norfolk (where he farmed for a time), the south of England and Flanders.[92] As published, though, Hutton's theory was little informed by such empirical material and, in this regard, failed to make a 'due impression on the world'.[93] Set against the powerful empiricist tradition that dominated English geological science in the first decades of the nineteenth century, Hutton was very much out of kilter. Nevertheless, his brilliance as a theorist was by no means completely unmatched by his facility as an accurate field observer.[94] He was always an avid consumer of geological facts.[95] The difficulty was that Hutton's science was 'anti-Baconian': for Hutton truth lay not in facts but in consistency of abstract ideas.[96]

James Hutton died in 1797 and most pundits of the time would likely have expected his theory of the earth to die with him. But one of his long-time friends, a mathematics professor at Edinburgh who was very much part of the city's intellectual ferment, set about revising Hutton's work and championing Huttonian theory. The man was John Playfair, whose *Illustrations of the Huttonian Theory of the Earth* of 1802 became justly celebrated. He produced a 'fresh and more lucid exposition of the theory'.[97] What is more, he managed to do so in a way that rendered it more palatable to evolving intellectual and social discourse. As geology, for example, was becoming more and more wedded to field evidence, to recovering the stratigraphic narrative, to constructing a chronology of earth history as revealed in the strata, so Playfair reconfigured Hutton's exposition in a way that *appeared* to reflect a similar pattern of enquiry.[98] Whereas Hutton produced an *ahistorical* account wrapped up in an old-style doctrine of final causes, Playfair presented the Hutton's story as if it was revealed in geological phenomena.[99] Whilst Hutton formulated his succession of dying worlds in the fashion of cosmological or abstract reasoning, looking for field evidence only rather incidentally, Playfair generated a picture of a succession of worlds drawn from material scientific evidence. In particular, he reinforced use of the concept of the geological unconformity as systematic proof of the Huttonian earth cycle. He also demonstrated, far more clearly than Hutton ever did, the landscape evidence for rivers as the chief agents of the earth's decay, generating a series of 'laws' of river systems.[100] Like Hutton, though, Playfair could not say how often the 'vicissitudes of decay and renovation' had been repeated.[101] There was no way of gauging future or past duration. The measure of human experience was evanescent by comparison, 'the momentary increment of a vast progression'.[102] Time merely performed 'the office of *integrating* the infinitesimal parts of which th[e] progression [was] made up'.[103] Here, then, was time as that horrifying abyss – deep time, as some later commentators described it.[104] Playfair continued by remarking that it was uncandid criticism for Hutton's detractors 'to load his theory with the reproach of atheism and impiety'.[105] The theory could claim a similar footing to that of Copernicus. Scriptural revelation could not 'furnish a standard of geological, any more than astronomical science'.[106] However, it was to take more than thirty years for this view finally to become widely accepted within the corridors of earth science. Moreover, in the minds of ordinary people, the Scriptures remained their standard gauge of time for much longer. They waited for revelation upon more populist commentators in the artisanal and gutter presses, and for champions like T. H. Huxley, 'Darwin's bulldog'.[107]

'Old Principles Working New Results' [108]

'Time, like an ever-rolling stream'

Church attendance in Britain remained high well into the nineteenth century. In a census taken on Sunday 30 March 1851, it was revealed that 7.25 million out of a total population of 18 million were present at some kind of service. Allowing for young children, invalids and those unable to leave their work, this was a remarkable tally, even if the religious authorities at the time were dismayed by the lowness of the figures.[109] It was while at worship, whether in church or in chapel, from pulpit or from lectern, that congregations were perpetually reminded of the story of the earth's beginning, as decreed in verse one of the Book of Genesis. How many of them noted the date of its creation, 4004 BC, shown in the margins of some bibles, is uncertain. But few would have failed to register the beginning of time, according to Holy writ. For British Protestantism was 'inveterately biblicist'.[110]

There were some features of church and chapel services, though, that might have sown doubts among more perceptive souls. The ever-popular hymn, 'O God our help in ages past', written by Isaac Watts in the mid-eighteenth century had 'time, like an ever-rolling stream' for the opening line of one of its verses, offering notions of infinitude.[111] The hymn's preoccupation with 'ages', by the 'thousand', held a clear resonance for early nineteenth-century engagements of geology and scripture. And Watts, although a Doctor of Divinity, was himself author of a book of instruction on astronomy and geography.[112] Other much-loved hymns afforded further uncomfortable reminders of new histories of the earth. In 'O Worship the King', written by Sir Robert Grant and published in 1833, at the height of the debates between geology and scripture, congregations found themselves reciting how the earth had a 'store [o]f wonders untold', established by a 'change-less decree'. Indeed, one could be forgiven for thinking that Grant's vocabulary was inspired by a reading of Hutton, Playfair or one of their later advocates: God's bountiful care streamed from the hills and descended to the plains, so the hymn continued; alongside, there was storm, dust and distillation.[113] It might easily have been a description of Hutton's Earth-Machine. Nor were Victorian hymn-writers slow to record the undermining of belief that such interpretations might signi-fy. Joseph Anstice's 'O Lord, how happy should we be', published in 1836, perhaps encapsulated the form; congregations found themselves opining that they could no longer trust Him as they should; theirs were 'faithless hearts', 'oft disturbed by anxious strife'.[114] It was left to hymn-writers like the redoubtable Mrs Alexander to try to stem the tide of unbelief. It was 'false science' that 'sang enchanted airs', so one of her 400 or so hymns recorded.[115]

The geological creed that, in the 1830s, came to haunt many a cler-gyman, not to mention many a thinking mind, was that contained in Lyell's *Principles of Geology*. Its first volume appeared in January 1830, the second in January 1832, and the third in May 1833, by which time the initial volumes had already run to new editions. It was Hutton and Playfair who provided Lyell's inspiration. The book's sub-title told all: it was an attempt to explain the former changes of the earth's surface by reference to causes now in operation.[116] And oppo-site the memorable frontispiece showing the Temple of Serapis at Puzzuoli, Lyell quoted from Playfair's *Illustrations of the Huttonian Theory*: 'Amid all the revolutions of the globe the economy of Nature has been uniform, and her laws are the only things that have resisted the general movement. The rivers and the rocks, the seas and the con-tinents, have been changed in all their parts; but the laws which direct those changes, and the rules to which they are subject, have remained invariably the same.'[117] Lyell's *Principles* was, in effect, one long argu-ment.[118] It was a text crafted in the manner of a lawyer's brief.[119] It was an elegant and passionate exposition of one viewpoint, 'hammered

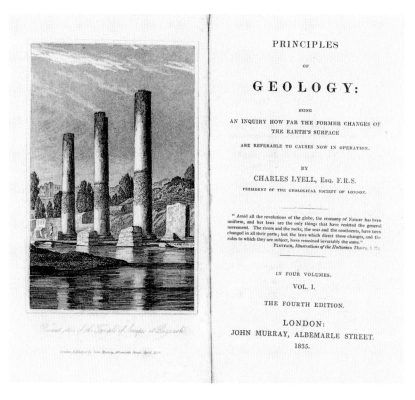

33 Title page and frontispiece of the fourth edition of Charles Lyell's *Principles* of 1835. The frontispiece, showing the ruins of a classical temple, added an air of respectability to what was otherwise a potentially heretical book.

home relentlessly'.[120] Lyell did not look for a beginning. For him, that was a metaphysical question, best left to philosophers and theologians.[121] What he sought were 'proofs of a progressive state of existence in the globe'.[122] Early geologists had misinterpreted signs of a succession of events, concluding that 'centuries were implied where the characters imported thousands of years, and thousands of years where the language of nature signified millions'.[123] Their view of geological time was imprisoned within the framework of Christian history.[124] They had consistently under-assessed and under-valued the quantity of past time. If one were to ascribe a single day for the erection of the Pyramids, one would have to infer some super-human power to be responsible.[125] If mountains were similarly viewed, they would have to be the result of some catastrophe or revolution of nature.[126] It was thus easy to think of the course of nature in earlier ages as widely different from that which Lyell himself sought to establish. The Creation and the Deluge fitted beautifully into such a scheme of things – events marked and measured in time in a way that human existence could easily apprehend. So the earth was born on the night preceding Sunday 23 October in the year 4004 BC, while Noah and his animals had joined the ark on Sunday 7 December in the year 2349 BC, the ark finally coming to rest on Mount Ararat on the morning of Wednesday 6 May the following year.[127] It was a wonderful and credible story.

One of Lyell's central difficulties in forwarding his vastly extended view of time was that the position of the observer was essentially unfavourable. All around, geologists saw not reproduction but decay: 'we know . . . that new deposits are annually formed in seas and lakes . . . but we cannot watch the progress of their formation'.[128] One could behold the transportation of rock material by a mountain torrent and so speculate, generally, about decomposition of rocks in the atmosphere. But there was no equivalent analogue when endeavouring to reason that many rocks had their origins beneath oceans.[129] The only way to study this process was through its frozen results – as evidenced in the geological strata. Another difficulty was the 'incalculable periods of time' that Lyell's thesis required for it to be sustained.[130] One needed to find a way of grasping the 'slow and insensible . . . lapse of ages'.[131] Unlike Hutton, though, Lyell had the benefit of the record of thirty more years of geological field investigation at his disposal.

34 Charles Lyell around the time of the publication of *Principles of Geology*

Sedimentary rocks in widely distant parts of Europe were found to have been formed simultaneously. The strata, sometimes wildly rich in fossils, were inexorably revealing a whole new time calendar.[132] This involved periods of vast duration. More and more, the evidence implied 'a uniformity in the laws regulating the changes of its [the earth's] surface'.[133] There was no validity in the idea that 'Nature had been at any former epoch parsimonious of time'.[134] Nor, by the same token, was there any substance in it having been 'prodigal of violence'.[135]

The reception of Lyell's *Principles* was much less condemnatory than the author himself had ever dared to imagine. Many geologists had remained fundamentally opposed to Huttonian theory, but Lyell's fluent prose and persuasively logical method attracted praise even from among those who disagreed with his views. The son of a Scottish landowner, he attended Oxford as an undergraduate and then trained for the law. However, because of poor eyesight, he chose to make geology his vocation. Soon he was at the pinnacle of the science, with its base in the salons and clubs of peers and gentlemen. How could one not listen to the ideas of an Oxford-educated lawyer who enjoyed such an elevated social position? Lyell benefited from being an accomplished field geologist, as his excursions on Wight have already shown. And unlike Hutton, he carefully sought to weave a seamless web between theory and illustration in his geological writings.[136] Despite all this, some critics still could not come to terms with the idea of limitless time. The geologist William Conybeare seriously doubted the erosional power of water, even if one threw all prejudice to the wind and allowed sufficient millions of years.[137] He mocked Lyell by suggesting that one should think not of millions of ages but 'infinit-illions' of ages in the nth power.[138] Since the Romans occupied the Vale of York, degradation had not even extended seven inches. How could raindrops and rills possibly account for the excavation of entire valleys?[139] Conybeare's beliefs lay in more catastrophic causes, involving violent currents and extended sheets of water.[140] Lyell did not himself reject the occurrence of such events, but they could be part of geological speculations of the past only as long as one did not imagine them to have been more frequent or general than in times to come.[141] They were merely part of James Hutton's terrestrial economy. The widely read *Gentleman's Magazine* was at a loss (in 1832) to understand what Lyell (in volume two of *Principles*) meant by an 'equal lapse of ages which immediately preceded our times'.[142] Its reviewer could not imagine how the elevation of strata on a continent-wide scale could occur other than by some great convulsion or catastrophe of nature. Later, though, in 1834, the same magazine, in reviewing Lyell's third volume, was more sanguine, conceding that there might be merit in the idea of 'constant and uniform change' through millions of years, even if this was 'upon a plan totally the reverse of the old geologists'.[143] The influential *Quarterly Review*, that pillar of the Anglican and Tory establishments, was (ironically) much more persuaded by Lyell's thesis. One could watch the destroying and renovating powers of the globe 'in their momentary operation' and then multiply them in the imagination 'by the effects of ages'.[144] The geologist could thus 'trace them equally on the grandest and the minutest scale – now rounding a pebble now laying the foundations of future continents'.[145] Gideon Mantell, in his *Geology of the South-East of England*, joined in Lyell's support. At Brighton, one could observe 'the ocean, silently, but incessantly . . . Carrying on the work of destruction'.[146] In a well in the lower rock gardens in the town, the molar tooth of an Asiatic elephant had been found embedded in a coarse beach material, a relic from some long-lost shore.[147]

In the event, Lyells's *Principles* proved to be wildly popular.[148] By 1840, it had gone through six editions and spawned a separate volume, *Elements of Geology* (1838), which itself went through numerous editions. The special appeal of *Principles* lay in its 'superabundance of details', so important at a period when field geology had become such a popular pastime, among women as well as men.[149] The details were also exceptionally carefully crafted, and as each new edition appeared, so new examples were incorporated and the old revised.[150] Lyell repeatedly sought to hammer home the progessive nature of the Huttonian creed. Swiss glaciers had their surfaces laden with sand and large stones, derived from the disintegration of the surrounding rocks as a result of frost action. These materials were often arranged in longitudinal ridges or mounds, sometimes thirty or forty feet high. Eventually the whole accumulation would be slowly conveyed to lower valleys where, once snow and ice had melted, the materials would be deposited or swept seaward by the meltwater streams.[151] On the tropical Indian sub-continent, Lyell calculated that the Ganges River discharged in a year enough mud to equal over three thousand million cubic feet of granite. This was more than equal in weight and bulk to forty-two of the Egyptian pyramids if they were considered as solid granite masses. Most of the mud was carried down in the four months of June to September. A glass of water extracted from the river in the flood season yielded mud to the proportion of one part in four.[152] Here,

35   Niagara, an imaginary view from the air, from Lyell's *Travels in North America*, 1845

then, was the manner in which new continents were made out of old. One could hardly have imagined a more vivid illustration. However, there was also the favourite example of Niagara, in all its subliminal energies. Here, so Lyell recorded, a spectacular sheet of water, split by a small island, is precipitated some 160 feet over a ledge of hard limestone, in horizontal strata, below which is a much greater thickness of soft shale that decayed away much more rapidly such that the limestone above periodically collapsed into the abyss below. Within the preceding forty years, so Lyell claimed, the line of the falls had receded by almost fifty yards. The logical inference was that the falls had once been located seven miles downstream, at Queenstown. This was perhaps 10,000 years before.[153] Equally, it was reasonable to speculate that, in the course of time, the falls would cut back to Lake Erie, twenty-five miles distant.[154]

If illustrations like the recession of Niagara quickly convinced readers of the destructive power of water, and if the Ganges River provided ample evidence of the way the foundation of new continents might be formed, there were still formidable difficulties in many minds as to the scale of time that the decay and reproduction of successive earths required – that is if the full force of Lyell's (and Hutton's) theory was to be fully accepted. In 1846, *Chambers's Journal*, the weekly popular magazine, priced 1.5*d*, remarked on the inadequacies of notions of this vast lapse of time. All that could be said, so it claimed, was that chalk strata were newer than coal strata and that they, in turn, were younger than mountain limestone.[155] Dickens's *Household Words* of 1851 tackled the issue in a rather different way – in a tale of a phantom ship on an Antediluvian cruise. Setting sail from London Bridge, leaving man behind, a thousand years were rolled back 'with every syllable uttered'. As the years receded, in their millions, so their dead inhabitants were temporarily restored: 'what is now continent has been sea before, as well as continent before, and will be sea as well as continent again'.[156] Earlier, in 1837, in the rogue 'ninth' of the famous Bridgwater Treatises,[157] which sought to reconcile science and religion, Charles Babbage had addressed the issue of the social construction of time. 'Time and change are great, only with reference to the faculties of the beings which note them. The insect of an hour, fluttering during its transient existence . . . would attribute unchanging duration to the beautiful flowers of the cistus, whose petals cover the dewey grass but a few hours after it has received the lifeless body of the gnat'.[158] Human appraisals of time constantly fell into the trap of applying human timescales to whatever was being observed. The testament of the rocks might, in general terms, provide a full 'monument of time', as the *Penny Cyclopaedia* recorded in 1838.[159] But there remained some difficulty in learning fully how to decipher that monument.

If, by 1850, there remained some for whom the idea of 'worlds without end' was untenable, they may yet have been persuaded by the poetry of Tennyson. For in the middle years of Victoria's reign, Tennyson became a critical focus of attitudes towards geology and of the engagement of science and religion. He was by then a practised field geologist and, upon taking up residence near Freshwater in the Isle of Wight late in 1853, the opportunities for geological observation were multiplied no end. The island had long provided a mecca for geological study, as already described in chapter one. Tennyson composed his first geological poem having read the account of the preliminary volume of Lyell's *Principles* in the *Quarterly Review* of 1830:

> The constant spirit of the world exults
>   In fertile change and wide variety,
> And this is my delight to live and see
>   Old principles still working new results.
> No thing is altogether old or new
>   Though all things in another form are cast

Here was the Lyellian creed in verse, 'old principles still working new results' (the uniformity of causes in time).[160] Nothing was old or new, echoing time's cycle, the idea that there was neither end nor beginning. Tennyson's most famous geological poem, however, was *In Memoriam*, composed over the sixteen years from 1833 to 1849 and published in 1850. Readers were astonished and intrigued by the ideas that it contained about the history of the earth, geologists not least among them.[161] In stanza XXXV, for example, Tennyson's thinking clearly reflected the ideas of Hutton:

> The moanings of the homeless sea,
>   The sound of streams that swift or slow
>   Draw down Aeonian hills, and sow
> The dust of continents to be;

These were themes that Lyell himself revived. But Tennyson could have read of Hutton's work from summaries in encyclopaedias.[162] He did not appear to have obtained any first-hand acquaintance with Lyell's *Principles* until late in 1836.[163] Subsequently, though, the book became a repeated source of inspiration, buttressed by a string of other contemporary publications on geology and on science and religion more widely, including the Bridgewater Treatises. From stanza LXIX, Tennyson recalled Lyell's ideas of climate change:

> I dreamed there would be spring no more,
>   That Nature's ancient power was lost:
>   The streets were black with smoke and frost,
> They chattered trifles at the door:

Here were worlds bound by fire and ice. Moreover, the theme continues in stanza CXIII: 'There rolls the deep where grew the tree'. And in the same stanza, Tennyson recalls once more the eternity of time:

> The hills are shadows, and they flow
>   From form to form, and nothing stands;
>   They melt like mist; the solid lands,
> Like clouds they shape themselves and go.

It was not Tennyson, though, but Dickens who, in the 1850s, arguably had the last word on uniformitarian time, or time as an ever-rolling stream. Time does not fly, so the writer in *Household Words* in 1853 reminded readers. 'Time is, and was, and will be the same – unchanged, unchangeable, immutable'.[164] And as if in specific illustration of Lyell's creed, time was 'calm, tranquil, unmoved by the course of centuries, and ages, and years'.[165]

'From Greenland's Icy Mountains'[166]

For the moving of large masses of rock, the most powerful engines without doubt which nature employs are glaciers.[167]

'Glacier Harbour' . . . [the] stupendous glacier, the accumulated snow and ice of ages, is about three miles from the ships; it extends for many miles along the supposed coast of Greenland in some parts in several thousand feet in thickness.[168]

The component of Lyell's theory of earth history in which it was claimed that climate had been modified in successive epochs had considerably greater contemporary resonance than one might at first imagine. It had been plain for some time that the coal-bearing Carboniferous strata were deposited at a time when the temperature of the northern hemisphere was very considerably higher than in the modern era.[169] Study of the fossil flora of coal seams had yielded 'the most extraordinary evidence of an extremely hot climate'.[170] Tree ferns had been discovered that grew to forty or fifty feet high. The petrified forms of the coal formations indicated species of a greater size than could be found in the hottest parts of the contemporary globe.[171] Lyell thus found freedom to speculate of an alternative time when much of the area of the globe, from the Poles to the 45 degree parallels, was frozen over.[172] The arctic wastes of Siberia would have extended into extensive tracts of the temperate zones, with surface ice and snow reflecting the sun's rays for the greater part of the year and so reinforcing the deep cold.[173]

36   HMS *Hecla* and *Griper*, from William Edward Parry's *Journal of a Voyage of Discovery of a North-west Passage from the Atlantic to the Pacific*, 1821

Not long after the publication of Lyell's *Principles*, there emerged within the study of geology a theory that argued that Britain and Europe had indeed once been buried beneath vast layers of land-ice. Its champion was Louis Agassiz, a Swiss palaeontologist who came to Britain in 1840 and expounded his ideas at the Glasgow meeting of the British Association for the Advancement of Science in September of that year.[174] He argued that in a period immediately preceding the 'present condition of the globe', in the Pleistocene, great masses of ice and, subsequently, glaciers, had covered much of Britain.[175] They could be traced through the occurrence of moraine deposits on valley sides and in valley floors. They were evidenced in rock striations, the result of the scouring action of hard fragments embedded in the bases of ice masses as they crept slowly forward. They also afforded a highly credible explanation for the great erratic boulders found on mountain tops, sometimes hundreds of miles from the rock formations of which they were once part. For those geologists who had remained sceptical of the erosive power of water action as proclaimed by Lyell, as well as by Hutton and Playfair before him, glacial theory came as an intriguing alternative. It was alternative, too, to the biblical flood. Agassiz was instead offering an altogether more powerful mode of landscape formation and, moreover, one that could stand *alongside* the work of water, even if it carried shades of catastrophism.[176] Glacial theory also had a more subtle and potentially much wider significance. It introduced the idea of glacial time, for some of those who followed Agassiz claimed not just one ice age but many.[177] The succession of former worlds of Huttonian theory thus encompassed various epochs when large parts of the globe were locked beneath huge ice masses. And these revelations came, by a strange as well as fortuitous coincidence, at a period when public interest in the Arctic regions of the modern world had never been greater. The Arctic became a rich source of the sublime, another of those worlds that 'puzzle, amaze, astound, enthral by their differences from our own world'.[178] Soaring ice-cliffs, ice-strewn waters and dazzling glaciers afforded one aspect of the imagery; another comprised 'mystery, coldness and vastness'.[179] Coleridge explored the theme in *The Rime of the Ancient Mariner* (1817). Although set in the Antarctic, the sources of the poet's imagery were mostly contemporary descriptions of the Arctic he had read.[180] Indeed, one of the most conspicuous characteristics of this part of the poem is its 'close adherence to actuality; and its power of striking through confused masses of recollections to the luminous point upon which they all converge';[181]

> And now there came both mist and snow,
> And it grew wondrous cold:
> And ice, mast-high, came floating by,
> As green as emerald[182]

Mary Shelley used the Arctic in *Frankenstein*, published in 1818. Indeed, the northern polar region 'frames the entire novel'.[183] In one sense, the spectacle of its frozen wastes surpassed even the most familiar beauties of nature; the human mind was exalted and the soul elevated.[184] However, the Arctic was in other ways a terrifying place, with extraordinary constrasts of darkness and light, strange and eerie visual effects and, most important of all, strange and unfamiliar sounds, all of which created a vivid backdrop to the central tale. As one of the narrators in Shelley's novel records, there was the thunder of a 'ground sea' as it rolled and swelled beneath the ice; then, 'with the mighty shock of an earthquake, it [the ice] split and cracked with a tremendous and overwhelming sound'.[185] All seemed preparation for a 'hideous death'.[186] The 'vast and irregular plains of ice' of the polar region seemed 'to have no end'.[187] Its frosts were seemingly 'eternal'. Here was a terrible abyss of nature, 'inhuman and infinite'.[188] The Arctic became a metaphor for deep time, a powerful example of space–time substitution. Human experience of its fascinations, its trials and its terrors made belief in former ice ages real beyond the most vivid of fables and fairy stories.

WINTER.	3. "Investigator."	7. House and Beacon on Whaler Point, to the right	9. "Enterprise."
	4. Snow Wall.	of which is seen a remarkable floating Iceberg.	10. Carrying Provisions to Whaler Point.
1. North East Cape.	5. Beacon erected on North East Cape.	7 a. Aurora Borealis.	11. Observatory.
2. Trapping White Foxes.	6. Leopold Harbour.	8. Captain Bird.	12. Cape Seppings.

37   'Winter', from Robert Burford's *Panorama of the Polar Regions*

Aside from 'gothick' creations like Frankenstein, the Arctic figured in the opening chapter of Charlotte Brontë's *Jane Eyre* (1847) as Jane let her imagination become lost in Bewick's description of the habitat of northern seabirds, providing respite from the taunts of her cousins at Gateshead Hall. Bewick's volume carried Jane into 'death-white realms', 'forlorn regions of dreary space', characterized by 'rigors of extreme cold'.[189] The *Penny Magazine* in March 1850 remarked on the 'extreme interest' still being attached to all subjects referring to Arctic voyages, and presented its readers with extracts from the description that accompanied Robert Burford's *Panorama of the Polar Regions*, a sequence of paintings that were exhibiting at London's Leicester Square at the time. Here desolation reigned 'triumphant'; all was 'wild order'. There were '[t]owering ice-bergs of gigantic size and the most fantastic shapes; immense hummocks; huge masses of ice formed by pressure; columns, pyramids and . . . brilliant icicles, exhibiting a thousand harmless effects of light and shade. . . . The prominent surfaces being tinged with vivid emerald and violet . . . whilst in clefts, crevices, and deep recesses [were] shades of the most intense blue'.[190] In 1860, Charles Dodgson (better known as Lewis Carroll) recorded going to see a series of 'dissolving view of the Arctic regions' whilst holidaying in Hastings. Such views were produced by the use of two lanterns, each containing a different image of the same object, and set out so that one image could gradually be substituted for the another. In effect, it was a primitive form of 'movie'.[191]

Charles Dickens's *Household Words* of 1856 included a short story set in the polar regions: *The Man on the Iceberg* told of a ship encountering a vast iceberg upon which a dead man could be seen. The berg was so large that it took the vessel five hours to circumnavigate it.[192] A few years earlier, the magazine had become embroiled in the debates about the fate of the expedition of Sir John Franklin to search for the Northwest Passage. Franklin had sailed in 1845 and was never seen again. In 1854, reports reached London that some of the crew's emaciated bodies had been discovered by eskimos. But more disturbing was that the state of the bodies suggested that survivors had in desperation resorted to cannibalism to prolong their lives.[193] Dickens refused to accept the truth of such accounts, preferring to glorify the expedition as proof of man's powers of endurance even in the face of overwhelming odds.[194]

In more comic vein, the magazine *Punch*, never one to overlook a public craze or preoccupation, speculated on climbing the North Pole and finding streets and courts formed of immense walls of ice.[195] *Punch* even had the railway mania sharing in the passion for the Arctic with a prospectus for a Great North Pole Railway. Its directors were Jack Frost Esq., Chairman of the North-west Passage and Baron Iceberg, Keeper of the Great Seal on the Northern Ocean.[196]

Fascination with Arctic regions had a much more familiar counterpart, of course, in the snow-covered Alps. The scenery of Europe's alpine regions had long been attracting travellers and tourists; indeed, it had become unambiguously part of many 'grand tours' by the late eighteenth century. It was not just the scale of the mountain summits and the stupendous form of the declivities that captured observers, but the tranquillity that could prevail, with vast snows untrod by human foot, as if in a giant frozen sleep.[197] By the early nineteenth century, though, it was alpine glaciers that were beginning to preoccupy many visitors and to offer an almost unending source of poetic and literary inspiration. In particular, it was the glacier's apparent capacity to appear static one moment and then roar into life the next that caught attention.[198] For Dorothy Wordsworth, in 1820, the Mer de Glace at Chamouni presented to her images of duration and decay, of eternity and perpetual wasting, that could have come straight out of Hutton's Earth-Machine.[199] The glacier had already been attracting artists, J. M. W. Turner among them, and poets such as Byron and Shelley were soon adding their own imaginative reactions. Shelley, especially, imbued the glaciers around Mont Blanc with a sinister, almost deadly presence. Here was a primaeval landscape in which glaciers crept like snakes. The slow-moving ice-wall, domed, pyramided and pinnacled, looking for all like a city of death, shattered and mangled all in its path. The dead as well as the living became its spoil. The ice-stream was perpetual, its melt-waters rolling down to the ocean waves.[200] Such frightening imagery, though, did not prevent travellers from making excursions on the ice, Dorothy Wordsworth among them. In parallel, Switzerland's glaciers increasingly became objects of study by scientists, including, of course, Louis Agassiz. The glacier bore upon its surface, as well as within it, a whole raft of clues about the history of the

earth, a history that transcended the memory of man: it was 'an endless scroll, a stream of time'.[201] And whilst the glacial ice might retreat
from whole valleys, the tell-tale signatures of their former presence
were everywhere to see, with Britain affording no exception.

Within a few years of Agassiz announcing his glacial theory in
Britain, some groups of geologists had apparently gone 'glacier-
mad'.[202] The geologist-editor of the *Scotsman* newspaper devoted a
string of articles to glacial landforms over two weeks in December
1840 and January 1841, all focused on the landforms around
Edinburgh as partly the creations of ice.[203] Some learned journals were
awash with geological papers on glacial topics. Geologists began turning Britain into a kind of ice-house.[204] Even Darwin caught the tide of
enthusiasm. Writing to Charles Lyell in 1841, he remarked what 'a
capital book' Agassiz's was.[205] He recalled his geological excursion in
North Wales with Adam Sedgwick in the summer of 1831. At the time,

38 The 'Mer de Glace', from
J. D. Forbes's *Travels through the
Alps of Savoy*, 1845

39   Snowdon, North Wales, from *Beauties of England and Wales*, 1812

neither of them noticed the glacial phenomena around them. But 'a house burnt down by fire did not tell its story more plainly', so he was later to reminisce.[206] Familiar valleys there had been fashioned not by the action of water but by vast masses of ice, thousands of feet thick. In the summer of 1842, Darwin spent some days based at Capel Curig, at the foot of Snowdon, examining the 'marks left by extinct glaciers'. Writing to the geologist W. H. Fitton, he observed that an extinct volcano could hardly have left more evident traces of its vast activity.[207] Glacial theory also captured the public imagination.[208] By 1850, for instance, *Chambers's Journal* was including 'The Glacial Theory' as part of its 'Popular Information on Science'.[209] It confidently asserted, contrary to some scientific opinion, that the land-ice theory explained the particular physiographic form of the Firth of Forth.[210] By 1855, the geologist John Phillips was treating readers of *Chambers's Journal* with an account of a 'glacial sea' in Yorkshire. He had icebergs floating over the whole of the Hull district. They carried with them blocks of stone from the Lake District 'now found perched on the limestone hills'.[211]

There were others, though, who were equivalently hostile. They could not comprehend the scale of Agassiz's Ice Age. It was quite inconceivable that mountain lakes could have been filled with ice to such depths. Sir Roderick Murchison, that pillar of the geological establishment, remarked, mockingly, how soon people would be claiming London's Highgate Hill as having been the seat of a glacier.[212] Even

Lyell himself wavered in acceptance of the full force of glacial or land-ice theory – perhaps because it appeared to hark back to catastrophic causes.[213]

Vestiges in Time

> The inhabitants of the globe, like all other parts of it, are subject to change. It is not only the individual that perishes, but whole species.[214]

> Everything is proved: by geology, you know. You see exactly how everything is made; how many worlds there have been; how long they have lasted; what went before, what comes next. We are a link in the chain, as inferior animals were that preceded us; we in turn shall be inferior; all that will remain of us will be some relics in a new red sandstone. This is development.[215]

The organic world was no less revealing of the enormity of time than the inorganic world of rocks. Tennyson remarked in his poem *In Memoriam* that Nature cared neither for the single life nor the type:

> 'So careful of the type?' but no.
>   From scarped cliff and quarried stone
>   She cries, 'A thousand types are gone:
> I care for nothing, all shall go.'[216]

The kingdoms of plants and animals were not guarded by some caring personality but were subject to all the 'mechanics of cycle and change' of Hutton's and Lyell's theories of the earth.[217] Organic nature, to trace Tennyson's lines further, battled only to: 'Be Blown about the desert dust/ Or sealed within the iron hills?'[218] Organisms became lost in the eternity of time, locked into the unending cycles of reproduction and decay.

Alongside this idea of extinction in time was another concept that involved species being *mutable* in time. Such notions developed a particular currency in France in the eighteenth century, although most of the men who advanced them had to contend with the wrath of orthodox theologians. There was the disturbing implication that not all plants and animals shared in the grace of Creation. One individual, Benoit de Maillet, had pointed to underlying similarities in the anatomies of different animals, suggesting that they might have changed 'one into another by specialisation of their parts during periods of time *immeasurably long*' (my emphasis).[219]

In England, Charles Darwin's grandfather, Erasmus, had also maintained a belief in the mutability of species.[220] The existence of vestigial organs in animals, without any function, implied a former function in predecessors. The existence of mutations, or 'sports', in both plant and animal kingdoms provided yet further illustration. Erasmus Darwin was roundly condemned by most of those around him. Not only were his views subversive of Christian teaching but they smacked of moral and social revolution. They appeared as contaminants from across the continent, uncomfortable reminders of the radicalism and terror of revolutionary France.[221] Such attitudes were reinforced by the work of another Frenchman, Lamarck, who became probably the first to construct a theory of evolution. It comprised a genealogical tree of the principal animal groups in which particular groups were displayed as 'branching off from others with which they shared common ancestors'.[222] In this evolutionary scheme, the simplest organisms were located at the bottom and the most complex, including man, at the top.

In the early part of the nineteenth century, despite more and more evidence coming to light indicating the possibility of species mutation, the idea, in general, remained hidden beneath a great raft of pessimism, suspicion and ridicule. It was enough to have to contend with new theories of the earth, let alone extend and explore such theories in relation to its organic kingdom. However, in October 1844, there was published by John Churchill in London a book that was, almost overnight, to transform this situation. *Vestiges of the Natural History of Creation*, from the anonymous pen of Robert Chambers, offered 'the first thorough-going presentation of evolutionary theory in England'.[223] It was not the work of a great mind, but that of a populist commentator. Robert Chambers and his older brother William founded the firm of W. and R. Chambers, an Edinburgh bookseller and publisher, in 1832. Its most important publication was probably *Chambers's Journal*, which is reputed to have sold 30,000 copies on first issue.[224] Robert Chambers's *Vestiges* was also an immediate best-seller. Within six months of publication, it had gone through four editions. After ten editions over ten years, it had sold close to 24,000 copies.[225] Ultimately, it ran to fourteen editions and almost 40,000 copies.[226] As for the number who read it, the figure is likely to have been many times more, for in the 1840s few people actually bought books, instead borrowing them from the many circulating libraries by then in existence.[227] Catalogues show that it was also found in many mechanics' institutes' libraries.[228] *Vestiges* was so successful partly because it was notorious. It was outwardly atheistical in creed, shallow and blundering in its science and, above all, socially dangerous. The spectre of Frankenstein loomed menacingly over it.[229] It could be allied to the 'infidel socialism' of the factory floor of the 'hungry 1840s'.[230] Its 'Principle of Development' gave hope to the under classes. By its creed, the aristocratic order appeared no longer divinely ordained.[231] Society could be 'driven' exactly like organic life – by laws and not by miracle.[232] However, for all the book's 'black materialism',[233] its bright red cloth binding like the cloak of the whore of Babylon,[234] it became so sought after for the plain fact that it was also a good read. One twentieth-century commentator has remarked how, today, few people bother to leaf through its pages, and yet it is actually 'a lively and energetic book', sometimes stiff in style, sometimes earnest, sometime naïve, but almost everywhere engaging.[235] Its carefully crafted narrative structure seemed to give the book a voice that was 'everywhere and nowhere simultaneously'.[236] In places, it read just like a novel.[237] Above all, though, it performed the act of reflecting a generation to itself.[238]

Chambers was a deist, much like James Hutton fifty years before. In a concluding note to the book, he records that it was his sincere desire 'to give the true view of the history of nature, with as little disturbance as possible to existing beliefs, whether philosophical or religious'.[239] He fully accepted that God created animated beings, but saw cause to reconsider how such creation was effected.[240] His argument was that organic creation was 'progressive through a long space of time'.[241]

The general point of departure for Chambers was provided by the new thinking on geology where, as already remarked, it had become clear (following Hutton, Playfair and Lyell) that 'the same laws and conditions of nature now apparent to us ha[d] existed throughout the whole time'.[242] Chambers transferred these ideas to organic nature. Systems of predation among animals, for instance, were much the same in the past as now. Adaptation among plants and animals in relation to their respective spheres of existence was likewise very similar. Chambers then reminded readers how species came to be withdrawn from the earth. The trilobite was found in only the earliest geological formations. Ammonites did not appear above the period of the great chalk seas. Hardly any species that flourished before the much more recent Tertiary geological epoch existed in present times. Setting these remarks alongside the knowledge of the gradual advance of species from the humblest to the highest, from the biologically simple to the biologically complex, Chambers determined to offer a novel mode by which 'the Divine Author' proceeded in the organic creation.[243] His argument alighted on the fact, already acknowledged by some, that although organic forms could be very different from each other, they nevertheless demonstrated 'variations of a fundamental plan'.[244] He drew attention to the way organs, although put to different uses in different animals, preserved a fundamental resemblance. Ribs in the serpent thus became a means of locomotion. He also remarked how analagous purposes were served in different animals by organs esentially different. Thus mammals breathed by lungs, fish by gills. Some

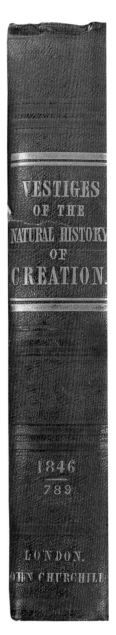

40 An example of *Vestiges of the Natural History of Creation* – the fifth edition of 1846, bound with scarlet boards

structures advanced in particular animals, even though they were not used. These tended to be more conspicuous in animals forming links between various classes.

Chambers went on to remark how 'the giraffe ha[d] in its tall neck the same number of bones with the pig, which scarcely appeare[d] to have a neck at all'.[245] Then there was the long-ridiculed idea of man having a tail: this was not in fact as silly as it appeared for the bones of such an extremity did actually exist in undeveloped state in the human subject.[246] Chambers's irresistible conclusion from this and other telling illustrations taken from the plant kingdom was that the organic creation was affected by a general law. Development was the principle by which the globe had become a repository of life as we know it: 'a process extending over a vast space of time, but which is nevertheless connected in character with the briefer process by which an individual is evoked from a simple germ'.[247] Here, then, were the same immensities of time required by the geological ideas of Hutton and his followers. Here, also, was the doctrine of macrocosm and microcosm. Drawing on a raft of sometimes disparate ideas and evidence, Chambers had produced a theory of evolution for a popular audience. Its ruling principle was 'everlasting, forward-moving change', with the stages of advance in all cases very small.[248] For Chambers, man became, in zoological terms, 'the type of all types of the animal kingdom'.[249] The inexorable logic was that 'the Adam of the human race was a baboon'.[250]

One delicious irony of *Vestiges* was that Chambers himself was a variation. Both he and his brother were hexadactyls, that is they sported six fingers on each hand and six toes on each foot.[251] Given the anonymity of the authorship of *Vestiges*, this fact never came to public view. Had it done so, one can be sure that the book would have been assailed by its critics with even greater vigour and even more sinister import than was actually the case. Even so, the still vast torrent of abuse that pursued *Vestiges* was, as Charles Darwin later observed, as good as praise for selling such a volume.[252] Disraeli had Lady Constance Rawleigh parodying it in his novel *Tancred* (1847): 'You know, all is development. The principle is perpetually going on. First there was nothing, then there was something; then, I forget the next, I think there were shells, then fishes; then we came'.[253] Adam Sedgwick spent ninety pages villifying *Vestiges* in the influential Whig journal *The Edinburgh Review*. Its philosophy was false and cast in serpent coils, so Sedgwick argued. It taught that the Bible was a fable. It annulled distinctions between the physical and the moral. The dead and the living became, by its example, a function of a 'rank, unbending, and degrading materialism'.[254] Over the years 1844, 1845 and 1846 alone, *Vestiges* attracted almost fifty separate reviews, ranging from the *Christian Examiner* and the *Methodist Quarterly* to the *North British Review*, the *Athenaeum*, *Fraser's Magazine* and the *Lancet*.[255] The Chambers brothers themselves added indirectly to the book's publicity by including in the *Journal* a whole string of articles under the headings: 'Sketches in Natural History' and 'Popular Information on Science'.[256] One of these dealt with the transmission of features from parents to their children, including six-fingeredness.[257] A later offering, in 1847, was entitled 'Nature at War' and invited readers to contemplate the 'system of reciprocative defensive and offensive warfare' in animate nature.[258] This orgy of 'confusion, anarchy, and mutual destruction' eventually gave way to equilibrium of species, 'the balance of creation'.[259]

Vestiges gave to its readers a new perspective on time, and one that, for all its extant relevance for the study of geology, found an audience that extended from the salons of the rich and famous to the reading rooms, lecture halls and taverns frequented by ordinary folk. Moreover, it is difficult not to read into *Vestiges* a 'certain supra-scientific intuitive validity'.[260] After all, why did so many reputable scientists expend so much time and energy on a book that was regarded with such derision? In its broadest meaning, *Vestiges* anticipated Darwin's *Origin*, and Darwin was not amongst those who pilloried it mercilessly – at the time that *Vestiges* appeared, he privately described the writing and arrangement of the book as admirable,[261] and in the third edition of the *Origin*, he remarked upon the 'powerful' and 'brilliant' style of *Vestiges*.[262] No matter that its method and underlying thesis were far from Darwinian, what it achieved, as Darwin himself recorded, was to call attention to the subject of evolutionary theory and to lay out the ground for 'the reception of analogous views'.[263] And as the various narratives of stratigraphy and palaeontology unfolded in the middle decades of the nineteenth century, so Darwin was able to gather more and more clues with which to refine his own ideas.

41 Advertisement for *Vestiges* from the *Quarterly Literary Advertizer* for 1849. Among John Churchill's other publications was Martin's account of the Isle of Wight Undercliff, re-emphasizing the prominence that the island held in contemporary discourse

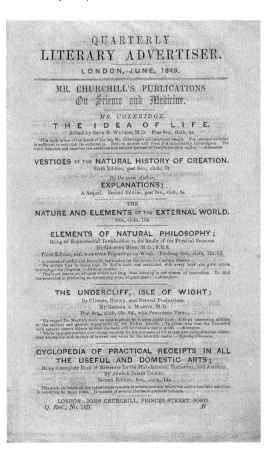

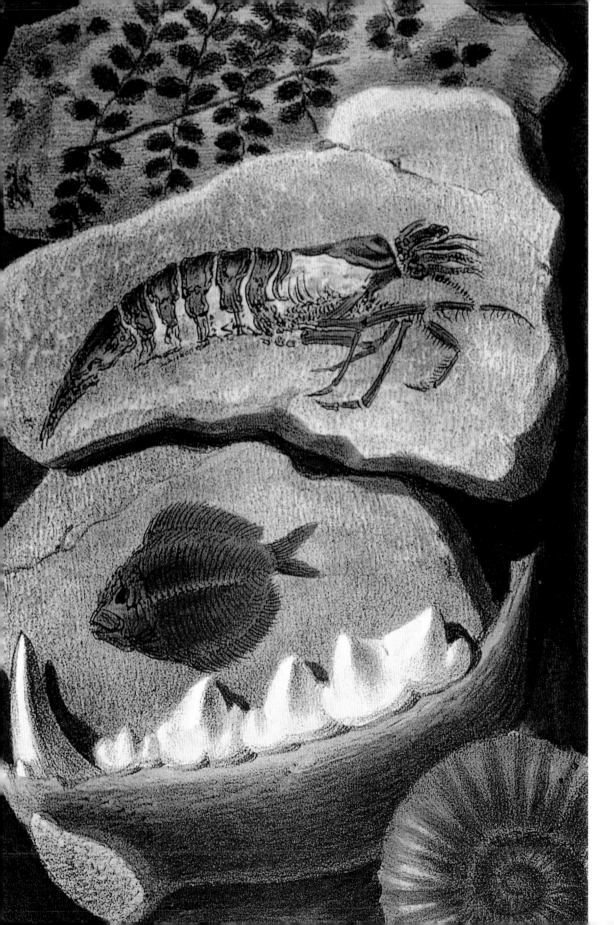

3
The Testimony of the Rocks

Ours is no coasting voyage by the sunny shores of some well-havened bay; we steer across the undiscovered oceans of truth, with compasses in need of correction, under the canopy of cloud and darkness which involves the origin of things.[1]

Discovering Eden

Nature is a vast tablet, inscribed with signs, each of which has its own significancy, and becomes poetry in the mind when read.[2]

The book in brown cloth bindings was number fifty-eight in the library of the United Presbyterian Church Literary Society. A tiny blue sticker attached to the bottom right-hand corner of the inside cover revealed that it had been purchased from a Bradford bookseller, one H. Gaskarth.[3] The title-page announced that the book was the 'thirty-ninth thousand', that is there had been thirty-eight thousand printed before it. And this was a volume that had first been published only ten years before. The book in question was *The Testimony of the Rocks* by Hugh Miller. Appearing first in 1857, it became the most popular geology text of the nineteenth century.[4] The edition belonging to the lending library of the United Presbyterian Church Literary Society had on its cover a small crest consisting of a cherub sitting astride a globe, geological hammer upraised in hand, the globe itself enveloped in a spangle of tiny stars. Hugh Miller was among the last of the line of 'scientist romantics'.[5] He developed his views out of a genuine love of nature, but was also a believer in divine providence, perhaps the final representative in a long line of scriptural or biblical geologists.[6] The crest on the book's cover reflected these features, and its subtitle, *Geology in its Bearings on the Two Theologies, Natural and Revealed,*

42 (*facing page*) The frontispiece to Gideon Mantell's *Medals of Creation*, 1844, including a coal-shale fern, a limestone fish, and part of the jaw of a hyena found in sandstone

offered a more specific guide to its contents. There were general chapters on fossil plants and on fossil animals, and more specific ones on the fossil floras of Scotland. However, the bulk of the book was given over to exploring geology in scripture. In this, Miller found no falsehoods propagated by geology in relation to the sacred text. Whether one focused on the Creation story or on the Deluge, on one phase of geological history over another, the overwhelming characteristic of both was progress. This was the essential point of union. Miller succeeded in sanitizing geological discovery for his god-fearing readers. At the same time, he turned geology into romance. Both features helped to account for the book's runaway success. No wonder some reviewers thought of it as a 'magnificent epic', the 'Principia of Geology'.[7] Even Sir Roderick Murchison, that pillar of the geological establishment, described Miller's work, generally, as 'worth a thousand didactic treatises'.[8] The creative vision of Moses was turned into high drama. Divine command saw lands rise from the deep, first as isolated reefs, then as scattered islets, later as extensive flat and marshy continents. One day the scene was of mighty forests of cone-bearing trees, in the recesses of which low, thick mists crept over dank marsh and along sluggish streams. Another day saw gigantic birds stalking sandy shores and monstrous armoured creatures heaving their bulk in clear blue seas.[9] Geology, in other words, afforded a 'sublime panorama of creation'.[10] And, finally, when morning broke on the last day of creation, there was revealed a landscape of man in which cattle and beasts of the field grazed on the plains.[11]

One of Hugh Miller's early books, *The Old Red Sandstone*, first published in 1841, went through twenty editions.[12] It told the story of a geological awakening in an Old Red Sandstone quarry in Cromarty in Scotland. Throwing aside their picks and levers, the quarrymen had resorted to gunpowder: 'not the united labours of a thousand men for more than a thousand years could have furnished a better section of the geology of the district', so Miller recorded having witnessed the blast.[13] The remains of the 'several creations' were presented for all to see, including fragments of almost every variety of rock: 'the young geologist, had he had all Europe before him, could hardly have chosen for himself a better field'.[14] Miller wrote how Nature was vast but knowledge of it limited. Thus no young man desirous of betterment could despair of not being able to add to the general fund of knowl-

43   Illustration of a fossil fish from Hugh Miller's *The Old Red Sandstone*, 1841

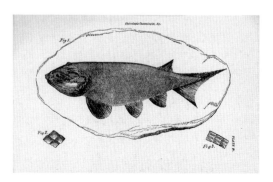

44   William Daniell, Loch Coruisk, with the Cuillins in the background, 1819

edge.[15] In fact, Miller's work on the Old Red Sandstone turned out to have singular importance in the history of British geology in the nineteenth century. He amassed a large collection of fossil fish, many of which were new to the science. Leading geologists of the day, Sedgwick, Lyell and Agassiz among them, were full of admiration for his collecting activity. Simultaneously, they could not help but marvel at the way he made his fossil fish swim and gambol as if they were living things.[16]

Not least of Miller's writings was his *Cruise of the Betsey* (1858), an account of a summer journey, largely by yacht, around the Hebrides.[17] He described the wildest and most desolate tracts of the island of Skye as if it was one of the painter John Martin's 'darker pictures': the pyramidal Cuillin Hills in a profile of black, the marshes below asleep in a blue vapour, the smooth sea reflecting the 'dusk twilight gleam to the north', the whole scene 'vast, dreamy, [and] obscure'.[18] Off the island of Eigg, through 'beautifully transparent' green water, he recorded pink sea-urchins 'warping' themselves up, green crabs 'stalking along the gravelly bottom', and jellyfish 'flapping their continuous fins of gelatine . . . a few inches under the surface'.[19] But this identical sea, 'heaving around stack and skerry', had once also thronged with reptiles – in shapes 'more strange than poet ever imagined', the dragons of fairytale and fable come to life.[20]

Hugh Miller had a famous co-worker in Philip Henry Gosse, the naturalist and Fellow of the Royal Society who sought to justify geology to godly readers of Genesis in his book *Omphalos*, published in 1857.[21] The book's basic notion was that all created things must bear some sign of a 'pre-created' state. Gosse saw the shoreline near his South Devon home, with its 'great prawns gliding like transparent launches' and its streams of seaweed fronds, as the Garden of Eden, as if 'Adam and Eve, stepping lightly down to bathe in the rainbow-coloured spray, would have seen the identical sights'.[22] But *Omphalos* offended everyone. The manner of its attempted reconciliation of geology and scripture, including its rejection of the mutability of species, was universally scorned; its thesis was untestable. Once a favourite of the press and the public for his lectures and his prolific writings on the natural world (especially on marine zoology[23]), Gosse found his nemesis. Atheists and Christians each looked at *Omphalos*, but then laughed and threw it away.[24] The man who T. H. Huxley once called the 'honest hodman of science', the expert collector and observer, author of poetic works on nature, in the end remained a humble slave to revelation.[25]

Both Miller and Gosse exemplified in a broader sense the way in which religion and science had remained deeply entangled for almost a century. As one modern commentator has put it, the difficulty as reflected in the scientific literature was 'one of religion . . . *in* science, rather one of religion *versus* science'.[26] More recently, it has been revealed that the relationship between religion and science in the early and mid-nineteenth century was both extraordinarily rich and complex.[27] It may even be possible to claim that the interaction between science and religion was actually advantageous to both.[28] From such distant perspectives, though, it is easy to lose sight of the degree to which nineteenth-century geology, particularly in its broader public face, 'remained embedded in controversy about Creation, the Fall and the Flood'.[29] William Buckland, founder of the so-called Oxford School of Geology, for a long while contended that 'geology evidenced divine design and a universal Deluge'.[30] Charles Lyell, though, sought vigorously to disconnect geology from scripture and was later to complain that Britain, generally, was more 'parson-ridden' than any European country apart from Spain.[31] However, there was no denying that the English clergy were at the heart of the establishment. They remained among the rulers of society. Church and State were deeply intertwined at all echelons of the institutional and social scale. The fact that so many members of the clergy themselves took an active interest in geological pursuits added to the brew. Then there was the residual belief among natural philosophers and early scientists in divine providence, in the doctrine of a prime cause. Both Hutton and Chambers, for all the atheistical labels attached to them, openly reflected such views. Lyell's *Principles* afforded 'new proofs of divine power and goodness', and, in this sense, was not without 'theological triumph'.[32] Even Darwin echoed the form, for example, in his capitalization of the word 'nature' in the *Origin of Species*.[33] The mid-Victorian realm, in other words, was still essentially a Christian culture.[34]

One feature of the emerging science of geology that placed its relationship with orthodox religion in a position of particular ambiguity was the use of labels from classical mythology. The followers of Abraham Werner, the distinguished Saxon mineralogist, subscribed to a theory of the earth that, in the words of one contemporary observer, claimed it to have once been covered with 'a sort of chaotic compost, holding in either solution or suspension the various rocks and strata which now present themselves as its exterior crust'.[35] They acquired the label 'Neptunists', after the Roman god of the sea. In opposition were a group of geologists known as 'Plutonists', after the Greek god of the underworld. They held a theory of the earth that saw subterranean fires, a species of vulcanism, as the overwhelming force that shaped it.[36] It was the action of heat that explained the consolidation of sediments at the bottom of oceans, and their subsequent uplift. Thus began the famous Huttonian cycle of restoration and decay. And Hutton's *Theory of the Earth* in effect became the focus for the 'Plutonist' school.[37]

Precisely when and how these particular labels came to be applied is not clear. In general terms, they reflected the significance of a classical education for many writers on natural philosophy and geology. But what they did was to offer reminders of false gods. No matter that Neptunism could be loosely allied to Noah's Flood, or that Plutonism could be allied to hell-fire judgement. And in an age of industrial revolution, such false gods readily found proxy in newly discovered industrial processes. Careful observation of these helped to underpin the possible scientific bases of the opposing sets of ideas. They also afforded no end of artistic inspiration, whether the apocalyptic scenes of John Martin's paintings and mezzotints, recalling the cyclical incidence of catastrophe in earth history, or startling canvases of industrial districts by night, in all their elaborate firework motifs, as found in the paintings of Phillipe de Loutherbourg. Travelling by night through the Black Country, for instance, brought the regions of Pluto to the imagination for some contemporary observers.[38] John

45   Detail of John Martin's *The Great Day of His Wrath*, 1852

Martin even used his local knowledge of the drift-tunnels of the Tyne's Allendale Gorge and the coal pits of Jarrow and Wallsend to set out the contours of the underworld, the hollow labyrinth of Hell.[39] More widely, storms, volcanic eruptions, earthquakes and other natural disasters 'swept like tidal waves through early nineteenth-century periodicals, broadsheets [and] panoramas'.[40] Catastrophic and apocalyptic visions acquired a remarkable common currency, the Malthusian spectre a constant reminder of the need for atonement.[41] For some onlookers, Martin's most famous canvases of divine revelation seemed simultaneously to encode new geological and astronomical truths. This was especially true of his *Deluge* (1826 and 1834),[42] and was even more powerfully demonstrated in *The Great Day of his Wrath* (1852), in which the Edinburgh of James Hutton, with its grand citadel, hilltop terraces and spectacular volcanic landscape, explodes outwards and appears suspended upside-down, flags still flying from its buildings, before crashing head-on into the valley below.[43] If this seemed a far cry from Hugh Miller's romantic prose, it remained true that Martin was equally adept in depicting visions of Eden. From his birthplace at Haydon Bridge in Northumberland, the Tyne River runs seaward through a broad, almost park-like valley, an earthly paradise that afforded the artist repeated inspiration for his historical paintings.[44] Just as Miller could sanitize geology to make it comfortable for Christian souls, so could Martin conjure from his native Northumberland Arcadian images of hope and renewal.

Reading the Earth

> It is the uniformity of superimposition, the invariable order of succession, sometime disturbed but never inverted, on which geology depends as a practical science.[45]

Biblical scholars like Ussher evolved their precise chronologies of earth history with the aid of great interpretative ingenuity.[46] When geologists began to try to do likewise with the structure of rocks, the task required them to be no less ingenious, but the difference was that they had material evidence all around them that could be repeatedly interrogated and compared. Miners and quarrymen had long recognized that rocks displayed some sort of uniformity, that there was a kind of order of succession. Early writers on natural history made similar observations. John Morton, for example, writing on Northamptonshire in 1712, detailed his method of studying the local ironstone, noting the order of the strata, 'beginning with the uppermost and so proceeding downwards to the bottom of the pit'.[47] Sometimes he observed disruptions to the strata, but before this 'they were continuous, their sites were often more uniform, eaven, and parallel to each other'.[48] Much later in the century, John Whitehurst appended to his *Inquiry* into the history of the earth an especially detailed account of the Derbyshire strata, much of it culled from experienced miners.[49] Here, stress lay on the *progressive* nature of the earth's formation and, once again, on the general *conformity* of strata.[50] This, too, had become the view of most members of the Geological Society of London, following its founding in 1807. Their mentor was Abraham Werner, whose continental teachings on the universal order of rock formations remained highly influential.[51] By the 1840s, G. F. Richardson, in his *Geology for*

COMPARATIVE   GEOLOGY.
Or a Familiar method of illustrating the Vertical and lateral Positions of Rocks.

46   A method of illustrating the order and disposition of rocks, the Bible used as keystone – from Granville Penn's *Conversations in Geology*, 1828

Beginners (1842), was persuaded to argue that 'the constant and unvaried sequence of the several geological formations' was one of the most important and useful lessons that the science was calculated to convey.[52]

One singular feature of the study of geology in England was that the country presented an 'epitome . . . of the greater part of the regular solid strata of the globe – a model in miniature nicely arranged and accurately defined, of the rocky masses, which in other countries spread over a large surface, or . . . raised into lofty chains of mountains'.[53] As travel in England became easier and more habitual, this feature became progressively more apparent, not least because of the way the lines of new turnpikes and railways could necessitate cutting through rock strata, as chapter one has already revealed. Phillips and Conybeare made much of this in their best-selling *Outlines of Geology*, published in 1822:

> If we suppose an intelligent traveller taking his departure from our metropolis, to make from that point several successive journies to various parts of the island, for instance, to South Wales, or to North Wales, or to Cumberland or to Northumberland, he cannot fail to notice (if he pays any attention to the physical geography of the country through which he passes) that before he arrives at the districts in which coal is found, he will first pass a tract of clay and sand; that he will observe next numerous quarries of the calcareous freestone employed in architectur; that he will afterwards pass a broad zone of red marly sand; and beyond this will find himself in

the midst of coal mines and iron furnaces. This order he will find to be invariably the same, whichever of the routes above indicated he pursues.[54]

The authors continued that, beyond the coal deposits, one typically came across formations of limestone hills, 'abounding in mines of lead and zinc'; and beyond these were mountainous tracts in which roofing slate, for instance, was all around.[55] Once one had generalized these various observations, the inevitable conclusion was that such 'concidences' could not have been 'casual'.[56] They indicated a 'regular succession and order in the arrangement of the mineral masses constituting the earth's surface'.[57]

The arch-exponent of this 'order' in Britain, of course, was the surveyor William Smith, whose practical work on canal and drainage projects, beginning in the 1790s, revealed much the same 'concidences' in the pattern of rocks across the country.[58] By 1800, his 'card' of the order of English strata had disseminated widely, including his particular system for identifying strata by the fossils they contained.[59] Smith's work had direct impact on the Board of Agriculture's new county surveys, with some of them – for example, John Farey's account of Derbyshire (1811–17) – forming model geological studies for their day.[60] Indeed, almost all of the first volume was taken up with rock strata and minerals.

The order of the strata was a function of the way rocks were formed by 'the motion of water, which arranges them in succession over each other in the same manner as the muddy waves of the ocean deposit their contents in regular layers upon the shore'. Or, at least, this is how Robert Bakewell's popular introduction to geology of 1813 described the process.[61] Moreover, there was no influence more legitimate than the one stating that 'the rock which supports another must be older than that which rests upon it, if their original position has not been changed'.[62] Chalk was the deposit of perhaps the deepest of the various oceans, traceable from the north of Ireland across to the Crimea and from Sweden down to southern France.[63]

Outlines of Geology's 'order of succession' became the object of part of 'A Geological Primer in Verse', addressed to the professors and students of the University of Oxford in 1820:

> When Nature was young, and the Earth in her prime,
> All the Rocks were invited with Neptune to dine.
> On his green bed of state he was gracefully seated,
> And each as they enter'd was civilly greeted.
>
> . . . Oolites, with sandstones, and sand red and green
> In a crowd, near the top of the table were seen.
> The last that were seated were Chalk-marl and Chalk.

They were placed close to Neptune to keep him in talk.
Now the God gave his orders, "If more guests should come,
let them dine with the lakes, in a separate room."[64]

In not dissimiliar manner, the satirical magazine *Punch* discovered in
the popularity of geology a never-ending source of copy and quickly
adapted the idea of the succession of rocks to the study of society. Thus
the 'primitive formation' provided the bases and support of all other
(higher) classes. They came in 'tag-rag' and 'bob-tail' varieties, 'distinguished by their ragged surface, and shocking bad hats'.[65]

However, the order of succession was not always quite what it
seemed. For George Greenough, a rank empiricist if ever there was
one,[66] stratification was a subject 'pregnant with controversy'.[67] It was
not enough, for instance, that a rock substance divided into parallel
planes. It had to do so as a function of the manner of its formation
from a fluid mass.[68] Otherwise one might claim that granite was stratified. Gideon Mantell, writing on the geology of south-east England,
commented how displacement and disintegration of strata made for
considerable difficulties of investigation.[69] Junction lines of strata were
also often covered by soil and vegetation. Charles Lyell, in his fieldwork on Wight in the summer of 1823, wrote to Mantell of the way
'almost the whole of the back of the Isle of Wight [was] in such a
dreadful state of ruin that . . . it ha[d] been the cause of much of the
confusion . . . with regard to your [i.e. Mantell's] Sussex beds.'[70]
John Stuart Mill saw a parallel here between the development of
geological and legal history:

> The deposits of each successive period [are] not substituted but
> superimposed on those of the preceding. And in the world of law
> no less than in the physical world, every commotion and conflict of
> the elements has left its mark behind in some irregularity of the
> strata: every struggle which ever rent the bosom of society is
> apparent in the disjointed condition of the part of the field of law
> which covers the spot; nay the very traps and pitfalls which one con-
> tending party set for another are still standing.[71]

For Mantell, where the relative position of rocks was involved in
obscurity, presenting 'insuperable obstacles' to accurate examination,
one could but proceed by replacing observation with 'induction and
analogy'.[72] Underpinning such an approach was the geological map
that effectively 'realized a *theory* of geological knowledge' that could
be predictive as well as interpretative.[73] William Smith was the pioneer
in such work, as chapter one has again shown,[74] but others soon added
to the framework and, later, the task became the preoccupation of the
Geological Survey. In the interim, however, there remained (perhaps
inevitably so) differences and conflicts of interpretation among leading

geological observers, in a few cases giving rise to major disputes, as in the case of the Great Devonian controversy. This centred on the so-called 'transition' rocks below the Carboniferous deposits which displayed little of the distinctive layering that was so prominent a characteristic of the great mass of Secondary strata above, the rock strata that had by then been so successfully mapped.[75] The controversy exercised the minds of the leading geologists of the day and exemplified the enormous intellectual puzzles that heavily folded and faulted rock strata, with relatively few or poorly preserved fossils, presented to those seeking to describe and interpret them.

The order of the strata, or stratification for that matter, was clearly not a concept that was relevant for rock groups deriving from the action of 'subterranean fires', or heat. Rocks like granite and basalt, although sometimes displaying spectacular structural properties, did not bear relationship to any process of marine or fluvial deposition. *The Geological Primer in Verse* of 1820 picked up the story in its further account of Neptune's dinner:

> . . . No rock from his seat
> Ever moved, or evinced the least wish to retreat;
> And old Neptune found out, as the wise ones aver,
> When the rocks are once seated, they love not to stir.
> So he rose unobserved, and began to retire;
> But 'tis whispered the Sea-God already smelt fire.
> Be this as it may – a deep hollow sound,
> Still nearer and nearer was heard under ground;
> 'Twas the chariot of Pluto, – in whirlwind flame
> Through a rent in the earth to the dinner he came' . . .
>
> Thrice he stamp'd in a rage, and with washes like thunder
> The earth open'd wide, and the rocks burst asunder
> And the red streaming lava flow'd over and under.[76]

Such convulsions were considered to be not unconnected with the great changes of levels that strata in general displayed. With marine remains discovered in the highest locations, for example, a fundamental problem of theoretical geology was to assign adequate causes for changes of level that allowed the sedimentary mass of an ocean's floor 'to rise in hills and mountains above its waves'.[77] A very different kind of convulsion was required in order to explain how, over all strata, there was spread, indiscriminately, a covering of gravel. This was 'seemingly formed by the action of a deluge which ha[d] detached and rounded by attrition fragments of the rock over which it swept'.[78] The same authority was so certain of the 'fact of an universal deluge at no very remote period' that the claim was advanced that, had geologists never heard of such an event in Scripture, they would have

required the 'assistance of some such catastrophe to explain the phenomena of diluvial action'.[79]

Artistic Visions

> A little hollow excavated round
> To a few fishing boats give anchorage ground
> Guarded with bristling rocks whose strata rise
> Like vitrified scoria to southern skies
> Called Lulworth cove. . . .[80]

While miners, quarrymen, natural historians and early geologists puz-
zled over the order of the different rock strata that comprised the surface of the earth, artists were drawing and painting them. They were
attracted by the idea of singularity in nature, and by 'accidents of
nature'.[81] This included the startling structural symmetry of certain
rock formations, the scale of rock structures, their sometimes striking
colouration, and the way rock masses could display extraordinary
patterns of folding or contortion. Such features helped to constitute
visions of the sublime and the picturesque, marking the beginning of a
'union between art and geology'.[82] It was not just coastal cliffs like
those revealed on Wight that answered to such visions. Inland, one
could find scarp and cliff faces, deeply incised valleys and gorges, and
cave systems, not to mention waterfalls, with their coincident focus on
differences or discontinuities of rock structure. Artists soon began to
register how the various rock types were associated with distinctive
sorts of scenery, whether in the form of the slates of the mountains of
North Wales or the volcanic basalts of the coasts and isles of western
Scotland. Just as William Smith had learned to pick out chalk by its
contour,[83] so painters' eyes began to register similar variations of landscape form. Nowhere was this more apparent than in the
Derbyshire peaks and dales. Visitors had long been attracted to this
region of Britain. Its scenery had a kind of alpine grandeur, littered
with cliffs and chasms, in places heavily wooded, and threaded by
mountain streams. There were also its famous underground caverns, a
perpetual fascination for tourists.[84] When the antiquarian and artist Sir
Richard Colt Hoare visited the district around Castleton in 1800, he
described the rocks around as 'grotesque and majestic' in form.[85] In
parts, they could almost have been the 'mighty ruins of some huge
fortress', erected for the purpose of guarding the pass he found himself
descending.[86] After visiting several of Castleton's caverns, Colt Hoare
remarked how their beauty was the greater the more candles there
were to illuminate them. In one particular cave, you embarked upon a

small boat, laid yourself down flat, and were pushed by a guide to larger and yet more spectacular interiors.[87] In the equally famous Peak Cavern, according to one account, gunpowder was sometimes used to replicate the sound of thunder, adding to the power of the sublime vision that greeted visitors.[88] The clue to such distinctive scenery was the mountain limestone from which it was created. Of all rock types, limestone is especially susceptible to solution weathering, that is the dissolving of its elements by rainwater. This occurs mainly along lines of jointing, helping to create a very distinctive surface appearance and leading, ultimately, to extensive underground drainage and the formation of subterranean streams. The cave systems were part of this network and the modern surface streams were but relics of formerly subterranean channels. Among the many artists who went there to paint was Joseph Wright of Derby, famous for his renderings of Vesuvius in eruption and for depictions of industry at work, not to

47   Joseph Wright of Derby's *Matlock Dale, c.* 1780–85

mention portraiture.[89] In the 1780s, he painted both Matlock Tor and Matlock Dale. Each picture offered a detailed and accurate portrayal in oil of the scale and form of the region's limestone strata, as well as conforming to ideas of what was then considered picturesque. In fact, Wright played specifically to the latter by also producing an oil painting of the Tor by moonlight. Some artists actually visited Derbyshire in the company of geologists. John Webber, for instance, painted Thurshouse Tor in Dovedale in 1789 whilst touring with the artist–geologist William Day.[90] The distinctive appearance of the limestone masses was keenly depicted, even if vertical scale was exaggerated to convey the legendary grandeur of dale scenery, everywhere 'wild and grotesque', according to one topographical commentator.[91] Dovedale also attracted Joseph Wright's artistic energies. At one point along the river valley, the rock pinnacles take on the appearance of medieval architecture, acquiring the name 'Tissington Spires'. Byron was later to write of Dovedale in tones that made it as noble as anything that one might see in Switzerland or in Greece.[92] It was no wonder that tourists flocked to view it. Wright's knowledge of the geology of Derbyshire's peaks and dales was almost certainly more than just the result of an artists's curiosity. For he knew John Whitehurst, author of *Inquiry into the Original State and Formation of the Earth* (1778), and painted him in a pose that showed the geologist looking over a section diagram of the strata at Matlock High Tor.[93] The portrait was completed around 1782–3 and Wright's various paintings of Derbyshire scenery mostly followed directly after.

Another tract of limestone country that held special appeal was Cheddar, in Somerset. Here, according to a commentary in the *Gentleman's Magazine*, was a sublime and tremendous scene. The vast opening of the rocky ribs of the Mendip Hills presented a striking combination of 'precipices, rocks, and caverns, of terrifying descent, fantastic form, and gloomy vacuity'.[94] The gorge itself was wonderfully captured on canvas by George Vincent in the early nineteenth century.[95] The picture forms one of the most accurate pieces of geological draughtsmanship that oil painting could provide. The rock masses are unmistakeably limestone from their massed jointings. Moreover, Vincent exploits the natural dip of the strata and the opposite downward slope of the roadway to give a diagonal focus for the eye. This focus takes the form of a sharp cleft in the left centre of the picture, a first glimpse of the giant gorge that lays just round the bend. Contemporary with this picture of Cheddar, the Bristol painter, Francis Danby also completed various views of the Avon Gorge, described by the artist Henry Fuseli as 'the finest scenery in the kingdom'.[96] The gorge was again chanelled out of limestone and demonstrated something of the blend of towering rock faces and luxuriant foliage that gave Derbyshire its special attraction. This was most apparent in Danby's *St. Vincent's Rocks and the Avon Gorge* (*c.* 1815), and also in *Hotwells in the Avon Gorge* (*c.* 1815). In the former picture, a combination of watercolour and pencil, the limestone massif made up two-thirds of the composition. In the latter picture, the eye focused more on the river, but to the left the rock bluff still towered over the scene. The two pictures actually formed a pair. The viewing point for each is identical, except that one looks down river, the other up.[97]

When painters took the mountains of Snowdonia as their subject, they were arguably more interested in scale, a primary requisite of the sublime, than in rock type and rock structure. However, the relationship between the two could not have been altogether lost on observers. The slate rocks of North Wales gave a distinctive line to the topographical form, not to mention the distinctive angularities arising from the erosive agency of glaciation. Phillipe de Loutherbourg's *Snowdon from Capel Curig* (1787) illustrated this powerfully, the more so given

48   George Vincent, *View of Cheddar Gorge*, n.d.

49   Phillipe de Loutherbourg, *Snowdon from Capel Curig*, 1787

the very large size of the canvas, almost seven feet wide. *Tryfan, Carnarvonshire* by George Fennel Robson (*c.* 1810–15) afforded another example. When Sir Richard Colt Hoare first saw the peak while journeying in North Wales in 1810, he was much struck by its curious form, the mountainsides fissured and barren of verdure and foliage.[98] G. F. Robson also applied his skills to the depiction of the Cuillin Hills of Skye in western Scotland in *Loch Coruisk, Skye – Dawn*, painted around 1826–32. Here the volcanic rocks again lend an alpine character (unusually so for Scotland) to the form of the mountainscape, a feature thrown into yet sharper relief by their black colouring and by the horizontal line of the loch that provides the other principal element of Robson's picture. The geologist, John MacCulloch had earlier been especially struck by the 'spiry and rugged' outline of the Cuillins and the way they had a uniform dark appearance, their summits enveloped in an almost perpetual mist.[99] Rather later in the nineteenth century, William Dyce registered how the mountains of Wales, often being composed of slate rock, generated 'more rugged, stony and precipitous' scenery.[100] The Welsh slate split and tumbled, leaving sharp and angular peaks, unlike the older rocks of much of Scotland which were 'rounded and their asperities

50   G. F. Robson, *Loch Coruisk, Skye – Dawn*, 1826–32

smoothed down'.[101] This was clearly displayed in Dyce's own picture of Tryfan (1860). Here the eye was immediately struck by the Swiss-style arrete that ran from left to right across the painting. And stories that travellers told of the perilous nature of the ascent of the mountain added to the fearful line of the view.[102]

Scale and form within mountainous districts also preoccupied J. M. W. Turner. But Turner sought to go further in order to bring into focus not just the physical masses but the immensity of nature's space. This involved the use of colouring and painting techniques that made such space atmospheric, even ethereal.[103] The object was to create skies of immeasurable depth, mountains that soared and, all told, a palpable sense of awe in the face of the viewer. *Dolbadern Castle, North Wales* (1800), located between the Llanberis Lakes, illustrated the phenomenon, as also did his earlier painting *Buttermere Lake* (1798).[104] Like some of John Martin's grand pictures, Turner sought suggestions of cosmic infinitude that were profoundly relevant to the contemporary geological imagination.[105]

On the west coast of Scotland, it was the remarkable structural symmetry of some rock masses that caught the eyes of artists, as much as the scale of the landscapes of which they were part. Probably the most

51 William Daniell, *Entrance to Fingal's Cave, Staffa*, 1817

prolific recorder of Scottish coastal scenes at this time was William
Daniell. For example, his eight-volume *Voyage Round Great Britain*
(1814–25) contained a total of 308 plates and Scotland's coasts figured
in four of the volumes.[106] Daniell was an accomplished artist and
engraver as well as a seasoned traveller both in Britain and abroad.
Some of the paintings that resulted from these excursions were exhibited at the Royal Academy, and in 1822 he became a full academician,
elected in that year in preference to John Constable.[107] The feature of
Scotland's west coast that had begun attracting travellers and artists
was the columnar basalt. It was displayed at its most spectacular on the
island of Staffa, first made famous by Joseph Banks after he had
explored the great sea cave there in 1772. Soon there was a string of
engravings of Staffa, especially of the 'Cave of Fingal', These were
often used as illustrations in tourist accounts or companions.[108] To
contemporary minds, the most striking feature of the basalt columns
of Staffa was their appearance of design. That is they accorded
with ideas of Natural Theology, with the notion of a divine creator.
Simultaneously, though, they tapped into debates about the origins of
the earth, and, in particular, about internal heat as a cause of landscape
formation and upheaval. In the opinion of some, the basalt of Staffa
and elsewhere were none other than lava remnants, the results of vulcanism. This is indeed their origin. At the time, though, there were also
those who claimed they were merely different forms of sediments. One
outcome of these various divine and scientific associations was that

52   Fingal's Cave as pictorial wonder, from the *Penny Magazine*, 1832

when artists came to depict subjects like Staffa they often resorted to techniques of 'artful construction',[109] overemphasizing symmetry, regularity and scale so as to produce a kind of pictorial wonder. Thus Fingal's Cave was transformed into a cathedral nave, with William Daniell initially among the exponents (1807).[110] However, the plates in Daniell's *Voyage* reveal rather less in the way of pictorial licence. Indeed, in *Entrance to Fingal's Cave, Staffa* (1817), Daniell almost conforms to being a geological draughtsman, the whole effect enhanced by the uncharacteristically calm sea. His attention to detail clearly reflected his reading of John MacCulloch's geological paper on Staffa published in the Geological Society's *Transactions* in 1814.[111] What attracted Daniell, in particular, was MacCulloch's plain and direct language, largely devoid of 'hyberbolic comparisons'.[112] The resulting plates seem to echo this feature. Although few of Daniell's other 300-odd plates matched the fine draughting of those of Staffa, his general sensitivity towards geological structure remained. On the north coast of Devon, for instance, the plates of Ilfracombe and Combmartin arrest the eye in their representation of the dip of the strata.[113] Off the coasts of western Scotland, the columnar basalt re-appears in plates of the Shiant Isles.[114] Northward still, *Entrance to the Cave of Smowe* and *Whiten-head, Loch Eribol* begin to help to illustrate the variety of Scottish coastal geology.[115]

The *Penny Magazine* described Staffa as 'one of the most wonderful natural curiosities of the world', the cave's interior supported on each

53 J. M. W. Turner's illustration of Fingal's Cave for *The Poetical Works of Walter Scott*, 10, 1834

side by the same perpendicular columns of basalt, extending 227 feet into the island's mass, its 'twilight gloom' and 'transparent green water' lending to the scene a 'profound and fairy solitude'. A 'mind gifted with any sense of beauty in art or in nature' could not fail to have been impressed.[116] Staffa, of course, was to become famous through the poetry of Walter Scott. His *The Lord of the Isles* (1815), for instance, incorporated a record of a journey along the west coast and islands in 1814. And Turner was later to visit the area (in 1831) to make drawings for an illustrated edition of Scott's poetry.[117] At the same time, he also completed the oil painting *Staffa, Fingal's Cave*, exhibited at the Royal Academy in 1832. The latter forms a powerfully atmospheric picture and the island is barely distinguishable. However, his illustration *Fingal's Cave*, for Scott's *The Complete Poetical Works* (1834), shows it from a previously unexplored angle, looking from the inside outward. The sides and roof of the cave are drawn with uncanny accuracy. At the same time, Turner demonstrates the effect of a strong head sea as it washes powerfully into the confined basin of the cavern.[118]

The singularity of sites like Staffa as natural phenomena was often more than matched by inland waterfalls, located not only where rock structures were presented bare to the observer's eye, but where the very existence of the fall could be directly related to differences or disjunctures in those rock structures. Nowhere is this more apparent than in John Pye's engraving of Turner's watercolour of *Hardraw Fall* (1818), prepared for Whitaker's *History of Richmondshire* (1823). The cove here displays a convex form, a function of the differential erosional capacities of the rock sequence. The beds nearest to the head of the fall are much harder than those underneath it, so that the latter are constantly being worn back, progressively widening the distance between the rock face and the fall of water, and thereby adding to the singularity of the display.[119] Landseer's similar engraving of Turner's watercolour *High Force or Fall of Tees* (1822), again for inclusion in Whitaker's work, shows a

54 (*left*) John
Pye, after
J. M. W. Turner,
*Hardraw Fall,
Richmondshire,*
1818, for
Whitaker's *History
of Richmondshire,*
1823

55 (*below*) John
Landseer, after
J. M. W. Turner,
Fall of the Tees,
1821, for
Whitaker's *History
of Richmondshire,*
1823

56 (*left*)   Charles Ray, *Blackgang Chine*, 1825

57 (*facing page*)   George Cooke, after J. M. W. Turner, *Lulworth Cove, Dorset*, 1812

closer view of the rock exposure over which the water plunges. The fall was supposedly one of the 'finest cataracts in the island', the sound of it audible long before it could be seen.[120] Both plates clearly reflect the cult of the sublime, especially in the diminutive scale of the human figures. But, as John Ruskin was later to remark, both also gave satisfaction to the geologist.[121] When Turner painted a further watercolour of High Force in the mid-1820s, he adopted a viewpoint much further away, leading to an emphasis on the scale and space of the mountain scenery. Ruskin was so struck by this particular rendering that he thought it was capable of forming a geological lecture. And this related not just to the rock formations but to the processes that had acted upon them in the past.[122] Very much earlier, Sir Richard Colt Hoare had viewed the same Fall on one of his journeys. He, too, had been struck by the perpendicular forms of the rocks and the way the strata revealed differences of colour, in some cases the more varied by virtue of the wild plants that had crept up between them.[123]

Waterfalls were not by any means confined to upland Britain. On Wight, for example, falls were found in a number of the chines that had proved so attractive to a whole array of artists. In *A Picturesque Tour through the Isle of Wight* (1825), Charles Ray included an aquatint of Blackgang. This particular chine had long fascinated

observers on account of its tumbled rock masses, 'the ruins of former worlds'.[124] But Ray gave special prominence to the waterfall at its mouth. Although the vertical drop was only forty feet, it had a 'most magnificent appearance' during the rainy season. He was also at pains to explain the stratification that his picture recorded. It consisted of alternate layers of calcareous rock (mainly chalk) and black argillaceous earth or clay.[125] Ray was obviously quite clear about the constituencies of interest that his volume addressed. Wight had already evolved a powerful pedigree in studies of the picturesque and the sublime, but simultaneously it was beginning to afford a remarkable geological laboratory.[126]

One geological form that held a special fascination for artists, and one that was most often displayed in coastal cliffs, was folding. Here, formerly horizontal sediments had been 'rucked up' to produce amazingly contorted sets of rock beds. Some of the best examples were found on the Dorset coast, particularly at Lulworth Cove. Turner sketched here in connection with W. B. Cooke's project, *Picturesque Views of the Southern Coast of England* (1814–26). From Turner's contemporary poetry, it is plain that he had more than an ordinary grasp of geology.[127] And the startling engraving of Lulworth (drawn by Turner) that Cooke published in 1812, is powerful testament. Not only does it afford a remarkable coastal panorama, but the viewpoint deliberately brings the tightly folded Jurassic beds to centre-stage. Here was Nature truly at its most singular. The observer was left to contemplate the apparently cosmic forces that had turned these strata from their recumbent horizontal state. Known today as the Lulworth Crumple,[128] these rock beds comprise mainly Portland and Purbeck limestones that form a natural rampart against the forces of the sea. The Cove was formed when these were finally breached and the sea began to erode the much softer rocks behind.[129] The blackness of the engraving around the 'crumple zone' was a stark reminder of all that was sublime. However, the Cove beyond is bathed in light. Sailing-boats lean to the breeze in the relative calm of the sea. Lulworth had long been a favourite of tourists and what Turner did was juxtapose geological allusion with scenic wonder. And many of those who visited Lulworth, having leafed through Cooke's volume, would have been interested in both. However, Turner's depiction of geology was rarely ever complete. In his 'finished' works, in common with many other painters, he avoided topographical 'mapping',[130] and geology was no exception. The Lulworth geology is accurate largely only within the confines of the central foreground of the picture. To have done otherwise would have disturbed the contrast between geology, rock massif and peaceful cove that the composition explores. As part of the same series, Turner drew the nearby coast at Bridport (engraved in 1820). Here the delineation of the stratigraphy east of West Bay is poor. Turner gives it a vertical emphasis, but this does not fully accord with what an observer actually sees. The East Cliff here is remarkable for its exposure of a superb horizontal sequence of Bridport Sands.[131] Turner misses this and picks up instead the much less sharp (vertical) camber jointing.[132] As a studio artist, Turner was revealing nothing inconsistent by working in such a manner.[133] It signalled the extraordinary depth of the imaginative process in his compositions.[134] Even so, as far as serious geological observers went, he still provided constant reminders of their stock-in-trade, whether it was the towering rock faces of the coast near Whitby in a storm scene, or Dover Straits with its undulating line of chalk cliffs.

Observation, Observation, Observation

'Persons have been called Geologists who, gifted with prolific imaginations have indulged in fanciful speculation concerning a former order of things. . . . Others, by careful, diligent, and extended observations of the present state of the earth's surface, have endeavoured in the path of induction, to trace the nature of the agents which have once been active, to ascertain how they are now operating, and to anticipate the results of their continuance'.

. . . These are really Geologists, and their aim, is not to imagine or suppose, but to discover. . . .[135]

This was how Brande began his Lectures on Geology, written not long after Wellington had achieved his triumph over Napoleon at Waterloo. Observation was to become the touchstone of geological investigations for the next forty or fifty years, and especially among English geologists. Working in the field quickly assumed 'the primary locus of encounter between the geologist and the phenomena of his science'.[136] Initially, the central figures in this fieldwork project were the 'gentlemanly specialists' who came to dominate the emerging subject of geology in the first half of the nineteenth century.[137] However, as the decades unfolded, the social profile of those who engaged in observation and collecting became broader. Thus while Henry De la Beche's manual, *How to Observe Geology*, was intended to 'afford assistance to the scientific traveller and student', it also presented the hope that the 'listless idler' might, too, be changed into 'an inquiring and useful observer', and would thus 'acquire the power of converting a dull and dreary road into a district teaming with interest and pleasure'.[138] Hugh Miller, in *The Old Red Sandstone*, issued the same challenge to young working men. The stonemason who spent so much of his time amid the rocks and quarries of widely separated localities was urged to acquaint himself with geology. Not for him the Chartist meeting with all its undercurrents of class contest; what mattered was study and getting ahead in intelligence.[139] How far these particular homilies were taken up is difficult to gauge, but there can be no doubting the measure by which the popular presses took up the cause of geology. Nor can one fail to register the popularity of the many geological manuals and guides published during the period. Gideon Mantell's *Wonders of Geology* (1838), with John Martin's famous frontispiece depicting a group of fighting prehistoric reptiles, went through six editions in ten years. *How to Observe Geology* (later re-titled *The Geological Observer*) was first published in 1835 and went through five editions by 1853. De la Beche's *Geological Manual* (1831) was re-issued in 1832 and 1833, possibly helped along its way by the runaway success of Charles Lyell's *Principles* published over those same years. Robert Bakewell's *Introduction to Geology*, which Lyell read as an undergraduate,[140] appeared first in 1813, had run to a second edition by 1815, and to a fifth edition by 1838. William Buckland's famous Bridgewater Treatise on geology and mineralogy went through four editions between 1836 and 1869.[141] And supplementing all of these various titles, of course, were the works of the wildly popular scientist romantics Hugh Miller and Phillip Henry Gosse, where some readers thought that geological chapters would be less dull if they 'left out the geology'.[142]

For gentleman specialist and for amateur alike, geological observation and the fieldwork it entailed had a further dimension. It held aesthetic and emotional experience: aesthetic in terms of the romantic movement's love of the 'natural'; emotional in terms of the way geologists so frequently used words like 'joy', 'pleasure' and 'delight' in accounts of their geological forays.[143] These features are powerfully demonstrated in a passage of Mantell's *Wonders of Geology* in which he describes a traveller's view of part of the country on the road between London and Brighton:

> At Sutton he ascends the chalk hills of Surrey, and travels along elevated masses of [an] ancient ocean-bed. . . . Arriving at the precipitous southern escarpment of the North Downs, a magnificent landscape, displaying the physical structure of the Weald, and its varied and picturesque scenery, suddenly bursts upon his view. At his feet lies the deep valley of Galt in which Reigate is situated, and immediately beyond the town is the elevated ridge of Shanklin Sand, which stretching towards the west, attains at Leith-hill an altitude of one thousand feet. . . .[144]

This was the man who, from the obituary written of him in the *Illustrated London News*, was the most attractive of lecturers, 'filling the listening ears of his audience with seductive imagery'.[145] However, Mantell could still display the qualities of a more strictly technical observer. On that same journey, the traveller would notice (so he went on to remark) that beyond Sutton the roads were made of broken chalk-flints, but at Reigate they were of cherty sandstone. Then, at Crawley, the material used was 'grit and stone, containing many fluviatile shells, bones and plants'.[146] The 'grit' was calciferous sandstone, according to Mantell. It formed excellent road material, with quarries to excavate it found on all the principal lines of communication between London and the south-east coast. This commentary was written just prior to the railway era and in Mantell's later book, *The Medals of Creation* (1844), he noted how the rapid transit afforded by the railroad on this routeway, for all its many advantages, had resulted in the loss of many road-coach services and had closed down the inns of country towns and villages where once 'the geological tourist always found a welcome reception'.[147] The possibilities of geological excursion did nevertheless enlarge enormously once the principal trunk railroads were in operation. Mantell's *Medals of Creation*, very appropriately subtitled *First Lessons in Geology*, had a whole section called 'List of Excursions'.[148] These embraced destinations all across the country, but the most attractive had to be that to Matlock in Derbyshire which, beginning at Euston station, could be reached by trains of the London and Birmingham, Midland Counties and North Midland railway companies. The journey was itself a geological excursion, of course. Once in Derbyshire, it was the 'mountain limestone' that came to view, with its 'glorious' cliffs and chasms, its precipices crested with forests and its rivers 'dashing' through ravines,[149] the very images that artists had been busy capturing over preceding decades. Mantell even advised his travellers what they needed in the way of equipment: a hammer, a leather or camlet bag, a paper for wrapping specimens, boxes, wadding, string, 'paper gummed on one side for labels, gloves, eye preservers, a measuring tape, compass, map, knives and chisels and a set of lenses'.[150] He gave detailed instructions on how to collect specimens and added how better it was to sketch cliffs or coast exposures in preference to writing a 'long memorandum'.[151]

Mantell clearly wrote for the more populist or amateur end of the 'gentlemanly' market. For 'gentleman specialists', geological observation and the fieldwork that it entailed had affinities with outdoor sporting pursuits and, more widely, with robust, manly endeavours.[152] Sir Roderick Murchison, who cracked the code of the ancient Silurian strata, perhaps formed the apotheosis of the breed, eventually tracing his 'Silurian system' into the heartlands of Russia.[153] Another contender was Adam Sedgwick, who developed a close friendship with Wordsworth and whose love of the great outdoors, particularly the Lake District and Cumbrian hills, ensured that he was never more at home than when leading geological expeditions.[154] William Buckland's zest for the field was never more apparent than in his investigations of the geology of Wight. And although Gideon Mantell wrote and lectured for a rather less adventurous audience, he was not himself any stranger to clambering over rocks in all weathers. Sometimes these intrepid fieldworkers had the company of their wives, particularly in summer seasons. This was true of both Buckland and Lyell. Otherwise, wives were left to lament their 'grass widowhood', as *Chambers's Journal* recorded in a poem of 1846, entitled *The Geologist's Wife*:

> To Her Husband setting off upon an excursion:
>
> 'Adieu, my dear, to the Highlands you go,
> Geology calls you, you must not say no;
> Alone in your absence I cannot but mourn,
> And yet it were selfish to wish your return.
>
> No, come not until you have searched through the gneiss,
> And marked on the Smoothings produced by the ice;
> O'er granite-filled chinks felt Huttonian joy,
> And measured the Parallel Roads of Glenroy'.[155]

This was clearly no idle piece of verse, for it embraced various of the geological controversies of the day, including Agassiz's idea of

PL. 3.

*Basaltic Columns on the
N. Side of Cader Idris
Sep: 9. 1812.*

58 The frontispiece to Robert Bakewell's
Introduction to Geology, 1813

a former Ice Age and the puzzle presented by the terraces or 'parallel roads' of Glen Roy which so excited Charles Darwin when he saw them in 1838.[156] These were supposedly relict shorelines arising from uplift of the land.[157] The poem is also interesting in the way it cast geology as a 'calling', offering shades of religious or spiritual awakening. In not unrelated vein was the *Quarterly Review*'s account of the first volume of Lyell's *Principles*, which referred to geology as a 'noble' object, a feature that was the more enhanced by the grand scenery within which it was so often transacted.[158] In similar mould was the idea that geological study bore parallels with the study of antiquities. Rock and fossil remnants were construed as pieces of artistry much as one would regard the ruined forms of earlier civilizations.[159] Geological investigation could also be entwined with the art of courtship, as, for example, in the fictional tale of 'Love and Geology' that *Blackwood's Magazine* printed in 1838. The story was also a gentle satire on the practice of rock and fossil collecting, with the characters sinking a borehole in order to aid their study of the strata, wondering whether or not they would manage to confirm the 'Mosaic theory'.[160] Geological study, though, did not always invite enthusiasm. It brought complaints of the decay of the picturesque for one contributor to *Blackwood's Magazine*. The quarries that supplied so many skeletons of giant reptiles were destroying some of England's finest natural scenery. Meanwhile, road-coaches 'bowled along in triumph' over some of the pulverized remains, 'sold by the boatload and measured by the ton'.[161]

The expeditions of gentleman specialists were sometimes captured in sketches or caricatures that the more artistically talented of their members made whilst in the field. The frontispiece to Robert Bakewell's *Introduction to Geology* (1813) showed one such specialist (perhaps the author in self-portrait) resting among the shattered basaltic columns on the north side of Cader Idris in North Wales in September 1812. Perhaps the most memorable caricature of contemporary field geology is Henry De la Beche's depiction of himself debating with Charles Lyell, drawn in about 1831.[162] It is set in mountainous terrain, but the most revealing aspect of the picture is the contrast that it presents between theoretical and practical geology. The former is represented in parody by the tailcoated Lyell, clutching behind his back a

59 (*above*) Henry De la Beche's sketch of himself (right) debating with Charles Lyell

60 (*right*)   Henry De la Beche's sketch of himself imprisoned indoors on account of inclement weather, unable to continue with his fieldwork

volume entitled *Theory of the Earth* and waving at his companion the 'tinted spectacles of theoretical supposition'. By contrast, De la Beche is presented as the practical man in working clothes, hammer in hand and collecting bag over his shoulder. De la Beche became founder of the Geological Survey of Great Britain and was a committed empiricist. He saw theoretical geologists as complacent, victims of a false sense of security. He did not object to theories as a means of promoting enquiry, but they were no substitute for the steady accumulation of facts.[163] And, in due course, this became the mode by which official knowledge of the country's geology was extended. The Survey's geologists were fundamentally firsthand collectors. They enjoyed a shared mission. Their 'boundary chasing and cask collecting' had a prevailing sense of direction. After the work of the Survey had been done in an area, what was left to be done became a kind of second-class science.[164] De la Beche had been one of the first professional geologists retained by the Ordnance authorities, beginning in 1832.[165] This had further distinguished him from Lyell and many other leading geologists of the day by his needing to carry out fieldwork to a large extent regardless of the weather or season of the year. It was in just such a context that he penned the caricature of himself in lodgings in Tregony in Cornwall in January 1837, peering out at the winter rain, unable to work.

Geological fieldwork could, of course, be a ready medium for the instruction of students of the subject, as William Buckland was quick to establish. The way one made the class remember the Kimmeridge Clay was to send them walking through its sticky mud.[166] Based in Oxford, Buckland had no shortage of field sites from which to instruct his students. One of the most frequented was Shotover Hill, south-east of the city. *Punch* later characterized such field excursions at British Association meetings as nothing but a species of junketing: geologists went on 'pic-nics with pickaxes',[167] the axes being most convenient for breaking the necks off champagne bottles.

Sketching in the field, for all its human interest, had a much more serious dimension, however. It was part of the emerging 'visual language for geological science'.[168] This does not, in the first instance, refer to the section drawings that geologists made in the field, the imaginary 'slices' through the earth's crust that were used to demonstrate, for example, the principle of superposition. These were critically important, but will be dealt with later. What sketching also encompassed was a system of topographical documentary, a way of recording landscape in a manner that enabled geologists better to conceive of the earth's 'former worlds'. At best, this became topographical draughtsmanship, something that the evolving techniques of line engraving helped readily to disseminate.[169] One arch-exponent was Thomas Webster, the Geological Society's official draughtsman, whose recordings of the coastal geology of Wight and of east Dorset were featured earlier in chapter one. As that chapter also revealed, though, there was no sharp dividing line between topographical draughting and artistic energies. The strata of Durlstone Bay in Dorset, 'bent and distorted in a very singular manner', according to Webster's own verbal account, became not only a geological pictorial but a piece of high visual drama, embracing imagery that satisfied senses both of the picturesque and the sublime.[170] And nor was Webster himself any stranger to the direct illustration of such sensitivities. A number of his drawings for Englefield's *Isle of Wight*, for example, were unambiguously romantic and antiquarian.[171] It would not be surprising, either, to have found gentleman geologists (not to mention their wives) as appreciative of such views as they were of those that conveyed plainer geological reference.

Webster, it will be recalled, made many of his coastal drawings of Wight and East Dorset from the vantage of a small sailing-boat in summer. This had initially been the intention of Richard Ayton and William Daniell when they set out in 1813 to record the 'ruggedness and sublimity' of Britain's coastal scenery at large in their *Voyage Round Great Britain*.[172] On less clement shores than southern England, though, winds and waves were often so intractable as to force them more often to 'sail' on horseback, or 'scud' in a gig.[173] Otherwise they faced a 'catalogue of horrors', comprising 'rapid tides, ground-swells, insurmountable surfs, and foul winds'.[174] *Voyage Round Great Britain* demonstrates much the same pictorial montage as Webster (and

61   William Daniell, *The Eligugstack, near St Gowan's Head, Pembrokeshire,*
1814

Englefield) had done for Wight and its adjacent coasts. In Ross-shire, Loch Duich provided a 'striking union of the sublime and the beautiful'.[175] Of Beaumaris, the authors remarked how dull and fatiguing watering-places could be in rainy weather when 'all the *readable* books of the Library [were] engaged'.[176] In general, though, they sought not to emphasize the coastal parts where sea-bathing frequented and where the adjacent country was 'divided into smooth paths for ladies to walk on'.[177] What preoccupied them instead was ruggedness, as in the coast east of Combe Martin where rocks appeared 'rising in fragments . . . like one tremendous ruin'.[178] It was here, as shown earlier, that Daniell, the artist and engraver for the project, revealed some sensitivity to geological interests. At St Gowan's Head in Pembrokeshire, he recorded the horizontal stratification of the principal stack, the accompanying text adding how some of the main cliffs showed ranges of nearly vertical strata surrounded on either side by strata that were completely at right angles.[179] In the Western Isles of Scotland, as already noted, the coastal and island geology took centre-stage, informed by readings of the Geological Society of London's own *Transactions*.[180] Fingal's Cave, in particular, became 'one of the infinite wonders by which the Omnipotent ha[d] manifested himself'.[181]

   Ayton and Daniell viewed the 'strangely rent and disordered' strata they saw on so many coasts as evidence of some 'great convulsion'.[182] Cliffs were observed that preserved 'deep traces of a violent shock'.[183]

Fingal's Cave, Staffa.   *The Scuir of Egg.*

62   Henry De la Beche's views of trap (volcanic) rocks in the Western Islands, 1830

They were not, in the authors' view, affected by 'any of the slow causes which are at present on the surface of the earth'.[184] Such rejection of Huttonian ideas of earth history in favour of catastrophic causes played admirably to the subliminal energies that Daniell's own draughtsmanship often sought to evoke. However, catastrophic views of the history of the earth were widely respected within geological circles at the time, so that one cannot lay the charge of romanticism upon Daniell's work. Not only could he be geologically informative, but, in cases (for example, Fingal's Cave), very accurately so.[185]

Topographical documentary may have been aided by use of the *camera lucida*, patented in 1807.[186] This was a development (by the geologist William Wollaston) of the *camera obscura*.[187] The Scottish geologist John MacCulloch may have employed this in some of his sketches used to illustrate the Geological Society of London's *Transactions*. This is suggested, for instance, by the incredible accuracy of his depiction of the 'parallel roads' of Glen Roy in the Scottish Highlands.[188] Gideon Mantell certainly used the device, for he recorded a field expedition to Mount Harry in August 1819 in which he drew outlines of the surrounding Sussex hills for his later monograph on Sussex geology.[189] In its earliest form, the *camera lucida* comprised a four-sided prism mounted on a small stand above a sheet of paper. By positioning the eye near to the prism's upper edge, the user was able to see on the paper a reflected image of an object situated in front of the prism. The image could then be traced and an accurate perspective drawing obtained.[190]

In the hands of some geological draughtsmen, or artist–geologists, the topographical documentary was adapted to the style of depiction

more recognizably akin to the modern visual language of geology.
There was an increasing formalization of structural features in draughting.[191] Henry De la Beche's illustrations in his *Sections and Views* of
1830 afford a superb example of the genre. As specimens of art, De la
Beche himself regarded them as valueless.[192] What they offered (in his
view) was the accuracy and detail of the painstaking eyewitness. They
were part of a science that was progressing on the basis of a hundred
facts not five.[193] De la Beche utilized both his own eyewitness
experiences and also those of others, including John MacCulloch's
depictions of volcanic rocks in the Western Isles, especially of the
Island of Staffa. *Sections and Views* is of separate interest in the fact of
employing a remarkable compound of visual languages for geology, as
the title itself implies. On the one hand, there were striking topographic panoramas that did indeed amount to pieces of artistry. On the
other, there were mining sections in all their engineering formality, as
well as coloured geological maps. In between, though, was a vast array
of traverses, some forming plain horizontal sections, but many presenting hybrids, half section and half topographical drawing. Such features recur in the Geological Society of London's *Transactions* for 1829
which incorporated in its end-binding a parallel array of types of geological illustration: maps, different kinds of section drawings, ordinary

63   Mary Buckland's sketch of the Great Bindon landslip, East Devon, 1839, drafted with an obvious
eye for the picturesque

topographic prints on which rock beds were identified by over-printed letters, as well as examples of topographical documentary with deliberate emphases on structural features.[194] The practice of combining 'sections' and 'views' was still current as late as 1840, as in the remarkable collection of illustrations used by William Buckland and William Daniel Conybeare to record the landslips on the East Devon coast between Axmouth and Lyme Regis that began on Boxing Day 1839.[195] The topographical draughting in this case was done by William Dawson (an Exeter surveyor and civil engineer) and by Mary Buckland (William Buckland's wife). Plate 2, showing a coastal panorama, was a typically hybrid illustration, partly a technical horizontal section (Conybeare supplying the stratigraphic detail) and Dawson sketching in the topography. Most of the remaining views, like that of the Great Bindon landslip, were straight topography and demonstrated a return to the pattern that had characterized some of Thomas Webster's drawings of Wight some twenty-five years before: they sought to provide a documentary of the landslips but did so with a clear eye to the picturesque, even given a list of subscribers that might have constituted its own geological 'who's who'. The final plate of the series (10) depicted relics of the landslip that had occurred between Beer Head and Branscombe in 1789–90. It was by Dawson and, with its peacefully grazing sheep and small boats under sail in the calm sea of the bay, might have been straight out of a collection of picturesque coastal views. The distinctions between formal geological draughting and artistic energies remained, in other words, highly fluid – even in the 1830s. While some geologists now sought to purge metaphor and hyperbole from their descriptive language, 'to dispense with the gauze of literary and picturesque associations',[196] the power of artistic vision as applied to features of the face of the earth was not easily erased or absorbed.

Brandishing the Torch of Science

> The advance of geological science has lately been so rapid, that it requires some exertion to keep pace with it.[197]

The adaptation of artistic visions for geological purposes clearly raised the profile of the subject in educated minds. But even more critical to this process was the agency provided by the British Association for the Advancement of Science.[198] Within the Association's annual meetings, the session devoted specifically to geology took centre-stage for several decades. As *Chambers's Journal* commented when the Association next met in Cambridge in August 1845, geology was allocated the 'largest and best room enjoyed by any of the sections' and

64 A session at the British Association for the Advancement of Science's annual meeting held at Cambridge, 1845, revealing a large number of women in attendance – from the *Illustrated London News*

was 'the most numerously attended'.[199] Geology was also about the only section 'enjoying feminine patronage',[200] mingling 'more pleasantries with science than any of the rest'.[201] Indeed, there seems to have come a point at which meetings, generally, were turned into a kind of rolling social pageant.[202] At the Manchester meeting of 1842, the Association's Treasurer had reported selling 331 'Ladies' tickets' out of a total ticket sale of a little over a thousand.[203] In 1859, the year of Darwin's *Origin*, the Association met in Aberdeen and it seems that geology's ascendancy was still unchallenged. The subject was 'surest of the largest audience'.[204] Moreover, for the meeting as a whole, some 2,500 people had applied to attend, one of them describing how he arrived in a train of thirty carriages and carrying above seven hundred passengers.[205] It was 'a frightful drag for the two engines'.[206] As well as indoor meetings and lectures, the Association's geology section invariably held field excursions in the local area. As recorded in chapter one, when the Association met in Southampton in September 1846, the visiting geologists spent a day circumnavigating the Isle of Wight in a steamer. Not only did seaward views of the island appeal to all the familiar senses of the picturesque and the sublime, but they simultaneously formed a geological spectacular.[207] Likewise, when the Association met in Oxford in 1832, William Buckland had no hesitation in regaling visitors with all the best geological prospects.[208] At the Newcastle meeting of 1838, Adam Sedgwick lead a coastal excursion that reputedly involved a crowd of 3,000 congregating on the seashore.[209] Those attending the Manchester meeting in 1842 had, as has been discussed already, the benefit of special trains on the Bolton Railway in order to view the fossil trees unearthed in the making of the line.[210]

The extent to which the annual British Association meetings came to occupy a position in the wider public consciousness is clearly revealed in the way popular magazines like *Chambers's Journal* chose to report them. But it was also reflected in the way copy-writers for *Punch* and *Household Words* repeatedly used them as a source of satire or parody. In 1842, when the Association was in Manchester, *Punch* described the organization as the 'British Association for the Advancement of Everything in General and nothing in particular'.[211] The Society was 'brandishing the torch of science in the eye of a flabbergasted world'.[212] *Punch* went on to parody its detailed proceedings, overseen by a battery of professors rejoicing in names such as Sniveldrivel and Leadenbrains.[213] The Association's support for the recording of railway cuttings came in for particular comment. As described in chapter one, several hundred pounds had been spent in making coloured drawings of them and one learned visitor desired to know what a sectional cutting meant.[214] Given the novelty of the section in the visual language of geology, this was perhaps not just a comic reference. *Punch* also gave space to the geological excursion in the field. Thus, so it was claimed, the proprietors of a local coalhole threw it open to the Association's visitors who subsequenly descended into it and duly reported their observations.[215] For Charles Dickens, it was the information conveyed in geological lectures that gave alarm. In a piece in *Household Words* for 1852, he remarked how unpleasant it would be if one had to face geological changes in the nineteenth century: 'what would people think . . . if a new chain of mountains were to rise up, one night, the whole length of Regent Street, London?'[216] The story may in part have been prompted by the coastal recession that Dickens had himself observed near Sheringham in Norfolk where groins were being constructed to combat it.[217] Much earlier, in 1836, Dickens had himself lampooned the British Association in his 'Mudfog' sketches.[218] Indeed, the inspiration for *Punch's* own treatment of 1842 may well have come from these. Mudfog was the town where the Association was to meet and the sketches took the form of a sequence of letters from a gentleman who was in attendance. The meeting's leading lights were Professors Snore, Doze and Wheezy.[219] Another celebrated participant was Mr Slug, a statistician with a dark purple complexion and a persistent habit of sighing.[220] Dickens went on to lampoon the various 'sections' of the Association's annual meetings, but he did not include geology among them.[221]

The British Association's provincial circuit was in part founded on a gathering array of philosophical and scientific societies and organizations that had grown up in the provinces, especially in developing industrial districts.[222] The Literary and Philosophical Society of Newcastle-upon-Tyne, founded in 1793, was one. Among the issues it addressed was local mine geology, both from the point of view of exploration and of safety.[223] The Society succeeded in gathering support within the local community and a number of the geological papers presented at its meetings were subsequently published in the Geological Society of London's *Transactions*.[224] One very early geological society in the provinces was that founded in Cornwall in 1814, again as a logical response to the geological interests among the county's mining concerns. The Royal Geological Society of Cornwall published its own *Transactions*, beginning in 1818, revealing a clear but far from exclusive emphasis on mining and mineralogy.[225] The first issue, for instance, included papers on coastal sandstone, granite veins, the structure of the Scilly Isles and the coast west of Penzance, and the geology of the Lizard peninsula. Like the Newcastle Society, the Cornish Society attracted wide support. Among its honorary members in 1818 were William Buckland, George Greenough, Robert Jameson and John Playfair.

In later decades, geological societies sprang up in many larger cities and even in some towns. Edinburgh had its own society by 1834 and Manchester had one by 1838. There was a society for Yorkshire by 1837. *Punch* parodied the phenomenon in the imaginary 'Geological Society of Hookham-cum-Snivey'.[226] It had its own little museum and published its own transactions. During summer months, its members were up to their knees 'in dirt and filth, gravel and gypsum, coal, clay and conglomerate'.[227]

The institutional apex of the emerging science of geology, of course, lay, in metropolitan institutions such as the Royal Society of Edinburgh, where James Hutton had figured so prominently, and the Geological Society of London, arrogant in its determination to forge an identity separate from the English capital's own Royal Society.[228] Although all of these organizations tended to be monopolized by affluent gentleman specialists, their significance was founded in the relative unity of the vision they displayed. In Edinburgh, as chapter two demonstrated, this was part of a broader intellectual project within which Hutton's *Theory of the Earth* was comfortably positioned. In London, it materialized as a grand stratigraphical enterprise, an exercise in empirical investigation, in the collection of facts about the geological strata and their subsequent classification. Whilst it was the Edinburgh 'school' that underpinned Lyell's hugely popular *Principles*, it was the London group that laid out the ground for launching geology in popular imagination,[229] for in its simplest sense, theirs was a research enterprise that all inquiring minds could follow. And many did.

Representing the Earth

> Without maps and sections of particular districts, representing the extent, thickness, and order of superposition of the several component rocks, the abstract truths of geology would never become of general interest or public value.[230]

In an article about the story of the Geological Survey, *Chambers's Journal* of 1861 suggested to its readers that the geological map was 'very suggestive of that grim anatomical figure styled an *écorchée*, which may often be seen in an artist's studio'.[231] It was a statuette of the human form 'with all the skin flayed off', with 'blue veins, gray muscles, and raw red flesh exposed'.[232] Geologists, when seeking to represent the earth, drew in a very similar style. The 'close-fitting carpet of green turf, purple heath', was rolled back, 'steeples, pretty villages and big black towns' were swept away and the underlying strata, 'the flesh and bones of the earth', were laid bare to view.[233] It was hopeless to expect to recognize a favourite landscape here. What one saw, instead, were the rich veins of minerals or the cumbersome masses of coal basins. The reader, though, was still left to wonder how geologists managed 'to see how the earth looks below the surface'.[234] They could not dismember it in the manner of an anatomist, for instance. What the map became, in effect, was 'a work of scientific deduction from a number of facts'. This needed 'keen eyes to collect and a well-stored head to comprehend'.[235] Geological mapping was thus as much a theoretical project as it was an empirical one. As one modern commentator has observed, the geological map, for all its outward empirical simplicity, is actually 'a highly complex, abstract and formalized kind of representation'.[236] And certainly this was the view of most of those who worked for the Geological Survey in the 1840s. Indeed, the Survey's approach to collecting was heavily influenced by theoretical views from an early stage.[237]

The science of geology was fortunate that its mapping project came at a time when the science or practice of mapping, generally, had shown rapid advance. It was not just the fact of the setting up of the Ordnance Survey in the 1790s that ultimately gave to the country the topographical map base that is so familiar a part of cartography today. What was also critical was the place of the map in evolving social and political discourse, and, more broadly, in evolving social and political formation.[238] The map had already become, over the eighteenth century, a vital prop to the emerging concept of private property in land. The profession of land surveyor had never been more productive, as William Smith registered from his apprenticeship days. Owners of landed estates, commissioners for enclosure and for the commutation of tithes, all needed accurate surveys and maps of the lands under their view. The map had become a vital support to the cash nexus of contemporary political economy. Simultaneously, the map also formed part of the agency of colonialism, an imprimatur of external political, social and economic control.[239] Within imperialism, the mere drawing up of maps sometimes came to substitute for conquest on the ground. Geology and geological mapping benefited from and reflected many, if not all, of these various features. One of William Smith's earliest attempts to plot outcrops was made in 1801 using John Cary's *New Map of England and Wales*, first published as seventy separate sheets in 1794.[240] Aside from the work of the Ordnance Survey, Cary was probably the most prominent map-maker of his day. Smith, though, had also used one of the many county maps that had been published in increasing numbers in the later decades of the eighteenth century, some of which were quite exceptional pieces of survey and draughtsmanship. Day and Masters's one inch to the mile map of Somerset formed the basis of his attempt at a coloured geological map of Bath in 1799.[241] In due course, as the national coverage of the Ordnance Survey's work of topographical mapping progressed, few geologists ventured into the field without the relevant sheets of the area they were investigating. As to the uses that the resulting geological maps could be put, the list was almost endless. In William Smith's printed list of his publications on English geology, which included not only maps but atlases, sections and tables, he noted how they could help in the improvement of landed property through drainage and irrigation, and through the finding of 'mineral manures', and road and building materials.[242] There was a sense in which geology became less about unravelling the history of the earth than about enhancing the capital value of land. Colouring the face of the map according to its underlying rock type also fitted beautifully into the nineteenth-century scientific preoccupation with classification.[243] The geological map thus became little different from the plethora of soil, vegetation and physical geographical maps that were later on in the century to feature in every commercial and school atlas. Geological mapping had affinities, too, with the mapping of empire.[244] Leading geologists, for instance, developed powerful senses of what was their own geological territory. Roderick Murchison's label as 'King of Siluria' was perhaps the apotheosis of the phenomenon,[245] but others, like Mantell and De la Beche, came to be closely identified with discrete territories – in their case with Sussex and the Weald, and with Devonshire. The various disputes over the identification of parts of the sedimentary rock series in the early to middle decades of the nineteenth century became almost as much about the integrity of personal territorial preserves as about the underlying confusions over rock classification that precipitated them.[246]

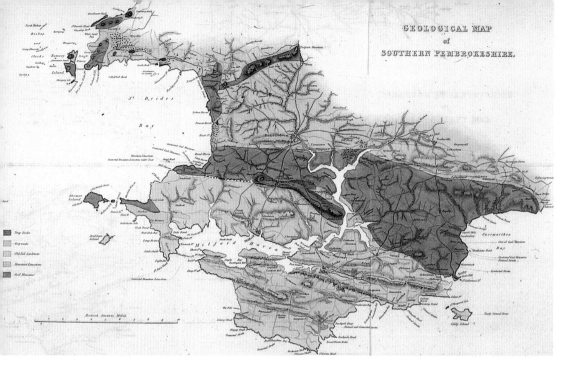

65 Map of the geology of Southern Pembrokeshire from *Transactions of the Geological Society of London* for 1829. The method of colouring outcrops remains largely the same today

When George Greenough wrote in 1840 of the geological map having no rival in conveying an idea of the visible forms of the earth, particularly against the consecutive form of words, he was doing so after four or five decades of trial and experiment among geologists in how to represent the earth as a three-dimensional structure within the two dimensions of the map.[247] Critical in this process was that only a limited amount of geological knowledge came from an objective recording of actual facts, as *Chambers's Journal* of 1861 recorded and as many subsequent writers have emphasized. So the map had to be as much inference as fact. William Smith registered this feature in the tentative way in which he coloured his first maps – usually by tinting each outcrop along its basal edge and fading the colour away towards the outcrop of the formation overlying.[248] Smith's 1799 map of Bath was coded in this way, as was his 1801 map of England and Wales. The reason for following this convention was that it made it easier to effect corrections and changes as more factual information came to light and as theory was refined or altered.[249] Smith's situation was perhaps distinct from that of some other of his fellow-workers (especially continental ones) in that, as remarked before, the English Secondary strata were arranged like overlapping and gently dipping 'slices of bread and butter', making for an elaborate surface sequence that was not always easily traced.[250] A variant on colouring the basal edge of the outcrop was to produce a 'cut' map, a primitive relief map, by scoring along the boundaries between successive strata and elevating the bases of the more resistant formations so as to replicate more clearly the three-dimensional form.[251] Neither of these particular techniques subsequently became the 'norm' in geological map-making, however ingenious they appeared at the time.[252] What was adopted instead was the method used by the Frenchmen, Cuvier and Brongniart, in their pioneering work on the geology of the Paris Basin. This was to colour whole areas of outcrops and to arrange a key to the colours as a vertical column at the side of the map that coincided with the vertical order of superimposition.[253] The result was a map that conveyed something of a three-dimensional structure. By the 1820s, the Geological Society of London's *Transactions* revealed this as having become the regular pattern.[254]

With different practitioners working in different parts of the country, including across the water in France, and given also the different sedimentary sequences that each was concerned in mapping, it was perhaps inevitable that there were initially no common conventions over colour coding. It was not just that William Smith sometimes used a weaker colour wash as one moved away from the base of an outcrop, it was also the case that different draughtsmen used completely different colours in identifying particular rock types. In Smith's 1801 map of England and Wales, chalk was shown as a bluish green and limestones as yellow. In John Farey's geological map of Derbyshire, only the Magnesian Limestone was shown as yellow, with the Lias Clay as blue and coal seams as green.[255] In the map of West Sussex in the Geological Society of London's *Transactions* of 1829, the Chalk was cream, while varieties of blue were used for the Gault and for Wealden clays. Sometimes the colours chosen reflected the colours of the rocks themselves, although even here there was scope for inconsistency given that rocks that were freshly exposed showed different hues to the same rocks after prolonged exposure to the elements. In the earliest geological maps, as well as in geological sections, it seems that colours were used largely as a means of differentiation *within* them, and not as part of any comparative geological project. Even in the two sections covering parts of the Devon/Dorset and Dorset/Hants coasts included in the Geological Society of London's *Transactions* of 1829, the colouring scheme did not match, except in the case of the pale crimson used for water-borne deposits, or 'diluvium'.[256] The British Association for the Advancement of Science, at a very early stage in its existence, proposed colour codes for geological mapping, but they were not widely taken up.[257] When the needs of comparison did press upon the science, so a more universal colouring scheme emerged. And for the work of the

Geological Survey, of course, such a consistent scheme of differentiation was a vital element of the project's success.

When the geological map had come of age, the vertical or columnar section that comprised its key, with the strata in their order of superimposition, had become a commonplace. But, like the map, the section actually had its own story of development.[258] The reason was that it derived from mining, quarry and water-well surveys. The recording of information on the stratification was generally for practical purposes only. Sometimes attempts were made to trace the section laterally when adits (horizontal excavations) had been driven by the miners in the process of prospecting or of 'breaking new ground'. However, rock structure was not generally considered if it was outside the limits of the mine workings.[259] William Smith's early geological sections were very much of this pattern, clearly connected to his experience of visiting collieries and also to his work as a canal surveyor. Thus his section of the strata sunk through for making the Wilts and Berkshire Canal Water Pit in 1816 focused on the particular engineering difficulties that supplying water to the canal presented.[260] The successive strata are detailed (and coloured) as well as the fossil bands. But no attempt is made to consider the strata beyond this narrow column. Eighteenth- and early nineteenth-century canal surveyors made hun-

66   Henry De la Beche's comparative vertical geological sections for Le Havre in France and Lyme Regis, Dorset, 1830

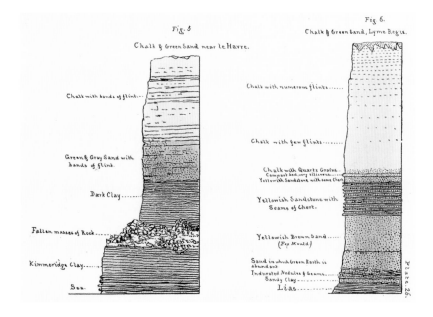

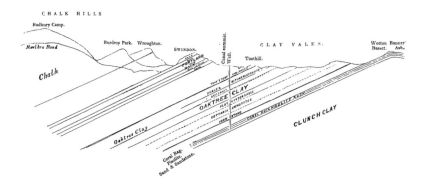

67   William Smith's traverse geological section for North Wiltshire, as published by John Phillips, 1817

dreds of similar sections in the practice of their professions. Unwittingly, many were laying the foundations for a much more abstract form of vertical section that generalized large tracts of country, even when those tracts were characterized by complex folding and faulting.[261] Henry De la Beche's *Sections and Views* of 1830 incorporated some singular examples, particularly the ability to 'facilitate stratigraphical correlation of the formations and to demonstrate the lateral variation of their thickness'.[262] Thus plate XVI of this volume compares sections for Freville on the Cotentin peninsula, for Le Havre and for Lyme Regis in Dorset, the intention being to try to illustrate the different development of the Chalk and Greensand in the three areas.[263] In the Cotentin, for instance, the Chalk was represented by a type of limestone rich in fossils.

The inevitable complement of the vertical or columnar section was the geological traverse, continuing in horizontal plane and on a much larger scale what the mine shaft or well-sinking revealed about an area's stratification. Geologists had long studied such sections in the natural form of coastal or inland cliffs. Wight held such a strong attraction for aspiring geologists because of its spectacular eastern and southern coastal cliffs which could be readily observed at close quarters on calm days. It was but a short step then to produce an imaginary or hypothetical traverse running across open country. Here, the geologist relied as much on theory as fact. Or, as William Smith recorded in 1839 shortly before his death, geology involved an experimental exercise, proceeding, like geometry, from point to point and from time to time.[264] By extrapolation and interpretation, the geologist was eventually able to produce a visual record or 'slice' through the earth's surface crust.[265] Where Smith produced such horizontal sections or traverses, they often had the appearance of engineers' drawings, showing differ-

South   North

68   Gideon Mantell's field sketch of the strata of a part of Sussex, showing the relation of the rock structure to the surface topography, 1822

ent strata as highly regular, parallel bands, a clear reflection of geology's origins within the practical context of mining and surveying.[266] His section of North Wiltshire lithographed by Phillips and sold in 1817 illustrated the pattern.[267] The first volume of the Royal Geological Society of Cornwall's *Transactions* (1818) incorporated a section diagram of the limestone at Veryan with divisions between strata marked out as parallel bands.[268] In due course, however, most traverses came to be drawn free-hand rather than ruled as a series of parallel lines.[269] The pattern seems to have been set by Cuvier and Brongniart's work on the Paris Basin.[270] However, the practice among English geologists of sketching the strata of coastal cliffs in the field (Wight affording one vivid example) was clearly not unrelated. It was here, too, that some traverses were drawn part as landscape and part as section, as described earlier. Among the plates in Henry De la Beche's *Sections and Views* is one that not only illustrated a whole range of traverses drawn by the free-hand method, but also demonstrated the significance of faults and flexures found in the strata. One showed a 'slice' through Somersetshire, and revealed striking examples of faulting, as well as other disturbances. As the author recorded: 'we here observe that the old red sandstone, carboniferous limestone, and coal measures were disturbed after deposition; that the red sandstone rocks, after filling up the inequalities which had been formed in the surface of the inferior strata, presented an horizontal or nearly horizontal surface, on which the oolite was deposited'.[271] It was after this that the entire consolidated mass was rent by fractures and faults.

When sketching such traverses, geologists were sometimes lead to imagine how strata might be extended to indicate the position of former land surfaces, part of the succession of James Huttons 'former worlds'. Thomas Webster did this in his drawings of the coastal geology of Wight and Dorset.[272] There is a similar sketch among the papers of

William Buckland.[273] Imaginary exercises of this kind perhaps found their most potent force in areas like the Weald of Kent and Sussex, where reconstruction of its 'dome' shape gave such stark visual meaning to the erosion story, to the vast depths of geological time.

Traverse sections were generally drawn to scale along the line that they represented. To do the same in their vertical dimension, though, was more problematic. Exaggeration was often necessary for simple reasons of clarity.[274] Questions of clarity also sometimes lead geologists to produce much more diagrammatic or schematic sections, including columnar ones. Indeed, there was a sense in which this was an inevitable part of a science that had to proceed as much by theory or interpretation as by fact.[275] Even William Smith followed such a pattern in his diagram of the succession of strata on his 1815 map of England and Wales.[276]

If section sketches and drawings gave geology its most powerful visual language beyond the basis provided by the coloured geology on a map, there was yet one other means by which the structure of the earth could be represented. This was to use a model, a miniaturized version of the real thing. The Polytechnic Institution in London had a model of the Isle of Wight, for Gideon Mantell records going to see it in December 1845.[277] This was probably the work of Ibbetson who had made 'beautiful models of the stratification of the Undercliff, and southern coast' of Wight, according to Mantell's guide to the island published in 1847.[278] In January 1846, Mantell also records taking 'a geological model of the Isle of Wight' to one of the Marquis of Northampton's 'geological' soirees in London where it was used in discussions with the Prince Consort.[279] This may have been one of the models that R. T. Wilde of Chancery Lane was then selling, priced from 5 shillings to 21 pounds and 2 shillings.[280] However, there was another proprietor, a Mr Sopwith, who made 'models of stratification etc' according to the list of dealers in fossils and minerals that Mantell had earlier appended to the second volume of his *Medals of Creation* of 1844.[281] Such dealers had also made models of the 'coal-trees' discovered in the excavations of the Manchester and Bolton railroad, so wide was the fascination with these relics of another world.[282]

How widely such geological models were disseminated is very difficult to gauge. If nothing else, they certainly equated with De la Beche's desire for 'miniature representations of nature', for a visual language for geology that was 'more conformable to nature' than hitherto.[283] For many fossilists or palaeontologists, though, conformity to nature became a critical difficulty of their pursuit. The reptile skeletons that they were soon finding in abundance in quarries and along coastal cliffs appeared to bear limited resemblance to anything then living. They were more like the dragons found in fable and in fairytale.

4

'Let there be Dragons'

And there was war in heaven: Michael and his angels fought against the dragon; and the dragon fought and his angels,

And prevailed not; neither was their place found any more in heaven.

And the great dragon was cast out, that old serpent called the Devil, and Satan, which deceiveth the whole world; he was cast out into the earth, and his angels were cast out with him.[1]

A Land with Real Monsters

Like Frankenstein, I was struck with astonishment at the enormous monster which my investigations had . . . called into existence.[2]

Dragons held a powerful place in the collective memories of those who lived in the Victorian age. Children learnt in the nursery and in the classroom of the legend of St George slaying the dragon. The archangel Michael was but one of many saints and heroes who appeared as a dragon-killer. According to chapter twelve of the Book of Revelation, the dragon and Satan were one and the same. John Milton, in Book IV of his epic poem *Paradise Lost* describes how the dragon 'came furious down to be revenged on men'.[3] There was a vast array of dragons in classical myth and in fairytale. Hydra, the water serpent with many heads, was destroyed by Hercules.[4] The ogres of English nursery tales were often synonymous with the dragons of gypsy tales of south-east Europe. The *Penny Magazine* of the Society for the Diffusion of Useful Knowledge described the dragon as 'by far the most poetical fabrication of antiquity'.[5] Consecrated by religions, objectified in mythology, 'it got mixed up with fable and poetry and history, till it was universally believed, and was to be found everywhere but in nature'.[6] But Nature was soon to yield a very different story. It began from a rather unlikely quarter: a watering place on the

69 (*facing page*) Thomas Jeavons, after J. M. W. Turner, *Lyme Regis, Dorset,* *c.* 1827–38, with the famous breakwater known as the Cobb to the extreme left, and the Blue Lias exposed on the near shore

Dorsetshire coast. This was Lyme. Its most important claim to fame
was as the landing-place of Monmouth's rebellion. First-time visitors
could not fail to register the massive curve of its sea-wall, the Cobb,
'the most beautiful sea-rampart on the south coast'.[7] Devotees of
Jane Austen would have read of Lyme in her novel *Persuasion* (1818).
The wide sweep of its bay, backed by dark cliffs, made it among
the most picturesque of settings for watching the flow of the tide or
'sitting in unwearied contemplation'.[8] Along the coast to the west
of the town was the famous undercliff, with its 'green chasms' and
'orchards of luxuriant growth',[9] a tumble of collapsed rock strata
much like that found west of Ventnor in Wight. Here, though, one also
got a sense of what some of the coastal strata of Lyme were soon
to reveal to an incredulous world. In high summer, the undercliff
west of Lyme was the nearest one came in England to a tropical
jungle.[10] It afforded a first glimpse of a primeval environment in which
pre-Adamite reptiles stalked a hot and humid earth and swam in its
warm seas.

The geology of this part of England's southern coast forms the base
of the Jurassic era, as chapter one has already described. Among its
limestones can be found a formation known as Blue Lias, 'lias' being a
corruption of 'layers' and referring to the alternate beds of limestone,
interspersed with clay or marl, which quarrymen over centuries had
been accustomed to extract for use in building. The Lyme quarrymen
gave the layers names: 'Table Ledge', 'Split Ledge', 'Glass Bottle',
'Rattle', 'Mongrel' and 'Speketty'.[11] What they also registered was that
the Lias abounded with what later became known as fossils. At the
time, they appeared as strange petrified creatures, or as the partial
remains of creatures. There were snake-stones (ammonites), devil's fingers (belemnites) and all kinds of shells. More puzzling were the bony
relics of what seemed like crocodiles, identified by their snouts and
great pointed teeth. The quarrymen collected the best of such relics and
took them home. Townsfolk accustomed to walking or scavenging the
beaches around Lyme did likewise. All sorts of speculations attached to
the petrified forms. For some, they were the gift of God, just part of the
interior ornament of the earth.[12] For others, they had supernatural
properties, curative of human ills on the one hand and illustrative of
the devil's visitations on the other.[13] For quarrymen and townsfolk
alike, what made Lyme so significant was the perpetual instability of its
coastal cliffs. It was not the Lias itself that was unstable but the Chalk
further above it. The permeable Chalk rests mostly on impermeable
Gault Clay and intermittently slips, fractures and tumbles.[14] The landslip between Beer Head and Branscombe, west of Lyme, in 1789–90,
afforded a spectacular example.[15] When coupled with the erosive work
of streams and the action of waves and tides, the outcome was that

Lyme and its surrounding cliffs and beaches were perpetually revealing fresh fossils. Indeed, Robert Bakewell, author of one of the earliest introductions to geology, considered the fossils of the Blue Lias to offer the most important geological character of any stratum of the British series.[16]

At Charmouth, on the coast two miles east of Lyme, where stage-coaches between London and Exeter made regular stops, a local man, William Lock, got into the habit of attending the coaches and offering for sale examples of fossils to interested passengers.[17] Described as 'curiosities'or 'curios', Lock (known locally as Captain Cury) used the common or vulgar names for the petrified specimens he sold.[18] He appears to have become rather successful as a fossil vendor, for by the very early nineteenth century, collectors were coming from London and elsewhere to deal with him.[19] One pioneer palaeontologist, William Cunnington, labelled him a 'confounded rogue' – not, apparently, to maintain that Lock was dishonest, but that he had a sharp nose for business and for the prices that the best curios could command.[20] At Lyme itself, Richard Anning, a local carpenter, followed Lock's example, combing the coast and displaying the fossils he had found on a table in front of his house.[21] Here, in season, they were seen by Lyme's increasing numbers of visitors (very probably including Janes Austen when her family stayed in the town in 1803 and 1804[22]), and Anning was thus able to augment the income of his otherwise impoverished family. In 1810, Anning died of consumption.[23] He left a widow and two young children, Joseph and Mary, with little means of support. However, all the family, it seems, had become accustomed, when Richard was alive, to search for fossils and this now became their principal occupation and their means of livelihood, albeit a precarious one.[24] At low water, the two children, in particular, would roam the beach and search the ledges of Blue Lias. In 1811, poking out of the marl below Black Ven, Joseph observed part of a 'crocodile'.[25] His sister subsequently traced the rest of the skeleton, all of seventeen feet long, and, helped by some of the local quarry-workers, the creature was disinterred and carried away. The local Lord of the Manor bought it for the sum of £23, intending it for his private collection. Later,

70 Head and eyes of an Ichthyosaurus, from William Buckland's *Geology and Mineralogy . . . ,* 1836

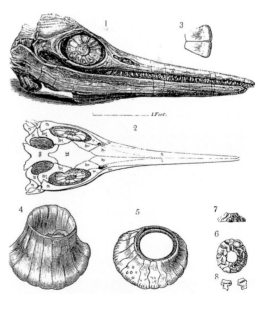

however, the fossil passed to Bullock's Museum in London where, in 1819, it was purchased by the British Museum for the much inflated sum of £47.5s.[26] The Annings's crocodile turned out to be 'a denizen of the new world that the embryonic science of palaeontology was beginning to reveal'.[27] The fossil came first to general notice in a short entry in the *Gentleman's Magazine* in 1813.[28] It gained the attention of the 'gentlemen of science' when Sir Everard Home published an account of it in 1814.[29] It was eventually named in 1817 as Ichthyosaurus, or fish-lizard, although its precise anatomy remained somewhat confused and waited upon the discovery and examination of further examples, one of which had been assembled by the Annings by September 1818.[30] By the time of the publication of Thomas Milner's *Gallery of Nature* in 1846, ten species of this fish-lizard had been enumerated ranging from five to over thirty feet in length. The animal had the 'snout of a porpoise, the head of a lizard, the teeth of a crocodile, the vertebrae of a fish . . . [and] the paddles of a whale'.[31] It appeared to unite in itself 'a combination of mechanical contrivances' found 'distributed among three distinct classes of the animal kingdom'.[32] Air-breathing, cold-blooded and carnivorous, it soon became known as the sea-dragon, a representative of a primeval, pre-Adamite world. In some eyes, it appeared to answer the words of Milton:

> With head uplift above the waves, and eyes
> That sparkling blazed, his other parts besides
> Prone on the flood, extended long and large
> Lay floating many a rood, in bulk as huge
> As whom the fables name of monstrous size,[33]

For others, such was this creature's conformity with dragons of antiquity, that one could almost be lead to suppose that 'among the buried learning of the earlier nations there lurked some knowledge of geology'.[34] Commentators soon marvelled in terrible fascination at certain of the animal's features. The eye, for instance, was enormous, the orbital cavity, in one species, being fourteen inches in its longest direction.[35] The eye also seemed to have a peculiar construction: 'to make it operate like a telescope and a microscope'.[36] Then there were the jaws, up to six feet long, filled with a vicious array of sharp, interlocking teeth.[37] Here, indeed, was a creature greater than any poet's fancy.

The Annings's talent for fossil-hunting was rewarded once more in the winter of 1823 when Mary disinterred at Black Ven a set of remains that bore no relation at all to anything so far seen. It was named in the following year as Plesiosaurus, and immediately took the palaeontological world by storm.[38] What was so startling about the nine-foot skeleton was its serpent-like neck, apparently as long as, or longer than, the body and tail combined. The neck seemed to have had

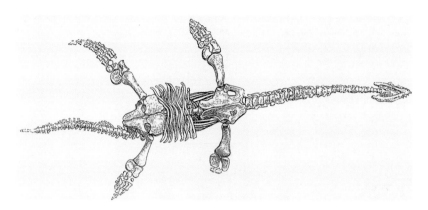

71   The skeleton of a Plesiosaurus, from William Buckland's *Geology and Mineralogy . . .* , 1836

thirty or forty vertebrae which confounded all extant animal anatomy. Even birds had up to only twenty-three vertebrae.[39] The creature was like a serpent threaded through the body of a turtle.[40] Here was a new sea-dragon more remarkable in its way than Ichthyosaurus.

There were some palaeontological authorities who initially thought the Plesiosaurus a forgery, given the measure of its anatomical deviation. But once the doubts were laid to rest, the Annings became recognized in palaeontological circles as reputable dealers in fossils and Mary herself became an object of interest.[41] Rather like William Smith, she lacked any formal scientific education and yet possessed not only great genius for discovery but great judgement and skill in extraction and development.[42] She became known by the scientific community as Miss Mary Anning and began supplying that community with drawings of fossil fragments.[43] According to one personal encounter, she had so thoroughly acquainted herself with the science of vertebrate anatomy that the moment she found bones in the Lias she seemed to know to what tribe they belonged.[44] The same authority also remarked how readily she was able to write to and talk with 'professors and other clever men on the subject'.[45] In a letter written to Charlotte Murchison in 1828, she remarked how much she enjoyed 'an opposition among the big-wigs'.[46] Some five years later she wrote again to Charlotte to say that she had not seen a geology professor for a year and thus knew no more of what was going on in geology than the man in the moon.[47] By 1840, however, the heaps of letters coming in from all over Britain and elsewhere made her 'almost ready to cry'.[48]

The Plesiosaurus and the Ichthyosaurus were inhabitants of primeval seas. But in 1828, Mary Anning uncovered the fossil remains of a dragon of the air, a flying fossil reptile. This was Pterodactyl, described

72   Cartoon sketch of
Mary Anning by
William Conybeare

by William Buckland as 'a monster resembling nothing that had been seen or heard of upon the earth, excepting the dragons of romance and heraldry'.[49] Its body and legs resembled that of a lizard, but it had claws like a bat and an elongated beak containing teeth like a crocodile.[50] What made the creature yet more improbable was that it had no surviving counterpart in the modern world. Whereas the Ichthyosaurus was extinct in its species, the genera survivied in the great Ganges crocodile. The flying dragon, by contrast, had 'passed utterly away'.[51] The discovery of the Pterodactyl at Lyme inspired, in 1829, a drawing that showed a great dragon 'spreading its wings over a storm-tossed, ship-filled and rocky Lyme seascape'.[52] The contrast with Turner's famous watercolour of Lyme Bay (1812), in all its romantic and picturesque imagery, could not have been starker.

The narrative presented by Lyme had a counterpart on the geologically similar Yorkshire coast.[53] Whitby, in particular, demonstrated rock strata much like those of Lyme. It soon had its own fossil dealers, with a local carpenter, Brown Marshall, dominant among them, much as Mary Anning dominated Lyme. In due course, fossil reptiles from the Yorkshire coast also found their way into museums up and down the land, specimens of the Ichthyosaurus and the Plesiosaurus prominent among them.

Mary Anning went on to discover many more new fossils, including 'fish more numerous than the present sea produces'.[54] She was soon so well-known by many of the scientific savants of her age that illustrious foreigners made pilgrimages to Lyme specifically to visit her.[55] The great English anatomist Richard Owen recalled a geological excursion at Lyme with Mary in November 1839 when they had almost been swamped by the tide.[56] William Buckland, who had been brought up at nearby Axminster, knew Lyme well and later used to bring his children to go 'fossilizing' with Mary, often ending up wading to his knees.[57] The dangers of the Annings' occupation cannot be underestimated. In 1833, Mary's dog was killed instantly when the cliff collapsed upon him.[58] On another occasion, she and her workman were almost drowned retrieving part of a Plesiosaurus; subsequently it made her cold to think of the incident.[59]

At much the same time that Mary Anning and her family were probing the Lias cliffs of Lyme for the remains of extinct creatures, Gideon Mantell was exploring the stone pits and quarries of the South Downs. Even though trained as a surgeon, and becoming at the age of twenty-one a partner in a busy medical practice in the town of his birth, Mantell still found the enthusiasm and energy rigorously to pursue his unravelling of the fossil and geological record hidden beneath his native Sussex landscape.[60] The task, however, proved no easy one. Whilst the chalk rock formations were readily distinguished, including the many invertebrate fossils of the former chalk sea, those rock beds in the area known as the Weald proved much more of a puzzle. And the puzzle deepened as Mantell's investigations proceeded. One quarry, at Whiteman's Green, at Cuckfield, became an especially important source of material, Mantell eventually forming an arrangement with the quarryman whereby fossils uncovered in the regular process of excavation were parcelled and dispatched to Lewes.[61]

As these consignments accumulated, it began to dawn on Mantell that the environment in which the fossilized remains (of plants and of animals) had originally thrived bore no relation to those of the chalk formations. There were fossil parts of plants and trees that differed fundamentally from the vegetation that clothed the Wealden countryside of Mantell's day. In its place there was a tropical flora. There were also vast numbers of bones and teeth of animals. But these remains were so damaged that it was exceptionally difficult to decipher the creatures to which they belonged. Mantell knew

73   The quarry at Whiteman's Green, Sussex, from Gideon Mantell's *Geology of the South-East of England*, 1833

of the Annings's discoveries at Lyme and thought at first that he had
uncovered the skeletons of a similar world. However, the bone fragments did not at all compare with those at Lyme. Their structure was
adapted not to surviving in the sea but to moving on land. They were
also a great deal larger. Mantell's conclusion was that he was likely
looking at a kind of crocodile, and the lush tropical vegetation was its
habitation. However, the quarries had not yet yielded a jaw, and some
of the teeth that had been discovered alongside the bones were not the
teeth of a carnivorous reptile. They had, instead, the characteristics of
a herbivore – similar to those of a modern mammal like a rhinoceros.
Mantell sought advice from various members of the scientific establishment of the day (of which he, as a provincial from a dissenting
background, without a university education, was excluded), but there
appeared to be little interest in his discoveries. William Buckland even
suggested that the bones came from the younger Tertiary epoch and
not from an epoch older than the Chalk.[62] After much persistence,
however, Mantell established similarities between some of the teeth he
had discovered and the modern Iguana. Initially labelling the creature
Iguanosaurus, he was subsequently persuaded to call it Iguanodon,
that is having the teeth of an iguana.[63] But what was by then beyond
dispute was the colossal size of this extinct reptile. Mantell reckoned
its length to have been up to seventy feet, fourteen times that of the
modern Iguana.[64] He arrived at the figure by comparing the sizes of the
bones and teeth of the Iguanodon with the Iguana.[65] And so it was that
Mantell cast himself as having given birth, so to speak, to a monster.
He was astonished at its potential magnitude, and sought constantly to
reduce its proportions.[66] Here, then, was a dragon of dry land, to be
set alongside the sea-dragons of Lyme.

G. F. Richardson, the curator of Mantell's Brighton museum when it
first opened in May 1836,[67] imparted the Iguanodon's story in verse:

> 'Tis indeed a world of wonder,
> Found within the earth and under;
> . . .
> Forms as wild as fancy wishes,
> Monster lizards, stony fishes;
> Fragments of the lost amphibia,
> Here a femur, – there a tibia; –
> Here the monster mammoth sleeping,
> Here the giant lizard creeping[68]

For Richardson, the Iguanodon realized the existence of 'monsters
wilder and more wondrous than ever Oriental fancy ha[d] portrayed!'[69] Buckland, addressing a meeting of the British Association,
suggested that such lizards were of so colossal a size that an elephant

would seem but a shrimp by comparison.[70] He added that as you journeyed across Sussex, the former 'Country of the Iguanodon', you would 'probably crush beneath your carriage-wheels the remains of creatures which, had you lived a hundred thousand years ago, might have turned the tables, and crushed you'.[71] Even if Buckland's time-scale was wrong, it was no wonder that Robert Bakewell was prompted to remark that Mantell would ride on the back of his Iguanodon 'into the temple of immortality'.[72]

Among the many bones and teeth that Mantell had assembled from Whiteman's Quarry at Cuckfield were some that did not belong to the Iguanodon. He initially thought they were part of a giant crocodile.[73] It later emerged, however, that they were

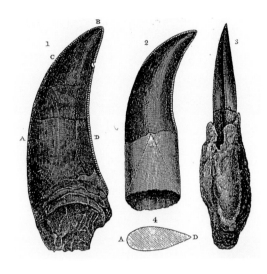

74 Teeth of a Megalosaurus, from William Buckland's *Geology and Mineralogy . . .* , 1836

very similar to the bones of a great carnivorous fossil lizard preserved in the Ashmolean Museum in Oxford.[74] The bones had been found in the famous Stonesfield quarries some miles to the north of the city. These same bones, in turn, had been observed to be similar to ones found in 1800 by Cuvier, the famous French comparative anatomist, at Honfleur.[75] What Mantell had stumbled upon in the Cuckfield quarry was the animal that became known as the Megalosaurus.[76] It was Buckland, though, who gained the credit for bringing it to public notice – at a meeting of the Geological Society in London in 1824,[77] the same one at which one of Mary Anning's later examples of a Plesiosaurus skeleton was reported upon and displayed.[78] The name Megalosaurus reflected the gigantic size of the Stonesfield fossil, thought by some authorities to have been up to forty feet long.[79] Moreover, Mantell could demonstrate from his Cuckfield specimens a set of bones belonging to the same kind of animal that were bigger still.[80] Buckland's Megalosaurus appeared to be amphibious, but not marine like Mary Anning's Ichthyosaurus or Plesiosaurus.[81] By contrast, Mantell regarded his Iguanodon as wholly a land-dwelling reptile.[82]

Like Mary Anning at Lyme, Mantell's wonder and satisfaction at his discovery was soon added to by the finding of another reptile specimen of an altogether different sort. In 1830, he had been much dismayed when visiting Whiteman's Quarry to learn that another collector had supplanted him and was obtaining all the best specimens from the quarrymen. It appeared to end all hopes of his ever obtaining a jaw of the Iguanodon.[83] However, when the quarry workers, two years later, were rock blasting and found a mass of bones, they agreed that Mantell was the cleverest man to examine them.[84] Having retrieved the best of the skeleton, Mantell managed to assemble a block four and a half feet long and two and a half wide. It became his 'celebrated new fossil'.[85] And what was most striking about the creature was a set of fifteen-inch spines that were exterior to the skeleton.[86] Soon Mantell registered that he had discovered a new reptile altogether, naming it Hylaeosaurus, that is, forest lizard.[87] In effect, he had found an armoured monster. He presented the news to a fascinated Geological Society in early December 1832.[88]

The Iguanodon story, though, had not yet run its course, for in 1834 Mantell received information from a quarry-owner just south-west of Maidstone in Kent of a mass of bones preserved in a block of Kentish ragstone. W. H. Bensted, the owner, had made drawings and had written descriptions of the fossil which he then sent to Mantell for comment.[89] The ragstone block turned out to contain the best specimen of the Iguanodon discovered up to that time. But unlike the Sussex Iguanodon, this animal was embedded not alongside freshwater fossils but marine ones.[90] The 'Maidstone Iguanodon', as it quickly became known, was soon secured for Mantell's collection and placed on public display in his new museum in Brighton.[91] Here it proceeded to attract much attention, including that of the artist John Martin. Mantell considered Martin's work to be 'the finest productions of modern art', and hoped that he could induce the artist to portray the 'Country of the Iguanodon'.[92] Martin does indeed appear to have been fascinated by Mantell's discoveries of an ante-diluvian age for he proceeded to execute a vast painting for which a mezzotint was later made to provide the frontispiece to Mantell's best-selling *Wonders of Geology* of 1838.[93] Martin depicted a landscape 'diversified by hill and dale, by streams and torrents, the tributaries of its mighty river'.[94] It was clothed with 'arborescent ferns, palms and yuccas', and inhabited by enormous reptiles.[95] The colos-

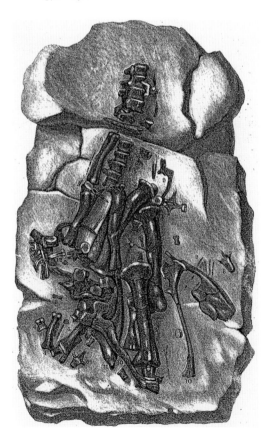

75   The skeletal remains of the Maidstone Iguanodon, from Gideon Mantell's *Wonders of Geology*, 1838

76   The famous mezzotint that John Martin created for the frontispiece of Gideon Mantell's *Wonders of Geology*, 1838

sal Iguanodon and the Megalosaurus were chief of these, but there were crocodiles, turtles, flying reptiles and birds, while the waters themselves 'teemed with lizards, fishes and mollusca'.[96] At the picture's centre, the herbivorous Iguanodon is being attacked by a Megalosaurus and a crocodile, a foretaste of an image, 'nature at war', that was later to send shockwaves through Victorian society with the publication, first, of Robert Chambers's *Vestiges*, and, later, Darwin's *Origin*.[97]

For those who never came to see Martin's renditions of the dragons of the primeval world, there was soon another reminder of these vicious creatures. This was the steam locomotive. As described in the prologue, Tennyson, in 1848, considered the best way to recall the lost world of monster lizards and giant ferns was by standing beside a railway line at night. The flaming eye of the locomotive's open firebox, the thunder of its iron tread, and the vomiting aloft of sparks and smoke, turned the engine into a 'great Ichthyosaurus'.[98] Benjamin Disraeli could still pick up the same theme in his novel *Lothair* some twenty years later, with Theodora complaining how difficult it was to read on a railroad journey 'amid the whirl and whistling, and the wild panting of the loosened Megatheria who drags us'.[99] The Megatherium was a creature resembling a giant sloth, a mammal that had once inhabited the snowy wastes of Northern Europe and Siberia. For a time, the uncovering of its fossil remains captured popular attention as much as the later reptile discoveries of Anning, Mantell and Buckland.

Recovering Former Life-Worlds

These remains of an earlier creation had long been known to the curious, and classed as freaks of nature, for so we find them described in the works of ancient philosophers who wrote on Natural History.[100]

In Cornwall, the tinners often found huge timber trees as they dug their excavations. On the coast at St Michael's Mount, at low water, you could find the roots and branches of trees embedded in turfs under the surface sand. Some of these trees still had fruits on them.[101] The earliest writers on natural history had long remarked on the frequency with which vegetable remains, especially timber trees like oak, were found at the bottom of lakes and ponds, and along the banks of rivers.[102] Canal excavation, from its very earliest years, had regularly disinterred the roots and trunks of trees.[103] The same was true of quarry and mine workings where, in the latter case, whole trees could sometimes slip down from above the working galleries once the seam of coal had been removed.[104] Serious injuries could be sustained by working colliers as a result. Among working men, generally, such subterranean wood was thought to have laid buried in the earth since the Flood.[105] Antiquaries likewise thought of the subterranean trees as having been lodged at the time of the Flood.[106] And so commonly were wood and vegetable remains found buried in England in this manner that the literal truth of the biblical account seemed to be constantly reinforced. Occasionally, though, difficult questions emerged. Why, for instance, were there fossil trees disposed in a vertical position. If they were the relics of a univer-

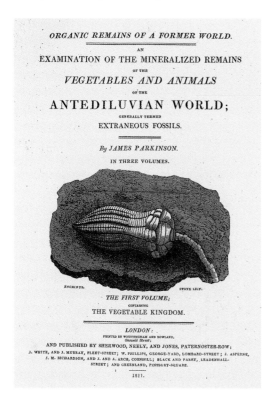

77   Title page of James Parkinson's *Organic Remains of a Former World*, the edition of 1811, illustrating a fossil lily

78   A fossil-hunter prospecting on the southern shores of the Isle of Wight, as recorded by Thomas Webster, *c.* 1812

sal flood, this was hardly consistent. Most trees would be expected to have been violently uprooted and to have floated away (in a roughly horizontal position) before being deposited in like manner once the flood waters receded. As writers on natural philosophy studied the various vegetable remains in more detail, it was also discovered that there were plants bearing no resemblance to any of the earth's known plants: 'hardly any agreement could be found between the fossil vegetable remains and those vegetables with which the earth is at present clothed.'[107] There was a similar 'want of agreement' between the fossil remains of some orders of animals and those currently existing.[108] In the minds of some commentators, the indications were of a world that pre-figured the Flood, what soon became know as an *ante-diluvian* world. As far as living forms went, though, there did not appear to be any line of separation between existing and extinct species, for remains of extinct species (and extinct genera) were often found with the remains of species 'very similar to, if not exactly agreeing with, species known in a recent state'.[109]

Another puzzle was that in some parts of the country there were hundreds of seashells and remnants of sea-fish buried in the earth, or 'enclosed in stone'.[110] Indeed, the teeth of sea-fish appeared more numerous than any other fossil bones.[111] As remarked in chapter one, when the Trent and Mersey Canal was being excavated in Staffordshire in the 1760s, the navvies were astonished to uncover the backbone and vertebra of a giant fish some fifteen feet beneath the surface.[112] Near Reading in Berkshire, naturalists had observed a whole bed of oyster shells lying in a stratum of greenish sand two feet thick. Some of the oysters had their valves or shells still intact.[113] In some parts of Suffolk, fossil shells were so numerous that they were dug up for manuring the fields.[114] Even in coastal cliffs, one faced the conundrum of finding marine shells that were either unknown in any living state or else native of far distant regions of the globe.[115] Where the remains of sea creatures were found in great variety and plenty 'at land', the inexorable logic was that 'the sea ha[d] formerly been there'.[116]

An equally perplexing phenomenon was the way sandstones displayed ripple marks. In the towns of Horsham and Crawley in Sussex, for example, slabs of local sandstone were used for paving. Their surface had the same appearance as sand along the sea-shore at low-water. And sometimes the stone was so rough that it formed ideal material for

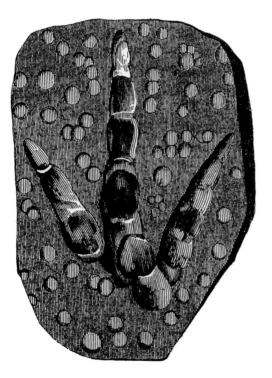

79 Animal footprint from the New Red Sandstone at Corn Cockle Muir in Dumfriesshire, from Thomas Milner's *Gallery of Nature*, 1846

use in stable yards, preventing the feet of animals from slipping.[117] In places, these former seashores displayed the footprints of animals. In 1828, an account was given to Edinburgh's Royal Society of the tracks left by an animal in a quarry in the New Red Sandstone at Corn Cockle Muir in Dumfriesshire. The tracks appeared in great abundance on many successive layers of stone. They seem to have come from a giant tortoise, and presented impressions of astonishing liveliness, recording the creatures as they walked to and fro at the recess of the tide.[118] In other sandstone quarries, slab surfaces had been observed to have a mottled appearance, with little circular and oral hollows. Ultimately, it was determined that these were the impressions produced by raindrops.[119]

It was in limestone strata, especially, that evidence of marine origins was most stark. Some beds, for example, contained shells, corals and exuviae of marine creatures in such abundance that they appeared to be composed of no other materials.[120] As earlier described, alongside the sea-dragons of the limestone cliffs at Lyme, Mary Anning

also found all kinds of fish as well as many sorts of marine inverte-
brates. Some parts of the Lias were even defined by their particular
fosilliferous concentrations.[121] The Chalk strata (chalk being an especially pure form of limestone) consisted of nothing but microscopic
marine life, each cubic inch containing 'upwards of a million of the
shells of these creatures'.[122] The chalk 'ocean' was of so vast a depth
and extent that it likely covered for many ages the greater part of
southern and central Europe.[123] It was a still rather than a stormy
ocean, however, for the Chalk revealed an almost entire absence of
water-worn or abraided pebbles and gravel, and, together with the perfect state in which its organic remains are found, the indications were
of deposition in the tranquil deep.[124] In Wight, even allowing for the
displacement of so much of its strata from the horizontal, the Chalk
demonstrated to its many Victorian observers a depth of over 900 feet,
all of this vast mass having once formed material at the bottom of the
ocean, yet now projecting way above its waves.[125]

In the younger Tertiary beds of Europe, the varied clays and sands
that cover somewhere over half of the land surface, marine shells, turtle carapaces and the skeletons of sea-fish were also found in abundance. Excavations of London's immediate substrata, for instance,
revealed that there was once sea there that teemed with marine life.[126]
However, such excavations also revealed fossils associated with freshwater. And among the Tertiary rocks, generally, there appeared to be a
mingling or alternating of freshwater and marine beds.[127] When the
shells were washed of the clay or sand in which they were preserved,
most had the appearance of being of very recent age, except for being
brittle and having lost their colour.[128] In the west of Wight, Headon
Hill had demonstrated vividly to early observers the 'alternation and
occasional intermixture of fresh and salt-water productions'.[129] One
had to imagine, it seems, that there were 'alternate inroads and retreats
of the sea, coupled with the occasional existence of freshwater lakes'.[130]
Such imaginings were rendered the more real, of course, by the evidence
of large-scale 'derangement' or faulting of the Tertiary strata, which
produced 'elevations and subsidences' of the type that the coastal cliffs
of Wight had so clearly depicted to its geological visitors.[131]

The inter-mixture of seawater and freshwater fossil life was a feature
not confined to the Tertiary beds. It was common, too, in parts of the
Secondary strata, and not unknown in even older rocks.[132] The various
limestone beds around Lyme revealed an alternation of marine
and freshwater conditions. The same was to some extent true of the
Weald of Kent. It was Mantell who first recognized the freshwater
character of some of its beds, most geologists having thought that they
were all of marine origin. Alternations of mud and sand, a general
absence of pebbles and shingle, demonstrated over parts of the area a

80   Fossil ferns from the Coal Measures of the Carboniferous, from James Parkinson's *Organic Remains of a Former World*, 1804

set of sediments forming the deposits of a river not of the sea.[133] Mantell discovered riverine shells, the bones of freshwater turtles and teeth of crocodiles, not to mention fossils of strange plants.[134]

The plants of the Wealden proved, as previously described, to be tropical. Mantell quickly established their broad identity by getting them compared with specimens in the British Museum and with living plants in hothouses such as the ones at Kew.[135] The plants included palms, arborescent ferns 'and the usual vegetable productions of equinoctial regions'.[136] All grew, it seems, upon the vast deltaic area of a wide tropical river inhabited by amphibious and terrestrial monsters. Here roamed the Iguanodon and the Megalosaurus. Turtles swam in its rivers and lakes, while enormous crocodiles basked among the ferns of its shallows.[137] One eminent geologist of the day compared the Wealden delta with the modern deltas of the Ganges and the Mississippi, once extending across much of south-east England and across the Channel into north-east France. Its deposits were up to 2,000 feet in thickness.[138] If Victorians had wanted a more exotic and sublime vision of a former world, the Weald would have been difficult to surpass. Evidence of former tropical conditions, though, could also be found in the much more recent strata of the Tertiary epoch. In the streets of London, for instance, workmen were frequently turning up the teeth and bones of elephants as they dug the foundations of the metropolis's new sewers. Occasionally, whole skulls, complete with tusks, would be unearthed.[139] These were not the elephants of the plains of Africa, but altogether different species. Other parallel discoveries included rhinoceroses, hyaenas and bears. All were clearly strangers to the recent climate, if not always to countries of lower latitudes.[140] In the wide sweep of the Thames estuary, the Isle of Sheppey gave prospecting fossilists a taste of the spice islands of the Moluccas, its beaches strewn with the remains of tropical plants and animals. One could find traces of palm-fruits, of plants allied to the bean and the cucumber. There were fragments of the cases of sea-tortoises or turtles, the vertebrae of crocodiles, the teeth of sharks and even the backbones of giant snakes. All of these in greater part belonged to species that no longer existed.[141] Some of the island's dwellers made a habit of collecting such 'curios' and sold them to visiting fossil-hunters much as the Annings had done at Lyme. Sheppey thus became another frequent point of excursion for all those who developed a fascination with natural history.[142] It was, according to one commentator, as if a magician had turned a spice-island to stone, complete with all its plants and animals, and transported it bodily to the mouth of the Thames.[143]

Tropical conditions were also revealed in geological eras long prior to the laying down of the Wealden beds. In the Coal Measures, for example, part of the Carboniferous series of rocks, dating from more than 280 million years ago, miners were continually discovering the impressions of ferns, flags and reeds, along with the trunks of large succulent plants. Many proved to be of unknown species, strangers to the present world.[144] These impressions were generally found not in the coal itself but in the intervening beds of clay and shale. All classes of vegetable matter could be recognized, including bark, leaves, seeds, stems, roots, fruit and flowers.[145] Some of the stems belonged to giant trees that seemed to have grown up to a hundred feet high, with bases twelve feet in circumference, later to be named Lepidodrendons.[146] At Killingworth colliery, on Tyneside, thirty trees of a similar scale were found above one of the coal seams in an area only fifty yards square.[147] They belonged to the family Sigillaria, the same type that were discovered by navvies excavating the Manchester and Bolton Railway in the 1830s.[148]

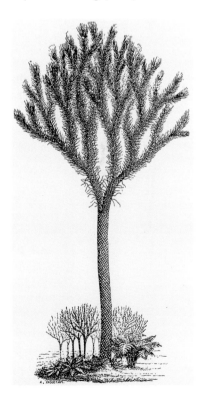

81 Reconstruction of a Lepidodrendron from the Coal Measures, from Louis Figuier's *The World before the Deluge*, 1867

Here, then, was what was once a 'coal forest'. Moreover, in many sandstone and limestone quarries of the Carboniferous series similar fossil trees were being uncovered.[149] Beneath the canopy of these coal forests grew reeds or rushes (Calamites) that sent up stems that sometimes reached forty feet.[150] They were actually the equivalent of the much more diminutive horsetails of the modern era, so common in ditches and in swamps.

To grasp the magnitude of such a forest realm, one had, according to Hugh Miller, to look at present-day ferns and club-mosses 'with the eye of some wandering traveller of Lilliput, lost amid their entanglement, like Gulliver among those fields of Brobignag.'[151] One also had to imagine an excessively hot and humid climate. How else could plants have attained such a scale? Coal was laid down in vast basins akin to the deltas and swamps of modern equatorial regions. Their warm waters swarmed with fishes and other life forms, whilst in the air dragonflies the size of seagulls hovered and darted.[152] In quarries and in mines, in canal trenches and in railway cuttings, Victorians and their immediate predecessors were perpetually having their eyes opened upon the world of the Carboniferous, a vast steamy swamp that by some apparent act of Providence had conferred upon them 'so great a variety of those mineral substances which minister to the necessities and comforts of life' and contributed to the making of so 'opulent and powerful a people'.[153]

Journeying the Vast Deserts of Space and Time

> Ages pass; the land rises slowly over the deep, terrace above terrace; the thermal line moves gradually north; the line of perpetual snow ascends beyond the mountain summits; the temperature increases; the ice disappears . . .[154]

The Victorians' increasing fascination with all aspects of the prehistoric world cannot be understood without grasping the particular ways in which that world was presented to them. Whilst some writers on geology, palaeontology and natural history wrote in a somewhat mundane, matter-of-fact or didactic style, others chose a narrative form to convey the startling realities of the worlds they sought readers to encounter.[155] The key to the success of the narrative style of *Vestiges* was that it involved writer and reader exploring nature together. Gideon Mantell, both in his lectures and in his writings, conveyed listeners and readers back into imaginary worlds full of seductive imagery, and left them openly amazed at the existence of so exhaustive a catalogue of prehistoric wonders.[156] William Buckland wrote his contribution to the Bridgewater Treatises in the form of a traveller's tale.[157] In part, such styles of presentation were necessary in order to 'meet with a sale, proportioned to the expense of the undertaking'. Or, at least, this is how James Parkinson justified his use of the epistolary form in writing on the natural world at the beginning of the nineteenth century.[158] He feared that a 'dry, strictly scientific work' might deter potential readers.[159] The book therefore began with a travelogue in which he first met with the petrified creatures of former worlds whilst taking tea in a wayside cottage.[160] The old woman who gave the travellers hospitality had her own treasured collection of 'curiosities'. The snake-stones (ammonites), it appears, were thought of as fairies turned first into snakes and then into stone on account of the crimes they had committed.[161] Parkinson began, in other words, with some fossil folklore, determined to adapt his work for readers in general.[162] Such pecuniary considerations were undoubtedly behind the style that Mantell chose for some of his geological writings, especially the later ones.[163] But Mantell (and others like him) was also tapping into a romantic stream of thought, the world of fairytale and legend, of dragons and monsters, and then using this as a springboard for populist writing. This was a momentum that became mutually reinforcing. Whilst in one sense there was an element of gothic horror about the discovery of monster reptiles from former worlds, the narratives into which they were soon woven gave them a much less threatening, even comfortable feel. Indeed, the writers of *Punch* found comedy in the genre, much as they had found comedy in geologists' field excursions and in the passion for collecting rock samples. By the late 1840s, *Punch* had the '*Nimrods* of the nursery' riding a Megatherium rather than a rocking-horse.[164] Meanwhile, the vast collections of fossils that were accumulating, particularly fossil reptiles, prompted *Punch* to speculate that Pterodactyls would soon be seen with their long beaks or snouts poking out of holes in the upper floors of houses.[165]

It was perhaps Mantell's *Wonders of Geology* of 1838 that set the narrative pattern. In the words of G. F. Richardson, writing in 1842, Mantell, 'adopting the image of an Arabian writer, introduced an imaginary being, endowed with superhuman longevity and power of observation, and in the position of this fictitious observer, describe[d] the chief geological mutations of our island in a style which combine[d] the most perfect eloquence, with the most accurate adherence to scientific fact'.[166] Thus did Mantell describe part of the 'Country of the Iguanodon':

Countless ages ere man was created . . . , I visited these regions of the earth, and beheld a beautiful country of vast extent, diversified by hill and dale, with its rivulets, streams, and mighty rivers, flowing through fertile plains. Groves of palms and ferns, and forests of coniferous trees, clothed its surface; and I saw monsters of the reptile tribe . . . basking on the banks of its rivers and roaming through its forests; while in its ferns and marshes, were sporting thousands of crocodiles. Winged reptiles of strange forms shared with birds in the dominion of the air, and the waters teemd with fish and crustacea. And after the lapse of many ages I again visited the earth; and the country, with its innumerable dragon-forms, and its tropical forests, all had disappeared, and an ocean had usurped their place. And its water teemed with nautili, ammonites, and other cephalopoda, of races now extinct. . . . And countless centuries rolled by, and I returned . . . the ocean was gone, and dry land again appeared, and it was covered with groves and forests, but these were wholly different in character from those of the vanished country of the Iguanodon. And I beheld, quietly browsing, herds of deer of enormous size, and groups of elephants, mastodons and other herbivorous animals of colossal magnitude. . . .[167]

Mantell, in other words, conjured up a picture of a country where once an Orinoco had 'rolled its waters'.[168] And when the great river had departed it was replaced by a vast 'chalk' sea. Thus did Gideon Mantell introduce readers to the ante-diluvian world.

Charles Dickens echoed Mantell's technique in a story in *Household Words* in 1851, though accorded his account a greater measure of intellectual sophistication. He related the tale of a phantom ship on an ante-diluvian cruise:

So we walk down Cheapside, bustle aboard at London Bridge, and sail out, leaving man behind us. . . . We have pased the Nore; we are on the ocean of a world which has not felt the footsteps of its master. Land ho! Then let us go ashore. This is some part of South America; there rolls a mighty river . . . we plunge into dense forests; let us now sit down under the trees and speculate upon the world, into which we spirits of the future have receded.[169]

Dickens had not just employed the idea of time-travel but he had substituted space for time by describing the landfall as forming some part of South America. Knowledge of such distant and exotic realms was increasingly being made available in illustrated books, the pictures far more numerous than in the past by virtue of cheap wood engraving. Londoners and those visiting the metropolis could also obtain vicarious experience of these realms by visiting Kew Gardens, particularly the hothouses, where the heat and the humidity of the tropics was reproduced. In parallel, the Zoological Gardens in Regent's Park gave many their first sight of hippopotamuses, rhinoceroses, elephants and other living creatures of the torrid zones.[170] However, Dickens's ante-diluvian cruise did not meet with such animals. Travellers instead came in sight of something altogether different. They observed a creature whose hind legs were 'three times more massive than an elephant's'.[171] The guess was that it was twenty feet long with a tremendous tail. Here was a 'ten-navvy power'.[172] Here was a monster of the prehistoric world. Trees and underwood gave way like grass at the approach of this mighty form.[173]

Readers of Hugh Miller's immensely popular writings on geology and the natural world also found themselves being taken on imaginary journeys back in time. Scotland's great Caledonian valley became an 'ocean-sound', the Pentland Hills submerged to half their height and the landmark of Arthur's Seat above Edinburgh a rock that appeared only at half tide.[174] All around there was drift ice, colonnaded in hoar-frost. Icebergs appeared like cathedrals complete with towers and spires. They gleamed in shades of emeralds and sapphire, 'the light polarized by a thousand cross-reflections'.[175] This was the Ice Age. Miller argued that modern science did not have to be 'adverse to the exercise and development of the imaginative faculty'.[176] It could, in his view, be as richly charged as 'those ancient tales of enchantment and faery which beguiled of old'.[177]

For Miller, the Jurassic epoch was an age of 'reptiles, reptiles, reptiles, – flying, swimming, waddling, walking; – the age of the cold-blooded, ungenial reptile'.[178] But the forests of the Jurassic were not planted for man, and as 'sinister eyes peere[d] inquiringly round', it was as well that the reader could return to the 'safer and better furnished world of the present time'.[179] No monsters reigned here. They were the victims of the 'great law of death that constantly pressed on the geologist'.[180] Such former life-forms lay 'as thickly on the surface of each of many thousand layers as leaves along the

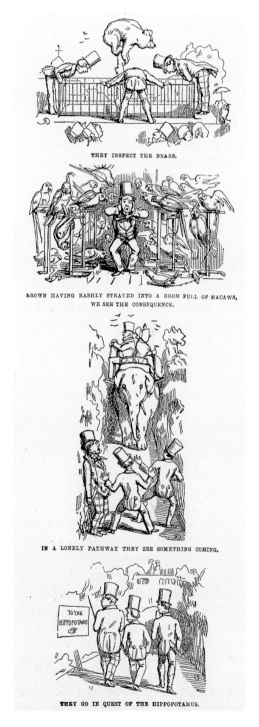

THEY INSPECT THE BEARS.

BROWN HAVING RASHLY STRAYED INTO A ROOM FULL OF MACAWS, WE SEE THE CONSEQUENCE.

IN A LONELY PATHWAY THEY SEE SOMETHING COMING.

THEY GO IN QUEST OF THE HIPPOPOTAMUS.

82 *Punch* visits the zoological gardens, 1850

forest glades on autumn'.[181] The whole incalculable past of the planet was one of production and decay going on inseparably.[182] When Miller was cruising around the Inner Hebrides in the Free Church yacht Betsey, he found in some of the coastal shales an abundance of reptilian bones of the Jurassic, apparently belonging to Plesiosaurs. He tried to explore the beds where they were exposed by the fall of the tide: 'fain would I have swam off . . . with my hammer in my teeth, and with shirt and drawers in my hat; but a tall brown forest of kelp and tangle . . . rose thick and perilous'.[183] Here Miller was not just exploring the 'wonders' of former worlds but turning it into an adventure. And to add further to the confection, there was the parallel account of the cruise itself, with the 'deep roar of the surf', and 'its undulating strip of white flickering under stack and cliff' as they ventured out into the Sound of Mull'.[184] Rounding the headland of Ardnamurchan, the passengers gazed back in the slant of the evening light at Mull's 'green acclivities', horizontal shadow lines marking the step by step rise of the terraces of volcanic rock.[185] The appeal of Hugh Miller's narrative was incontestable.

The famous French writer Jules Verne was shortly to carry this kind of narrative a stage further in his fictional thriller *Journey to the Centre of the Earth* (1864). The book cleverly integrated knowledge and discovery with science and adventure.[186] Beginning in the crater of an extinct volcano in Iceland, the travellers of the story descend into the bowels of the earth through one of the old lava 'chimneys'.[187] For the narrator, the fascination of the void is irresistible. His hair stood on end and vertigo rose to his head like intoxication. Through the Pliocene, the Miocene, the Eocene, the Cretaceous and the Jurassic geological eras they descended, then on and on to the 'primordial stratum'. After ten hours, they were ten thousand feet below the level of the sea, and found themselves in a gallery lined with lavas studded with gleaming crystals of opaque quartz. The gallery lead them off some miles to the south. Eventually, the coating of lava on the walls and floors gave way to solid rock. They were back in the Secondary Epoch of limestones and shales. They began to discern the impressions of primitive organisms in the rock walls. As they walked further, they found themselves surrounded by coal. But then they came upon a new tunnel that plunged back into the bowels of the earth. After dicing with death, the travellers reached an underground sea. It had the 'capricious contour of earthly shores', but was 'utterly deserted and wild in appearance'.[188] They were in a huge cavern, the overhead part of which was composed of huge clouds, a mass of 'shifting and changing vapours'.[189] As the travellers began walking again, their eyes cast upon a tall dense forest, the trees 'shaped like parasols, with sharp geometric outlines'.[190] Coming closer, it was easy to recognize that these were the lowly shrubs of the earth grown to phenomenal dimensions. It was as if the

83   Reconstruction of an Ichthyosaurus and a Plesiosaurus in combat, from
Louis Figuier, *The World before the Deluge*, 1867

scientists had constructed a huge hothouse of ante-diluvian plants. And
when they looked upon the floor, their eyes cast upon bones, unmistakeably those of ante-diluvian reptiles. Eager to explore further, the
travellers took to a raft and set sail on the underground sea. The story's
narrator found his imagination carried away. He had a prehistoric daydream. The whole fossil world of extinct animals came vividly to life.
As the travellers voyaged on, the narrator began to 'gaze in terror' at
the sea, expecting one of the monsters of the Jurassic to leap into
sight.[191] In a few hours, they were indeed awoken by a violent shock.
The raft had been lifted clear of the water. They were amidst a herd of
sea monsters. The reptiles seemed to move 'at a speed greater than any
express train'.[192] Soon they witnessed two of the creatures engaging in
mortal combat. They were none other than Ichthyosaurus and
Plesiosaurus. They attacked each other 'with indescribable fury', 'raising mountainous waves which rolled as far as the raft'.[193] For two
hours the contest raged until both dived, leaving a whirlpool in the
water. After some minutes, an enormous head shot out of the water. It
was the Plesiosaurus in its death throes. It had been cut in two, its head
severed from its body.

Journey to the Centre of the Earth has been described as 'a poetic
elaboration of the prosaic facts of scientific geology'.[194] Indeed, it seems
to have been planned 'as a geological epic'.[195] Verne was assiduous in
ensuring that his information was up-to-date. Not only was he well-
read in geology and in palaeontology, but he was also conversant with
some of the latest ideas and theories about the earth's interior.[196] The

book also explores the dissonance between the Biblical account of Creation and the scientific accounts that decreed millions of years of earth history, including the gradual evolution of plant and animal life.[197] In Verne's story, the days occupied in the descent to the centre of the earth are equated with the successive geological epochs, mirroring some of the contemporary attempts to reconcile scripture and geology. The heroes of the story thus appear directly to experience the days of Biblical Creation.[198] Verne also achieves a very clever play on the relations of space and time. The descent in space is simultaneously a passage back in time. The discrete geological layers are testimony to this journey.[199] But they indicate, too, that time's components ultimately break down – matter is turned to dust, much as Hutton and Lyell described the process of earth decay. Such is the 'verve' of Verne's narrative skill, that the journey back in time is seamless, continuous, like a film run backwards.[200] When the travellers reach the underground cavern, it is not a dead world that they witness but one that is vividly alive: 'past time is literally there to see'.[201] The interlude of the story of the prehistoric daydream forms a moment when the narrative is actually suspended and forward progress interrupted, and the narrator speculates on nature and on existence.[202]

Jules Verne's novels were immensely popular in Britain. First translated into English in 1872,[203] Verne took the astonishing discoveries of the prehistoric world and converted them into high drama. But he also engineered a voyage 'to the centre of the self', recalling the all too familiar doubts and unbeliefs that Victorians had been forced to contemplate.

The Craze for Natural History

> How lovely are the shells – how symmetrical – how beautiful! How vivid their colouring; how elegant their form; their convolutions, how delicate.[204]

According to *Chambers's Journal* of 1855, the great business of seaside resorts, next to bathing, had become 'the collection of shells and weeds, and creeping things'.[205] Since the appearance of the natural history books of Philip Henry Gosse and Charles Kingsley, everybody seemed to have one or more of those glass tanks called vivariums or aquariums.[206] Victorian dressing-rooms became the repositories of such strange receptacles, while bathrooms in turn became homes for the items the tanks could not hold.[207] Instead of spending six weeks in summer at a watering place, making bad sketches, walking up one parade and down another, reading the silliest of novels, and staring out the window with a telescope at an empty horizon, Kingsley argued that you try to discover the 'Wonders of the Shore'.[208] His book of that name began life as an article in the *North British Review* of November 1854. For Kingsley, you would, at every step, find phenomena 'stranger than any opium-eater dreamed'.[209]

In the twenty or thirty years preceding, natural history, generally, had passed from being honourable to being fashionable to being a popular pastime.[210] High society began to purchase natural history specimens as objects of art.[211] Books on natural history were finding their way more and more into drawing rooms and schoolrooms. One of the most popular works of the 1840s was Thomas Milner's *Gallery of Nature* (1846).[212] The thirst for such knowledge seemed unquenchable. Daughters were seized by 'Pteridomania', the collecting of ferns. Instead of 'crochet, Berlin wool and "fancy-work"', they contemplated the 'real beauties' of nature.[213] Members of London's burgeoning clerical classes could be found, by night, wandering in Epping Forest with lantern, jar of strange sweet compound, and pocketfuls of pillboxes. They were not poaching or deer-stealing but searching for moths. Back in their Pooter-ish terraced dwellings, they would show you 'glazed and cork drawers' full of the delicate insects.[214]

The *Penny Magazine*, as early as the mid-1830s, had tapped into the popularity of natural history. Over 1833 and 1834, it ran a whole sequence of articles on the 'Mineral Kingdom'. From Section 8 onwards, it dealt with 'Organic Remains', reminding readers how fossils revealed 'the important and wonderful fact that the Author of Nature had created different species of animals and plants, at successive and widely distant intervals of time, and that many of those that existed in the earlier ages of the globe had become totally extinct . . .'.[215] Although the remains of 'saurians', or 'dragons', featured prominently,[216] the various articles took in fossil plants, including those of the coal period,[217] as well as fossil animals of the most recent epoch, among them the famous Irish Elk. The bones of this particular animal were found in such abundance in the bogs and marl-pits of that island that they had long ceased to be regarded as curiosities by local inhabitants. Even so, no reader could fail to be astonished by the Elk's size, the specimen preserved in the museum of the Royal Dublin Society having a spine nearly eleven feet long and horns or antlers measuring nine feet between their tips.[218] Other articles in the magazine examined 'Curiosities of Natural History', focusing on plants and animals still living, very much in the manner of Gosse and Kingsley's later writings on the 'wonders of the shore'.

By 1856, *Chambers's Journal* claimed that almost every county had its own natural history society.[219] Sometimes these were founded alongside, or as part of, museums of natural history. At Ipswich,

Ransomes, the farm implement manufacturers, founded such a museum in 1848. The firm's owners hoped their own workmen would attend the lectures provided. It was also hoped that the hundreds who 'now sauntere[d] in the fields, uninterested and without object', might turn into 'cultivators of flowers, rearers of pigeons, or collectors of insects'.[220] Mechanics' Institutes, for all their implied emphasis on engineering and technology, were no strangers to the craze for natural history, either. Edwin Sidney, rector of Little Cornard in Suffolk and renowned popular lecturer, was a frequent guest of the Sudbury Literary and Mechanics' Institution, where, as well as lecturing to audiences on geology and on soils, he also spoke on the fossil animals of the Tertiary epoch, including various species of fossil elephant found whilst navvies were excavating the railroad between Sudbury and Bury.[221] In Mrs Humphry Ward's novel *Robert Elsmere* (1888), Robert, as the parish priest in the village of Murewell, ran a natural history club, based in the Workmen's Institute. Glazed compartments had been fitted into the windowsills and were filled with specimens: eggs, butterflies, moths, beetles and fossils. There were two large books in which members were encouraged to record their observations. Jars full of strange creatures stood on the tables. The club had turned loutish youths and common ploughboys into amateur naturalists.[222] In north-west England in the early to middle nineteenth century, interest in entomology, zoology and botany was considerable among artisan groups. Natural history became a communal pursuit among such men, involving a social network that was based around oral exchange, its values tied to the artisanal values of the workplace.[223]

Specifically among the middle classes, Charles Dickens memorably parodied the craze for natural history in his sketch of the Gradgrind family in *Hard Times* (1854). The younger Gradgrinds knew nothing of nursery rhymes or nursery tales, but they had 'cabinets in various departments of science', including ones for fossils. And all had been lectured on these subjects from the very tenderest of years.[224] *Blackwood's Magazine* had touched much the same mocking theme in 1837 in 'Notes of a Naturalist'. These told of a white and pink spotted caterpillar that had been spied on a 'broom-shaped vegetable'. Its excited collector spent a whole evening devising a suitable abode: 'an old lozenge-box with six holes bored in the lid with a pin'.[225] Some twenty years later, the same magazine began a sequence of 'Sea-side studies' that parodied (among others) Kingsley's *Wonders of the Shore*. The 'proper equipment for a day's hunting' included baskets, nets, wide-mouthed phials, wooden-capped bottles and an oyster knife.[226]

In chapter four of his *Geology for Beginners* (1842), George Richardson offered a complete guide for novices to rock and fossil collecting. He incorporated lists of dealers, of societies and of lectures, not to mention recommendations on the best books and guides to read.[227] Fossil botany, he wrote, 'enables us to restore the vegetation and temperature of the primeval earth at a period, when the English vales were rich savannahs, thick and matted jungles, or rank and swampy marshes, abounding in gigantic mosses, colossal herds, or huge aquatic plants'.[228] The study of fossil shells led one to imagine the nautilus and the ammonite:

> Each sent to float, in its tiny boat,
> On the wide wild sea of life!
>
> For each could swim on the ocean's brim,
> And when wearied its sail could furl;
> And sink to sleep in the great sea-deep . . .[229]

When a traveller brought home to you a madrepore or 'brain-stone' from a pacific coral reef to display on your mantlepiece, you came to learn that its first cousins were the soft, slimy sea-anemones seen in every rock pool around the coast, but without a trace of bone or stone in them.[230] In observing and seeking to analyse the natural world of the present, you were inexorably drawn into the natural world of the past. The identification of new species was impossible without it. And alighting upon new species (and their naming) became one of the Victorian naturalist's central pleasures and delights.[231] You might go down to the seashore after a gale and pick up a couple of delicate sea-ferns. Under a good pocket magnifier they seemed nearly identical. In fact, to much surprise, you learn that the 'two species of animal which have formed them are at least as far apart in the scale of creation as a quadruped is from a fish'.[232] Such was the popularity of books on the seashore that Philip Henry Gosse's son, Edmund, was later led to remark how armies of collectors ultimately helped to destroy the marine paradise that first inspired their composition. The rock basins of the shore became 'profaned and emptied, vulgarized'.[233] A 'fairy paradise' was 'crushed under the rough paw of well-meaning, idle-minded curiosity'.[234]

One vital agency in this process was the increasing facility for cheap illustration in books. Charles Kingsley's *Wonders of the Shore*, in the fourth edition of 1859, incorporated a fine set of coloured lithographs that conveyed more than any lines of prose the startling shapes and colours of submarine life in the shallows. Woodcut engraving had already revolutionized book illustration and the new steam presses of the 1830s permitted drastic reductions in the price of such printed media.

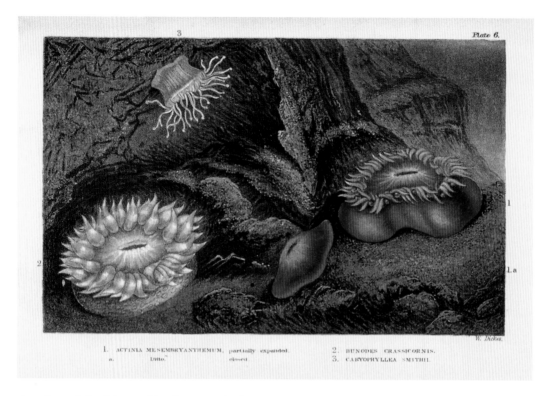

84   One of the lithographic plates in Charles Kingsley's *Glaucus, or Wonders of the Shore*, 1859

Charles Knight's famous *Penny Magazine* was founded on the back of these techniques. Lithography, however, did not involve any re-working of an artist's drawing. It was a medium that could handle anything from a pencil or chalk sketch to a detailed painting in oil. With its introduction, the exact repetition of pictorial statements became possible.[235]

Gideon Mantell's fossil museum in Lewes (later removed to Brighton) was, in a sense, a progenitur of the popular volumes on natural history that began to flood the cheap book market from the 1840s. Visitors to the museum first found themselves facing a large drawing that represented a restoration of the Sussex Weald. At its centre was a vast river, 'some mighty Nile, or still mightier Mississippi', which, it was conceived, had flowed through the district. Around its streams and banks were the strange, singular forms, both animal and vegetable, that had 'peopled' it.[236] It was this scene that partly inspired John Martin's painting *The Country of the Iguanodon*. Once visitors had taken in the scene, they then entered the apartments containing the fossil relics, all arranged in cases in their due order of succession.[237]

The drawing that graced Mantell's museum may have afforded part of the inspiration for the 'extinct animal park' that was made at Sydenham in the early 1850s as part of the project to re-erect the Crystal Palace there.[238] Mantell had been consulted about the scheme, but he was by then very ill and did not survive to see its fruition.[239] His role as an advisor was assumed by the anatomist Richard Owen, formerly an archrival. And Owen, in collaboration with Benjamin Waterhouse Hawkins, set about the task of re-creating, in life-size, the dragons of the primeval world.[240] They chose to set them on islands in a 'tidal' pool or lake. As the *Illustrated London News* recorded, 'at a convenient distance from the spectators, islands of irregular shapes will be placed, and covered with luxuriant vegetation. On one of these islands will be placed, in natural attitudes, and amid appropriate vegetation, animals of the secondary, and others of the tertiary period'.[241] It was nothing less than the restoration of a lost world. Engineers were to arrange the waters of the pool to give a five-foot rise and fall, having the effect of partially submerging the amphibious inmates, just as if it was an actual tide.[242] Visitors would see the Iguanodon pausing amongst the rushes, the Megatherium would appear 'in the act of climbing an ante-diluvian tree', whilst giant turtles, jaws gaping, would bask upon the pool's banks.[243] The pool was thirty acres in extent and was designed to be viewed from ground forty feet in height.[244] The task

85   A wood-cut from Richard Owen's guide to the extinct animal park at Sydenham, the re-erected Crystal Palace in the background, the restorations of giant extinct reptiles around the tidal pool, 1854

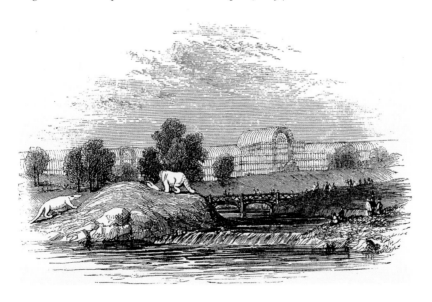

of actually constructing the extinct animals became as absorbing a spectacle as did their permanent display. From miniature models in clay, Hawkins and his workmen, advised by Richard Owen, began constructing in a workshop at Sydenham the set of life-size animals in concrete. The Iguanodon, the largest of them, amounted to a scale of project 'not less than building a house on four columns'.[245] The central iron supports were nine feet long and seven inches in diameter. To provide the bones, sinews and muscles, Hawkins used 100 feet of iron hooping and twenty feet of cube-inch bar iron. The flesh and skin of the animal were then made from 600 bricks, 650 two-inch half-round drain tiles, 900 plain tiles and 650 bushels of artificial stone (i.e. concrete).[246] As a publicity stunt, Hawkins then organized the famous banquet of New Year's Eve, 1853, when the inside of the Iguanodon was converted into a *salle à manger* in which distinguished guests wined and dined to the memory of the famous men whose investigations of geology and the natural world had made the whole enterprise possible.[247]

The press did not hesitate in rising to the occasion. The *Illustrated London News*, for instance, included a startlingly clear engraving of the 'Extinct Animals' model-room at Sydenham, with the Iguanodon occupying the centre of the picture. Alongside it was the armoured bulk of the Hylaeosaurus, together with various other fiercesome-looking creatures.[248] Simultaneously, *Chambers's Journal* was recording the restorations of this lost world. It related how visitors would see glaring at them from the island in the middle of the lake a set of gigantic diluvian animals 'of a period unknown for its remoteness'.[249] Their creators had, according to the account, 'rendered the Park a great lecture-room on geology'.[250] What is more, no effort was spared to ensure that there was easy public access to it. The Crystal Palace Company engaged with the Brighton and Croydon Railway Company to carry a branch in to the Park itself. Passengers leaving from London would be able to purchase a return fare to Sydenham that would include 'admission to all that the Crystal Palace and its grounds [could] afford'.[251]

The satirical magazine *Punch* quickly discovered in the 'extinct animal' park a repeated source of copy. Children were depicted being lead unwillingly around it to improve their minds.[252] Other visitors were presented as having terrible dreams as an outcome of the monsters they saw.[253] The Park at Sydenham became, in effect, a sensation. Hundreds of thousands of people flocked to see it. Richard Owen wrote a guide-book to it.[254] It was no matter that the life-size models were actually rather inaccurate representations of the fossil structures that Mantell and others had discovered.[255] Their scale and the savagery they portrayed were enough to fire popular imaginations already familiar with

THE ANTEDILUVIAN REPTILES AT SYDENHAM—MASTER TOM STRONGLY OBJECTS TO HAVING HIS MIND IMPROVED.

86   *Punch* offers a caricature of the extinct animal park at Sydenham

tales of supernatural dragons. Here, according to a writer for *Blackwood's Magazine*, one came face to face with the 'first primeval earth', an original world of dazzling 'sunbright seas', 'palmy verdure' and 'gorgeous unknown flowers'. In its midst, though, giant scaly monsters wallowed in 'pure mud', jaws gaping and with 'eyes in which no speculation dwells'.[256]

The Sydenham spectacle was actually a form of narrative, for as well as distinguishing animals of the different geological epochs, it also afforded visitors 'fully proportioned representations of the strata in which the remains of the beasts were found.[257] In a sense, as one proceeded through the Park, one became a time-traveller, exploring the unfathomable abyss that the early geologists like Hutton and Lyell had found themselves forced to invoke in order to make coherent their ideas about earth history.[258]

5

'Washing Away a World'

In the six hundredth year of Noah's life, in the second month, the seventeenth day of the month, the same day were all the fountains of the great deep broken up, and the windows of heaven were opened. And the rain was upon the earth forty days and forty nights.[1]

Prelude

Of all the stories recounted in the Book of Genesis, the Flood was the one that resonated most with Victorian senses of the sublime. It was not the ark that was important, but the overwhelming of a landscape by a great deluge. Among commonplace perspectives on earth history, the Flood had, as previously described, become a vital point of discontinuity. It marked the substitution of old world by new. Everything that was part of the old order became ante-diluvian, thought of as belonging to an era before the Flood. For devout Christians, of course, the catastrophe of the Flood afforded a vivid reminder of man's capacity for sin, and this was reinforced in the more apocalyptic-sounding Deluge. For some natural philosophers and geologists, though, the Flood had a vital *material* significance. It helped to explain the shape and form of the surface of the earth. It also made more credible the vast array of dragon-like creatures that had once inhabited its climes. They were *ante-diluvians*, rendered suddenly extinct by the great inundation of the waters. The biblical flood provided a repeated source of imaginative inspiration over the early to mid-nineteenth century. It is the story of an extraordinary engagement of art, religion and science, one in which seemingly opposing conceptions of truth seemed to sit comfortably side by side in men's minds.

Re-imagining the Deluge

> Ye wilds that look eternal, where shall we fly?
> Not to the mountains high,

87 *(facing page)* John Martin, *The Bard*, 1817 – man pitted against the wild forces of nature, the setting based on the Allendale Gorge in Martin's native Northumberland, a setting that he was later to re-use in *The Deluge*, 1826

For now their torrents rush with double roar
To meet the ocean.[2]

The British have never been strangers to great floods, even in
everyday life. On an island with a predominantly westerly air-stream
tracking across 3,000 miles of ocean, rain and wet weather have long
been inscribed in national character, quite apart from the role of rainfall in sculpting the landscape. During the early nineteenth century,
however, the effects of floods attracted a spate of books, pamphlets and
articles. Often these were not merely factual reports, but accounts that
turned such events into dramas.

In 1818, William Garret published a booklet that recorded the great
floods that affected the rivers of Northumberland and Durham in the
years 1771 and 1815. Horses, cattle and sheep had perished in their
hundreds in the raging torrents; coffins were torn out of churchyards;
the living and the dead clashed indiscriminately; river bridges and
whole sides of houses collapsed into the cascading water.[3] In 1830,
likewise, readers of *Blackwood's Magazine* were treated to extracts
from Sir Thomas Dick Lauder's account of the great floods that had
affected Morayshire in August 1829. They had been signalled, so Dick
argued, by the *aurora borealis*, or northern lights, appearing with
'uncommon brilliancy' at the beginning of the preceding month.[4] As
flood levels rose, trees and other wrecked detritus floated past like
straws. At times the force of the flow sounded like gunpowder ignited
within the confined tube of a cannon.[5] Lauder's book, which went into
a second edition almost immediately after publication, turned a natural disaster into a spectacle, a powerful illustration of the sublime, but
the more vivid because it could be enjoyed in the safety of the drawing-
room or the parlour.[6] As Garret remarked of the Tyne floods, 'the
shrieks of women and children with all the agonies of despair, will better be conceived than described'.[7]

By 1838, subscribers to the *Gentleman's Magazine* were reading of
the very destructive floods that had affected the principal rivers of
Yorkshire in December 1837. In Leeds, the water levels exceeded any
known since 1775. A steam boiler weighing four tons was wrenched
from its location on a river-bank. There was a great loss of property
throughout the county and three of Bradford's inhabitants were
drowned.[8] Nor were other river basins strangers to such inundations.
In January 1834, Charles Darwin's sister, Catherine, had written to
him from the family home in Shrewsbury of the constant deluges of
rain that they had experienced that winter. There had been 'three
Floods of the Severn': no frost or snow, just constant pouring rain –
more than anyone could remember.[9]

For those poor souls who shared in the anguish and terror, and for
the farmers and householders who lost possessions, such events could

not fail to recall to mind the story of the biblical flood, or deluge. Indeed, Sir Thomas Dick Lauder, in the conclusion to his account of the Moray floods, seemed unambiguous about the force of such association. The deluge poured down the mountainsides 'in a thousand cataracts, like the mighty host of God's destroying angels'. All was 'sudden dismay, clamour and dread'. The living, together with all their material goods, were left 'suspended over the depths of eternity'.[10] However, Moses' famous account of the event traditionally relied for its impact on the thought and fear that it invoked in men's minds. It forced them to recall their potential for wickedness and their fragile tenure of the occupancy of the earth. Despairing of the capacity for sin that man had revealed, God had used one of the elements of Creation, water, to visit upon the living earth a terrible cataclysm. Only a chosen few, namely Noah and his family, were to survive the Deluge.

The story in the Book of Genesis was perpetually re-told in the home and in the classroom, whether as part of nightly bible-readings, or as exercises in learning the art of reading itself. In George Eliot's *The Mill on the Floss* (1860), Maggie Tulliver found herself recalling the biblical catastrophe as she was swept downriver in the rising flood-waters that had followed days of incessant rain: she remembered 'that awful visitation of God which her father used to talk of – which had made the nightmare of her childish dreams'.[11] From Anglican pulpits, parsons had long recounted the story of Noah and the Ark and of its extraordinary survival in the face of the 'fountains of the deep'.[12] There were psalms that told related tales: of the heads of dragons being broken in the mighty flood (psalm 74); of 'water-floods' that threatened to swallow up the fallen (psalm 69). Among conscientious souls, the Deluge was a perpetual reminder of the need to achieve absolute purity and self-denial. For some non-conformist brethren, the Deluge became one of the symbols of eternal punishment – to be set alongside the awful spectre of the Day of Judgement as a means of inculcating obedience and humility in wayward souls. The Lord God was an all-consuming, jealous God. The Deluge became like the everlasting fire that wiped out Satan and all his angels. It helped to configure in men's minds a truly morbid sense of sin.

The overwhelming spirit of eighteenth-century enlightenment, of course, was to believe and hope only in the material and the practical.[13] In this sense, the familiar but invisible biblical horizons were slowly shrinking and did so apace from the mid-nineteenth century.[14] And the Deluge, for all its occasional material reminders in the inhabited world, was no exception. What happened instead is that it was swept from the world of religious faith and life into the world of the artistic imagination. For according to one writer in *Blackwood's Magazine*, all that was *ante-diluvian* had become poetical – from the eve of the Deluge right back to the Creation and to the Fall.[15] It was Byron's *Heaven and Earth* (1823) that provided one cue for the magazine's writer. The poem was 'solemn, lofty, fearful, wild, wicked, tumultuous and shadowed all over with the darkness of a dreadful disaster'.[16] At the approach of the flood, Byron described the deep sounds that howled from the mountain's bosom. There was no breath of wind yet leaves quivered and every blossom dropped as the earth groaned. When the waters rose, men flew in search of safety, most to be overwhelmed by the roaring torrents. Only the ark, floating in the distance, provided hope.

Byron's poem was tapping into a centuries-long tradition of artistic renderings of the Deluge in poetry and in plays.[17] But during the 1820s and 1830s, the genre attained an unrivalled display. In 1818, Pleydell Wilton published his *Geology and other Poems*, focusing on the 'submarine commotions' of the Deluge and the sedimentary processes that were the result. In 1826, the Reverend William Bassett published his *Molech; or The Approach of the Deluge*, a sacred drama in which the uncorrupted sons of God became ante-diluvian heroes. Bassett was not especially concerned as to the truth of this viewpoint, but it was sufficient for the purposes of his poetry.[18] Thirteen years later, John Reade penned a Deluge drama, in twelve scenes. When the fiercesome winds awoke the lightnings, every mountain, every crag, every cave and every deep were for one moment ablaze in the most intense light, only to be swallowed the next moment 'in Darkness as a grave!'[19] Then the flood followed:

> Through Earth's rent sides the waters of the Deep
> O'er the low plains deliriously sweep,
> In waves like rolling mountains,
> And towers of men are borne before the floods;
> Or, Crushed in one enormous mass. . . .[20]

Reade's drama appeared in the same year as a poem by McHenry: *The Antediluvians; or, The World Destroyed*.[21] It was reviewed at great length in *Blackwood's Magazine*, even if the writer of the review seemed uncertain about its qualities of the sublime. But since he (or she) claimed to be less affected by Byron's *Heaven and Earth* than by a paragraph in the *Westmorland Gazette* telling of the loss of a post-chaise, horses, driver and a pretty girl in the quicksands between Lancaster and Ulverston, perhaps this was a rather jaundiced view.[22]

The Deluge was just one of many episodes of the supernatural in scripture that had been catapulted by poets and writers into the realm of the artistic imagination. The great revival of interest in Milton's epic poem *Paradise Lost* had started the process in the later part of the eighteenth century. This had even extended to pocketsize abridgements for mass consumption, turned out by none other than the Methodist leader John Wesley.[23] And in the 1830s, Charles Darwin even had a pocket version with him while circumnavigating the world on HMS *Beagle*.[24] Solomon Gessner's poem *The Death of Abel*, translated into English in 1761, went through countless editions over almost three quarters of a century. Gessner began where Milton left off, producing a work that united the epic, the tragic and the pastoral.[25] The death of Abel was cast as the most remarkable event recorded in sacred history between the Fall and the Deluge. Here, as in *Paradise Lost*, the romantic and the sublime were helping to re-validate the supernatural. The scriptures were being theatricalized, cast as works of mystery and imagination.[26] Byron's *Heaven and Earth* was subtitled 'a mystery'.

In the early nineteenth century, the Milton revival acquired a new champion in the painter John Martin. Indeed, Robert Montgomery, author of the poem *The World before the Flood*, regarded Martin as a second Milton.[27] Martin was the child of a fiercely Protestant upbringing and, throughout his life, the stories of the Bible, especially the Old Testament, were to preoccupy him.[28] Soon these were being translated on to canvas, producing panoramas that embodied a stupendous sense of theatre. Hilltop castles, precipitous crags and swirling elements were combined to create renewed sensations of the sublime, the whole awash with references to good and evil, the terrestrial and the eternal.[29] Martin produced his first 'Miltonic' picture in 1813.[30] And in 1823 he received a commission to illustrate *Paradise Lost* in twenty-four mezzotints. The resulting engravings proved a runaway success. Four differently priced sets of Milton's epic were produced and there were many subsequent editions, the last in 1876.[31] Even among the critics, there were few who did not recognize the grandeur and mystery of Martin's designs for *Paradise Lost*.[32] He set out the contours of Heaven and Hell in a manner that turned them into high drama. The fame of the mezzotints, with their startling chiaroscuro, contrasts of light and darkness, spread far and wide.[33]

It was inevitable that, of all the stories of the Old Testament, the Deluge would attract Martin's genius. In fact, he was working on a large painting of it, almost nine feet long and over five feet high, at much the same time that he was engraving the plates for *Paradise Lost*.[34] Of all the scenes from Genesis, the Deluge had come to enjoy an extraordinary vogue among artists. For those in Britain, this apparently sprang from

88   Nicolas Poussin's *Winter, or the Deluge*, 1660–64

a single prototype: Nicolas Poussin's painting *Winter, or the Deluge*, one of a series of pictures depicting the seasons, executed by the artist between 1660 and 1664.[35] Until Poussin's work, the traditional manner of pictorial treatments of the Deluge was to place Noah and the Ark as central components. Poussin halted this pattern by making the Deluge into a landscape; 'an atmospheric landscape with jagged rocks, convoluted trees, and water moving turbulently in the center of the painting'.[36] The colouration of the painting added enormously to the effect. According to John Opie (1809), there was 'neither white, nor black, nor blue, nor red, nor yellow'; the whole mass was a 'sombre grey'.[37] Nature seemed to have become 'half-dissolved', 'verging on annihilation'.[38] When a British observer, Henry Milton, first saw the painting in the Louvre in Paris in 1815, he wrote how 'The effect produced by the mere colouring [was] most singular and powerful. It convey[ed] to the mind such an image of the destroying element, as no exposition of its actual effects could have produced'.[39] Henry Milton went on to remark how the picture seemed to reveal a quality of air 'burdened and heavy with water' with 'the very light of heaven . . . absorbed and lost'.[40] Never before, he argued, had a picture been more in accordance with its subject.[41]

It was little wonder, then, that it inspired a host of similar scenes by later artists, not to mention direct copies of the design. The picture also spawned at least twelve engravings.[42] Of similar scenes, no less than eight Deluge paintings were exhibited in England between 1770 and 1800, including one by Phillipe de Loutherbourg, an adaptation of his earlier shipwreck scenes, which focused on the suffering of three central figures.[43] Executed in 1789–90, it was later engraved for *Macklin's Bible* and enjoyed considerable popularity.[44] Some English critics, though, found Poussin's picture too tame as an inspiration: it was merely the inundation of a valley, according to one authority.[45] Thus nineteenth-century scenes of the Deluge revealed a preoccupation with the depiction of tempest and disaster. These were themes that had already become a kind of stock-in-trade for painters by that time.[46] What then followed were 'some of the sublimest of all romantic paintings'.[47] These were the Deluge pictures by Turner, Danby and, of course, Martin.

J. M. W. Turner completed his first version of the Deluge around 1804–5, apparently prompted by a visit to the Louvre in 1802 when he studied Poussin's picture.[48] Like Poussin, Turner chose to focus on Nature rather than on people. But he departed from Poussin in his desire to depict the cataclysmic violence of the Flood. Thus whilst the lower portion of the canvas is filled with people in varying states of drowning, the overwhelming impact of the picture comes from the palpable energy of the elements. To the right, a vast tidal wave is surging

89   J. M. W. Turner's *Deluge*, 1804–5

across the landscape, part of a fierce storm that has already flattened trees in the violence of the winds and swamped a boat full of people. The distant horizon has escaped the storm for the present and in its centre floats the ark, presumably with Noah and the animals aboard. Whereas Poussin's picture was buried in a striking uniform colour of gloom, Turner chose rich colours for his picture, in the manner of Titian and Veronese.[49] And when the painting was exhibited at the Royal Academy in 1813, the catalogue entry was accompanied by lines from Milton's *Paradise Lost*:

> Meanwhile the south wind rose, and with black wings
> Wide hovering, all the clouds together drove
> From under heaven . . .
> . . . The thicken'd sky
> Like a dark cieling [*sic*] stood, down rush'd the rain
> Impetuous, and continued till the earth
> No more was seen.[50]

Turner was plainly eager to remind those who saw the painting that the biblical story was the ultimate source of his inspiration. The landscape in the left middle distance appears to be the Garden of Eden, and the rich colouration of the painting may in part stem from that arcadian image.[51] Milton's poem also goes on to describe how the Garden of Eden was washed away in the Flood. Turner's Deluge was subsequently engraved, in a modified design, by I. P. Quilley. The large print appeared in 1828 and presented the picture to a wider audience.

The most famous and widely known rendering of the Deluge was undeniably John Martin's. And there was a sense in which, of all his immediate contemporaries, Martin was uniquely placed to explore the point of convergence between nature, catastrophe and apocalyptic sublime.[52] Even before the success of the mezzotints for *Paradise Lost*, Martin had been attracting public attention with his grandiose and theatrical paintings. In 1816, Princess Charlotte and Prince Leopold appointed him their Historic Landscape Painter in the wake of the Royal Academy's showing of *Joshua Commanding the Sun to Stand Still upon Gibeon*, in which the first of Martin's visionary cities makes its appearance amidst a vast mountainous landscape where weather effects have an almost palpable presence and human figures are reduced to the most diminutive of scales.[53] The painting that really set Martin's fame, though, was *Belshazzar's Feast* of 1821. Exhibited at the British Institution, it attracted so large a viewing public that it had to be roped off and the exhibition itself extended by three weeks.[54] Struck by its public attraction, Martin set about engraving the picture, but his work on the illustrations for *Paradise Lost* delayed publication until 1826. It was then an immediate best-seller, unlettered proofs tripling in price in three years, with Martin having to engrave a second plate in 1832.[55] In the meantime, Martin had been working on his painting of the Deluge, exhibited at the British Institution in 1826. Initially, the picture attracted limited interest and was unsold. It was returned to his studio to be 'worked up', later finding its way to exhibitions in Edinburgh.[56] Here it attracted the attention of one of the correspondents of *Blackwood's Magazine*. At first, it seems an incomprehensible picture, but the longer one gazes at it the more intelligible it becomes.[57] Another observer, Edwin Atherston, this time for the *Edinburgh Review*, praised the artist for 'his power in depicting the Vast, the Magnificent, the Terrible, the Brilliant, the Obscure, the Supernatural and, sometimes, the Beautiful'. No painter, according to Atherston, had succeeded so well in representing the 'immensity of space'.[58] Martin produced a four-page descriptive pamphlet to accompany the painting, including the relevant lines from Byron's *Heaven and Earth* and from the Book of Genesis, inviting observers of the picture to exercise their own imagination when responding to it. It was an example of art that did not live by methods of painting alone, as one later viewer commented.[59]

Martin always considered the Deluge to be among the most important of all his works,[60] and so it comes as no surprise to find that he

90   John Martin's mezzotint of 1828, based on his *Deluge* painting of 1826

quickly produced an engraving of it – in 1828. The resulting mezzotint was a roaring success. Some 600 copies were printed, most of them selling at two and a half guineas apiece.[61] Even after Martin's death, the plate was still being reprinted.[62] Although the engraving and the painting were not identical in their pictorial composition and although painting as well as engraving plate were re-worked (the latter necessarily through wear), the broad features of the composition remained the same. The focus is overwhelmingly upon Nature, whether the power of its dynamic forces or the character of darkness and light. The setting for the picture is a vast rocky gorge that is about the be engulfed by water coming not down the gorge but from above its precipitous sides. As the immense force of water is unleashed, so rock masses on the mountain sides are seen being cast into the gorge below. Martin gave the vertical scale of the gorge as 15,000 feet,[63] in other words a stupendous natural rift more akin to the Alps than to northern Europe. In its broad shape, the gorge clearly mimics the gorge of his earlier painting *The Bard* (1817).[64] Both actually took structural inspiration from the Allendale Gorge in Martin's native Northumberland which he had explored many times as a boy. Here the prospects were indeed 'alpine' in their grandeur, the ravine of the River Allen 'wild and romantic'.[65] All around were 'jutting crags, covered with trees, loose masses of rock, and lofty precipices'.[66] Legend had it that over one of these precipices, near Staward Peel, a horse-stealer had once leapt to avoid capture, escaping with only a damaged clog.[67] Above the point where the East and West Allen rivers unite, the water gathers force as a succession of smaller streams tumble down from the surrounding moors. In summer, severe convectional storms often transformed the Allen and its tributaries into raging torrents. A better setting for the Deluge could hardly have been imagined.

Martin's Deluge is simultaneously suffused by darkness, not the darkness of heavy storm clouds but the darkness associated with a solar eclipse. What light there is becomes incandescent as an outcome, picking out the terrified poses of humans as they perceive the magnitude of the approaching cataclysm, illuminating in sharp relief the tumbling rocks on the side of the gorge, and reflecting on the roaring surf of the tidal vortex. The contrast of light and darkness was revealed at its starkest in the mezzotint. The artist's son, Leopold, described the quality of the picture's light as 'awful'.[68] It derived from a very careful use of ink, the ink blended from burnt oil, whiting and Frankfort black.[69] Soon after the first mezzotints had been engraved and offered for sale, Martin produced an eight-page pamphlet about them,[70] a narrative guide much like the shorter one that he had produced for the oil painting. In it, he incorporated stanzas from Bernard Barton's poem, written shortly after the mezzotint was released.[71] They encapsulate contemporary reaction to it:

> The Awful vision haunts me still!
> In thoughts by day, in dreams by night;
> So well had art's creative skill
> There shown its fearful might.
>
> The flood-gates of the foaming deep,
> By power supreme asunder riven;
> Heaven's opened windows, – and the sweep
> Of clouds by tempest's driven; –
>
> The beetling crags which, on the right,
> Menace swift ruin in their fall;
> Yet rise on memory's wistful sight,
> And Memory's dreams appal.[72]

Aside from its focus on Nature, Martin's Deluge was above all a representation of a biblical story and both of the descriptive pamphlets helped to guide observers in their readings and interpretations. The location of the ark, for instance, is identified, high on a rock outcrop where it would be removed from the shocks and heavings of the earth and from the breaking waters.[73] The old man in the bottom right of the picture is Methuselah, whom Martin had ingeniously calculated to have been alive at the time of the Deluge. Meanwhile, the man whose mouth is being covered by his wife's hand is a blasphemer, an illustration of the ante-diluvian wickedness that the flood was intended to punish.[74] As previously noted, the thrust of enlightenment thought was increasingly to cast doubt on the literal and historical truth of the biblical account of the earth's history, but Martin seemed eager to reconcile science and religion in his picture. Familiar with the relationship of the moon to the ebb and flow of the earth's tides, he linked the Deluge to a celestial conjunction: the coming together of sun, moon and comet, or, as one commentator describing this aspect of the painting wrote: 'a fearsome moon, a'drenched in blood, in conjunction with a fiery comet'.[75] The gravitational pull of the comet was what drew forth the 'fountains of the deep' as recalled in Genesis.[76] And, interestingly, there were theories of the earth then current among natural philosophers that argued just such a view.[77] Martin made further scientific allusions in terms of the cave that he pictured in the bottom right of the picture. It is most clearly delineated in the earliest mezzotints and at the cave's entrance there stands a hyena. Martin added to this in his descriptive pamphlet, referring to the existence of other dens of ferocious animals. And in the mezzotint, in particular, a whole array of ante-diluvian creatures, including a herd of mastodon, can be seen. These references clearly related to the discoveries by the Oxford geologist William Buckland of the skeletal remains of hyenas and other tropical animals in a cave in Kirkdale in Yorkshire.[78]

There was clearly a sense in which Martin was offering onlookers a kind of encyclopaedic art.[79] And this feature was demonstrated further when some fourteen years after completing the original painting, he made two more paintings in order to amplify the record. These were *The Eve of the Deluge* and *The Assuaging of the Waters*, both exhibited in 1840. Martin himself stated that he had found it impossible to execute the epoch of the deluge in one design and so resorted to three separate scenes.[80] He was encouraged to make the new pictures by Prince Albert, even if it is plain that the ideas had been conceived some years before.[81] *The Eve of the Deluge* depicts a scene from nature: a broad valley, bounded by mountains, of a style Martin had used in many earlier compositions. The distant ground of the picture is bathed in twilight. In the darkening sky to the left, comet, moon and sun are revealed at an earlier point in their coming conjunction. In the left foreground, Methuselah, surrounded by Noah's family, is reading from a scroll on which the impending catastrophe is prophesied. In the valley floor, under trees, revellers of the ante-diluvian age appear unaware of the fate that awaits them. In the far distance, on the right, is the ark, ready to rescue Noah, his family and the animals.[82] As with the *Deluge*, Martin published a pamphlet to accompany the picture in which he set out a detailed description of the scene. He was at pains to cite his theological sources: *The Book of Enoch*, preserved by the Ethiopians; *Hebrews* and *Jude* from the New Testament; and Cartland's *Josephus' Antiquities*.[83] Martin's third deluge painting, *The Assuaging of the Waters*, is ostensibly set within the same landscape as the second but now drowned beneath a vast expanse of water, save for a rocky promontory in the left foreground. The overwhelming 'motif' of the picture is rebirth, even including the frame that Martin designed specially for it. The sun is just beginning to break through from behind a passing mass of cloud, casting a bright glow upon the seascape and highlighting the presence of a white dove plucking an olive leaf to take back to Noah to indicate the presence of dry land. The flowers and leaves of the framing mimicked this part of the biblical story. Near to the dove, a raven rests on a tree branch around which a drowned serpent is entangled. The serpent, according to the biblical story, had caused the catastrophe by tempting man to sin.

Both of these additional 'Deluge' pictures, as well as reflecting Martin's desire to extend his conception, can also be related to changing Victorian taste. *The Eve of the Deluge* arguably fitted a more melodramatic preference, and a reviewer in *Blackwood's Magazine* complained that Martin had failed sufficiently to represent awe in the picture. The colours were 'gay' rather than 'ominous'. There was nothing to convey 'wickedness on the earth'.[84] *The Assuaging of the Waters* tied in with the evolving appeal of sentimentality in the arts.[85] And once again, Blackwood's reviewer disliked the violence of the colour.[86] It remained true, however, that the symbolic significance of the extinction of a world, as recounted in the deluge story, was still powerful amidst the backcloth of depression in the early 1840s and the social turmoil that accompanied it.

The appearance of Martin's new deluge paintings in 1840 seems to have prompted Turner to revisit the subject.[87] The outcome was two new pictures, both exhibited at the Royal Academy in 1843. However, Turner used the occasion to treat the deluge in an altogether different manner, subordinating individual details to overall effects.[88] Colour was one aspect of this, the artist using the work to explore Goethe's controversial ideas about colour (1810) which ran counter to Newton's optical experiments. In the first painting, *Shade and Darkness – The Evening of the Deluge*, the dominant colour is a chill blue. In the second, *Light and Colour (Goethe's Theory) – The Morning After the*

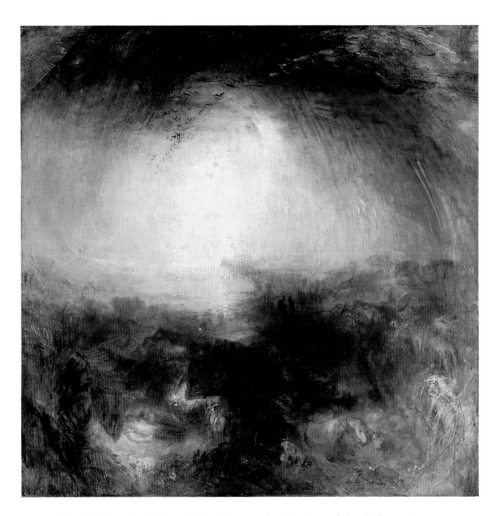

91   J. M. W. Turner's *Shade and Darkness – the Evening of the Deluge*, 1843

Deluge – Moses Writing the Book of Genesis, it is a lively yellow that dominates.[89]

Another aspect of the subordination of detail was the way Turner completed the pictures as octagons within which were formed circular outlines.[90] This alluded to the idea of cycle, whether the succession of night by day that the two paintings evoke, the hydrological cycle in which floods are merely one component in the system of condensation, rainfall, run-off and evaporation, or, more widely, the idea of rebirth from elemental inundation, part of James Hutton's succession of worlds.[91]

The idea of cycle, of course, was not without biblical allusion, particularly in relation to salvation. And it was here that Turner sought to reassert the 'purely human and mythical status' of the Flood.[92] In the centre of the second painting, Moses is seated on a cloud-wreathed mountain top, recording the cataclysm for posterity. The first painting has the ark in its centre background. As such details are few in both paintings, their presence takes on particular significance.[93] In fact, they afford a striking counterpoint to the colour spectrum represented in the portrayal of Goethe's theory. The two paintings are thus readily telescoped into one, in a manner that distinguishes them from Martin's pair of deluge paintings. In turn, the contrast with Poussin's *Deluge*, the prototype for depictions of the Flood among British artists, could hardly have been starker. One reviewer thought *The Evening of the Deluge* 'the strangest of things', unable to make out what was shade and what was darkness. *The Morning After* was similarly 'quite hopeless to make out'.[94] For all Turner's mastery of his craft, the message of these two pictures proved obscure for some viewers.

So frequent a subject had the Deluge become among leading painters in the nineteenth century that any one of them who had not yet portrayed it seemed compelled to do so as a matter of maintaining status and reputation. Among this group was the Bristol painter Francis Danby. Danby seems first to have toyed with the idea in the mid-1820s,[95] but after making studies for the painting, he abandoned the project. The success of Martin's *Deluge* (particularly his profitable and acclaimed engraving of it) was one reason, but Danby also considered that Martin's painting was a plagiarism on his own planned composition, even if the substance of this charge appears arcane.[96] Ten years later, though, Danby's patron and picture-dealer commissioned him to undertake a deluge painting. The bargain involved a canvas of vast size, ten feet by fifteen, larger than any of the deluge pictures made by Martin and by Turner.[97] The picture was completed in 1840 and exhibited privately in London. Why it was not exhibited at the Royal Academy is unclear. The *Blackwood's* reviewer certainly thought that it would have created a sensation there.[98] The idea for so large a picture seems partly to have been inspired by Danby's studied observation of Géricault's famous *Raft of the Medusa* (1819) which he had viewed many times whilst living in France during the 1830s.[99] It is known, too, that Danby admired Poussin's *Deluge* and had actually copied it for a client in 1837.[100]

Like Turner's first Deluge and like Martin's Deluges, Danby's own picture centres on a representation of Nature and the cataclysm that the flood visited upon it. Many of the familiar elements of the other paintings are there; the ark, the serpent and the cowering human figures. There is no tidal vortex or tidal wave, but there are great black cataracts of water enveloping the struggling souls. And following Poussin, Danby drowns the entire scene in a 'cadaverous mono-

92   Francis Danby's *Deluge*, 1840

chrome',[101] within which there are found deep contrasts, giving to the picture a striking luminosity. The unambiguous centrepiece of the composition is the steep rock mass up which desperate souls are seen trying to clamber to escape the rising waters. In this respect, Danby's *Deluge* is a more realist and less complex narration than either Martin's or Turner's paintings; '[t]here is no delirious panorama and no accumulation of marvels, only what creditably could be seen at one moment'.[102]

Danby's *Deluge* was especially admired by Thackeray. In a commentary in *Fraser's Magazine* he regarded it to be *the* picture of the biblical catastrophe.[103] Ultimately, though, it was Martin's *Deluge* that became most impressed on the nineteenth-century imagination. One critic of the 'British School of Painting', having bemoaned the school's comparative mediocrity alongside Milton's 'historical pictures' in *Paradise Lost*, Byron's 'brightest colours' in his 'Oriental Fictions' and Scott's 'burning' of the 'soul of painting' into his poetry and prose, maintained that Martin's first *Deluge* deserved unqualified admiration. The picture was wild and riotous in its luxuriance, reflecting the 'boundless power of a magician, than the faithful chronicler of things'.[104] If there were faults of colour, technique and conception in

Martin's compositions, few disputed the power and brilliance of effect
that he achieved. It seems he was capable of being rivalled only by himself.[105] By the 1830s, Martin's fame was such that inferior copies of his
mezzotints, the *Deluge* among them, were being made to the detriment
of his own printing and publishing business. However, at the Brontë
parsonage at Haworth in Yorkshire, it was Martin's original mezzotints that graced the walls of that austere home. Among them were
Joshua, *Belshazzar's Feast* and, of course, *The Deluge*.[106] Patrick
Brontë delighted in explaining to visitors the subjects they depicted.
Martin's influence was thus not just confined to the salons of the rich
and famous. It extended deep beyond even the provinces, to a rugged,
forbidding rain and windswept landscape that the artist himself would
have delighted in. Danby, though, was always somewhere in the background. In 1860, for instance, a commentator in the *Westminster
Review* likened Currer Bell's (Charlotte Brontë's) descriptions of nature
to the 'loaded atmosphere of a picture by Danby'. One saw the Pennine
moors through 'a haze, glorious . . . and beautiful'. In fact, Emily
Brontë's (Ellis Bell's) *Wuthering Heights* was probably even more
Danby-esque than anything her elder sister wrote. Thus images of the
Deluge ranged far and wide. Even as late as 1855, *Punch* could not
resist its comic appeal.[107]

The Deluge as Geological Witness

> Theobald walked over to the Rectory one Sunday morning early in
> December [1825] – a few weeks only after having been ordained.
> He had taken a great deal of pains with his sermon, which was on
> the subject of geology. . . .[108]

Re-imagining the Deluge was not just the province of artists or poets.
Geologists also participated in this extraordinary process, as already
described in chapter two. The position they occupied was the more
intriguing by virtue of many of their number being clergymen.
Alexander Catcott, vicar of Temple Church in Bristol, was among the
earliest of them. His *Treatise on the Deluge*, first published in 1761 and
already in a second, enlarged edition by 1768, produced a whole series
of 'natural proofs' of the scriptural account.[109] The 'circular shell' of
the earth contained a huge reservoir of water that was more than capable of producing a universal flood to the height represented in scripture.[110] Catcott argued that the entire solid structure of the earth was
dissolved when this mass of water welled up from its subterranean
abyss. The present solid structure of the earth was then formed when
this fluid body subsided 'together with the animal vegetable bodies

inclosed within it'.[111] Catcott claimed that the conical shapes and regular slopes of hills and mountains were further proofs of one universal deluge,[112] and much the same applied to the observed regularity of stream networks.[113] The former were features that Catcott had noted in the limestone country of the Mendip Hills, south-west of his native Bristol. Moreover, some of the valleys here were 'dry', that is bereft of rivers. It was thus not altogether illogical to think of such structures as having been formed by vast torrents rather than normal river systems. But surface streams were absent here largely because drainage had, over time, become subterranean – a common feature of limestone strata on account of its susceptibility to solution weathering. Thus whilst certain of Catcott's observations were quite astute, he was flawed in his identification of cause.[114]

Catcott's treatise, in any case, was later plainly contradicted by that of James Hutton. His self-regulating Earth-Machine, operating over unimaginable antiquity, had no place for the flood.[115] But to impugn the Mosaic account was, as already noted, to invite charges not only of atheism but of immorality, and Hutton's ideas were quickly derided. His principal antagonist, as described in chapter two, was Richard Kirwan, one of the first honorary members of London's new Geological Society.[116] Kirwan determined to rehabilitate the Deluge with an enthusiasm that was all his own. He did so using not the language of science but that of religious dogmatism, even though his scientific credentials had previously won him election to the Royal Society.[117] He ridiculed Hutton's prolix literary style and repeatedly misrepresented his ideas. He was so convinced of the absurdities of Huttonian theory that he did not consider it necessary to read more than a few chapters of Hutton's book.[118] Kirwan's own geological justification for the Deluge consisted of 'wonderfully imaginative speculations'.[119] They involved the vast waters of the southern ocean sweeping across Asia and North America, helping to explain the fossilized remains of tropical plants and animals found thousands of miles from their natural habitats.[120] For Kirwan, the account of the Deluge in the Book of Genesis was a 'central article of faith'.[121] All geological reasoning had to be subordinate to it. Despite making numerous geological excursions in his native Ireland, he appeared incapable of engaging observation and theory.[122]

Another of Hutton's antagonists was the Swiss geologist Jean-Andre de Luc. As also described in chapter two, de Luc was happy to accept that the six days of creation in Genesis were allegorical, in other words allowing a far greater age to be ascribed to the earth. However, he remained convinced of the literal and historical significance of the Flood, asserting that the catastrophe was caused by the sudden collapse of the crusts of the ancient continents which allowed vast incursions of the sea. In the process, most life-forms were extinguished.[123] De Luc's ideas first came to notice in Britain in 1793,[124] but his *Lettres Physique et Morales sur l'Histoire de la Terre et de l'Homme*, published in the early 1790s, appeared in a new (English) edition only as late as 1831, fourteen years after his death.[125] De Luc in fact spent much of his later life in England, finding royal favour with Queen Charlotte, including a house at Windsor.[126] He thus became an establishment figure whose 'proofs' of the Mosaic deluge sat comfortably with such a conservative social identity.

The work of Kirwan and de Luc in reinstating the Deluge in earth history soon found a much more convincing champion. It came in the person of the French comparative anatomist Georges Cuvier, famous for his studies of fossil reptiles unearthed in the sediments of northern Europe. These discoveries led him to argue that lands that were once dry had been successively inundated by flood waters. Sea and lake levels had thus been subject to constant fluctuations. Moreover, such fluctuations were, in his view, sudden rather than gradual, attested by the way geological strata were so often dislocated and overturned. In these frightful cataclysms, all life on dry land was engulfed and destroyed. And when the waters retreated, the same fate befell all life in lakes and seas.[127] Thus the Chalk of the Paris Basin, once conceived as modern in origin, was now conceived as related to one of these earlier revolutions.[128] Here, then, was the explanation for the whole skeletons that Mary Anning had uncovered on the beaches at Lyme. Here, too, were explanations for the broken relics that Gideon Mantell had found in the quarries of mid-Sussex. Cuvier's work was published in France in 1812 and the prefatory portion of it, a discourse on the history of the earth, translated into English in the following year by Robert Kerr, with mineralogical notes and preface by the geologist Robert Jameson. By 1829, the book had gone through five editions. Cuvier saw the biblical deluge as the record of an actual historical event, although not the only such cataclysm to affect the earth. He did not date it in line with Bishop Ussher's calculation of 2349 B.C., but placed it roughly around the time Ussher had identified for the Creation.[129]

For the more educated members of the English clergy, Cuvier seemed to have produced an ingenious compromise; the Bible remained as a record of mankind's origins.[130] Moreover, Robert Jameson added to the book's impact by suggesting in his separate preface that the Deluge was a supernatural phenomenon, despite Cuvier himself making no such suggestion.[131] Among the book's readers was Lord Byron, who admitted to adopting 'the notion of Cuvier, that the world had been destroyed several times before the creation of man'.[132] And Pleydell Wilton, writing in 1818, looked to English geology to find just such a champion as Cuvier. Wilton had no difficulty with the idea that, during the 150 days of the Flood, existing strata might quickly have been 'overspread with fresh masses of shells irresistibly driven by some new submarine commotion'.[133]

Cuvier's studies were indeed quickly exploited by a host of leading English geologists. They were opposed to Hutton's theory that present-day valleys had been excavated by the rivers that still flowed through them. In other words, they were convinced that they were formed by causes that no longer operated.[134] And what better cause than some kind of deluge. This was the view of William Buckland, holder of the first Oxford teaching post in geology. His many pupils were soon to dominate the ranks of English geologists and Cuvier's doctrine became central to their interpretations. Buckland readily conceded that the present surface of the planet was built upon the 'wreck and ruins of one more ancient'.[135] However, directly quoting Cuvier, he was clear that 'the present order of things' could not 'be dated at a very remote period'.[136] This order, moreover, could be understood only in relation to the occurrence of some kind of universal deluge. And so as to emphasize his scientific credentials as a geologist, Buckland insisted that evidence of such a deluge was found in the record of the rocks not in the record of the sacred text.[137]

93 William Buckland, with a drawing of the Kirkdale Cave hanging on the wall behind him and holding a specimen taken from the cave's assortment of animal bones

Within a few years, Buckland was presented with evidence that seemed to him to offer incontrovertible proofs of such a cataclysmic event in the earth's recent history. It came in the form of the vast assortment of animal bones that Yorkshire miners had found in a large limestone cave at Kirkdale in the Vale of Pickering.[138] The bones belonged to a whole range of species, but included hyenas, lions, tigers and other creatures that inhabited tropical climes. Buckland found some of the bones so perfectly preserved in mud and silt that he could draw no other conclusion than that they had been deposited there in a catastrophic flood. His investigations won immediate and widespread attention, helped by reports in the *Gentleman's Magazine*.[139] Soon geologists were excitedly exploring caves up and down the country. Coupled with the evidence of 'erratic' rock boulders found high up mountains, hundreds of miles from their origins, and of beds of clay

94   Entrance to the Kirkdale Cave in Yorkshire

and gravel over extensive tracts of the lowlands, proof of the Deluge appeared incontestable.

Buckland assembled all his views and observations in *Reliquiae Diluvianae*, published in 1823.[140] Here he recorded evidence from not just the Kirkdale cave, but a host of other cave investigations in Britain and on the continent. He also rehearsed at length the various features of the landscape that offered proofs of a universal deluge, echoing the ideas of Alexander Catcott, some sixty years before. Indeed, Catcott's work soon experienced something of a revival. The bones he had found in caves in Somerset went on display in Bristol City Library and his *Treatise*, 'after slumbering in dust upon the shelves of the Bibliomaniast', became 'very scarce'.[141] For Buckland, like Catcott, the shape and position of hills and valleys revealed conformities of slope and angle that could be understood only in terms of the universal action of flood waters.[142] Much the same conclusion followed from the universal confluence and successive 'incosculations' of minor valleys with each other, and their final termination in a trunk valley leading seaward.[143] Buckland rejected the view that any of such features could be related to the effects of ancient or modern rivers.[144] Some of the best examples of valleys produced by the action of the Deluge could be found in the various valleys falling into the bay of Charmouth in Dorset,[145] and Buckland incorporated coast sections in the book to display these.[146]

Buckland's *Reliquiae Diluvianae* was widely praised by leading commentators of the day. In particular, it was acclaimed for the way it gave a new and illuminating account of the past fauna of the European continent.[147] But more important in the eyes of the Anglican-dominated intellectual establishment was that the work attested to the action of a universal deluge. A reviewer in the *Gentleman's Magazine* saw Buckland as having corroborated religious creed.[148] In Oxford itself, this was no small consideration and Buckland's inaugural lecture of 1819 had taken pains to demonstrate geology's relationship to natural theology. Quoting Paley, he remarked that if there was one train of thinking more desirable than another, it was that which regards the various phenomena of nature with constant regard to a 'supreme intelligent Author'.[149] Why else could coal strata in England be found in an inclined position so allowing a far greater facility of excavation than if they were found in an horizontal position.[150]

At Cambridge, Adam Sedgwick, as newly appointed Professor of Geology, echoed much the same themes.[151] There were few who could deny the relationships and coincidences of sacred and material worlds. One writer to the *Gentleman's Magazine* described days wandering amidst the rocks and tors of Dartmoor, 'a grand representation of the wreck of the ante-diluvian world' following Noah's great flood.[152] Another related details of the organic remains of rhinoceroses, hippopotamuses, elephants and hyenas found at Kent's Cavern at nearby Torquay, all relics of the world before the Deluge.[153] Indeed, during the 1820s there was hardly a month when the *Gentleman's Magazine* did not include some kind of reference to geology and the Deluge. Not only were Buckland's publications reviewed at some length, but there were reports of new cave investigations and all kinds of speculations on the physical causes of the Deluge.[154]

Buckland's 'diluvial' geology, though, did not long survive the rapidly evolving scientific discipline that geology became over the 1820s and 1830s. As chapter two revealed, Lyell's *Principles of Geology* (1830–33), taking its cue from the work of James Hutton, argued that hill and valley, mountain and ravine, had been sculpted by the slow action of forces still in operation, but extending over unimaginable aeons of past time. More specifically, Lyell denied that the Deluge recounted by Moses had left any geological monument.[155] Buckland himself was by then also revealing doubts. These centred most of all around the absence of human skeletal remains in the detritus of the Deluge. Adam Sedgwick was very quick to abandon the Mosaic Deluge in the face of such a glaring inconsistency and performed a kind of ritual recantation in his last address as president of the Geological Society in 1831.[156] Buckland, though, still clung generally to the idea of reconciling geology with sacred history, revealing his colours in chapter two of his Bridgewater Treatise of 1836, a work that was otherwise a standard and competent geological text, with one volume devoted entirely to illustrative plates, including many of fossils.[157] Indeed, it proved a best-seller, the entire first printing of 5,000 copies effectively selling out before the book was properly on the market.[158] When the Treatise was reviewed for the *Gentleman's Magazine* in 1837, there seemed no hesitation in claiming that Buckland had established geology 'as an effectual auxiliary and hand-maid of Religion'. The evidences of Nature supported all the arguments of the divine, making 'doubt absurd, and atheism ridiculous'.[159] What Buckland did concede was that geological deluges may not have been of the 'same denomination' as Noah's deluge, but he urged for there to be more detailed investigations.[160]

Buckland's predispositions may well have been reinforced by the continung artistic preoccupation with the biblical catastrophe. It was also the case, as remarked before, that leading artists such as Turner and Martin were far from unfamiliar with contemporary geological discourse and with its points of engagement with the sacred story. Turner, in particular, had geologists (if not Buckland) among his close friends from the first decades of the nineteenth century,[161] and his tour of western Scotland in 1831, which generated so many drawings and paintings, including Fingal's Cave, was in part prompted by his reading of the first volume of the Geological Society's *Transactions*.[162] More widely, Turner had a keen eye for the structure of the earth, already revealed in some of his 'picturesque' representations of English coasts.[163] It was quite logical, therefore, for Turner to offer in his pictures allusions to some of the theoretical positions that character-ized contemporary thought and debate. This was expressed most plain-ly in an early, unexhibited version of *The Eve of the Deluge* that incorporated what looked like an Ichthyosaurus in the bottom right-hand corner of the painting. The precise identification of the creature has been the subject of dispute.[164] In some ways, it more resembles a crocodile, even allowing for the fact that the first skulls to be found of the Ichthyosaurus in Dorset were regarded by local people as belonging to the crocodile family. However, the central allusion for the painting was the issue of the continuity of life-forms. The Deluge sup-posedly destroyed all earlier life-forms, save for those assembled in the Ark. But there remained the difficulty of the similarity between some ante-diluvian remains and creatures that now inhabited the earth. The exhibited version of this picture erased the reference to the Ichthyosaurus, perhaps indicating acceptance of the Deluge as no more than sacred myth.[165] In doing so, though, Turner was effectively rein-forcing the imaginative power of the picture, even more so given the impressionist style in which it (and its pair) was painted.

Geologists themselves were not in any way remote from such imag-inative power, for when Turner's geologist friend John MacCulloch produced a description of Loch Coriskin (or Coruisk), he followed the style of a poet. First sight of the valley appeared to transport the viewer into the magical wilds of an arabian tale. The valley's giant yet simple proportions and absence of ordinary vegetation made humans seem insect-like.[166] It might have been a 'habitation of Genii among the mysterious recesses of the Caucasus'.[167] And when Turner was prompted to paint the loch around 1832, the resulting picture had shades of a deluge composition about it.[168]

John Martin also had geologists among his friends, including Gideon Mantell for whom he produced some memorable book illustrations in the mid-1830s.[169] Only a few years earlier, Cuvier had paid a visit to Martin's London studio. He came when the artist was absent but Martin's son, Leopold, recorded how the Frenchman had been fascinated when his eyes alighted on the painting of the Deluge, still unsold and having been much worked over since its original public showing.[170] Martin was much taken with Cuvier's approval of the picture, especially in terms of its points of scientific reference, and he had no hesitation in relating the incident to friends in subsequent years.[171] Buckland himself, as Cuvier's leading apologist in England, may possi-

95   John Martin's second version of his first 'Deluge' painting, completed in 1834

bly have counted among Martin's acquaintance in the 1820s. Certainly, Buckland's work on the Kirkdale cave was known to the artist, as already described, even if clear proof of personal contact does not appear until 1845 when Buckland had left Oxford to become Dean of Westminster.[172] Martin always thought of the Deluge as his favourite picture and it was the one case where he painted a second version of one of his major works, apparently because the original had become lost.[173] This second painting was exhibited in Paris in 1834, proving a sensation for some critics and observers, but disappointing others in seeming not to conform to be a painting at all.[174] In 1837, it was exhibited in London at the Royal Academy, despite the veto on pictures that had previously been exhibited elsewhere.[175] Once more there were those who were enthralled by its ghastly and sublime quality, while others thought it vulgar.[176] For Martin, the Deluge became ultimately less a human tragedy than an elemental upheaval in which the human race had had the misfortune to become involved.[177] Martin's perspective was very much that of James Hutton's: an earth millions of years old that had, over the ages, supported progressively higher forms of life, man being the most recent.[178]

As geologists in the country's leading academic institutions increasingly distanced their conceptions of earth history from Genesis, they found themselves opposed by another group who remained insistent on the need for literal adherence to the biblical account in understanding the earth. They got out their own 'fantastic' geologies.[179] They mocked the geologists' 'ephemeral and half-digested theories, and bubble-blown hypotheses'.[180] Not for them the 'insuation' [sic], following Lyell, that the globe they now inhabited was 'to endure to all eternity'.[181] Not for them the idea that all the causes of degradation, destruction, elevation and depression could be observed in the present state of the world.[182] George Fairholme asserted in 1833 that '[g]eologists, without any knowledge of the original text' and 'learned men without any knowledge of geology' had unintentionally achieved a species of coalition such as to strike deep at the foundations of Christian belief.[183] He confidently asserted that 'all the appearances of the surface of the earth' could be accounted for by 'an *attentive*, an *unprejudiced*, and, above all, a *docile* consideration' of the Creation and the Deluge. Thus coal was diluvial in its origin, traceable to the 'ruins of the whole vegetable world' at the period of the Deluge.[184] In short, the earth that now existed was wholly distinct from 'the earth that then was',[185] with the Deluge as the primary point of reference. Men such as Buckland and Cuvier had merely proceeded from 'doubt to infidelity', launching their students into a 'sea of clouds and thick darkness'.[186] It was almost as if they had been launched into one of Martin's very own paintings. The *Gentleman's Magazine* detailed Fairholme's arguments at some length, under the title: 'Fairholme on the Mosaic Deluge'. This included his fierce denial of the views of geologists that there had been several deluges and that these may only have been partial. He also denied Lyell's idea that the earth was immensely remote in age.[187]

Granville Penn attempted to combine Genesis and Newtonian physics, claiming that God had depressed portions of the earth and filled them with sea, only later to re-elevate these portions. The effect of such uplift was to swamp what had before been dry land. This was the Deluge.[188] Penn's work proved highly popular, not least because he was an Alpinist and seemed adept at conveying the romance of geological investigation in the field. He even produced a book for children: a series of 'geological conversations' between mother and children.[189] Other commentators, among them George Bugg, reiterated the adequacy of the Genesis narrative in all the features of the earth that geologists sought to explain. If there were elements that were incompatible, it was merely that God had forgotten to include them in the story.[190]

John Martin's brother, William, was another ally. His pamphlet on the Flood, published in 1834, maintained that the Deluge was the 'Proper Cause' of all the different geological strata.[191] A reviewer for the *Gentleman's Magazine* of 1831 summed up a range of reactions: the early geologists had 'wrecked their successive theories against the sublime simplicity of the Mosaic statement'; meanwhile, later geologists 'had failed in the other extreme – in an attempt to patronize the inspired author of Genesis', showing how he might be 'reconciled with scientific ambition'.[192] In 1844, at the annual meeting of the British Association, the Dean of York fired what was to be one of the most memorable last salvos in defence of the biblical Deluge. He not only spoke at the geology section, but caused a sensation by publishing his address as a pamphlet the next day.[193]

Another defence came in Hugh Miller's wildly popular *Testimony of the Rocks* (1861) in which two whole chapters were devoted to 'The Noachian Deluge'.[194] A great flood, even if 'restricted and partial', might yet have been sufficient 'to destroy the human race in an early age'.[195] Miller refused to concede that it was 'an incredible event'. Incredibility was found in the 'mere glosses and misinterpretations in which its history ha[d] been enveloped'.[196] It thus remained one of the vital 'evidences of faith'.[197] Miller was supported in his efforts by a former editor of the *North British Review*, the Reverend John Duns, whose *Biblical Natural Science* sought to reconcile holy scripture not just with geology, but with botany and zoology, including Darwin's theory.[198]

By the 1860s, however, adherence to a literal reading of the biblical account of earth history was being jettisoned even by the

96   The Asiatic Deluge from Louis Figuier's *The World before the Deluge*, 1867

Church of England. The findings of geological science, as summarized in an address to the Geological Society of Edinburgh, were accepted. The earth was 'a work of unknown beginning, gradual . . . development, and still in progress'.[199] The Deluge came to be regarded as an event restricted to a limited area of the earth's surface.[200] John Pye Smith, an academic theologian, had begun the trend as early as 1839. The alleged discrepancy between the Scriptures and the discoveries of science were 'in semblance only' and 'not in reality'. The Bible, 'fairly interpreted', was not adverse to a belief in 'an immeasurably high antiquity of the earth', nor of the possibility of the Deluge having had 'limited existence'.[201] Eventually, the position was fully laid out in *The Bible Commentary* of 1871.[202] Even so, this did not overnight erase the force of the Deluge either in the poetic, artistic or literary imaginations. John Linnell painted a Deluge as late as 1880.[203] In France, Louis Figuier published *La Terre avant le Déluge* in 1863, to considerable popular acclaim, and within two years, an English version, *The World Before the Deluge* had been translated from the fourth French edition. The translation, somewhat ironically, was done by a member of the Geological Society of London.[204] As a piece of popular science, Figuier's book was heavily illustrated. Essentially, it offered a record of

the succession of geological epochs, reconstructing the plant and animal life of each. The book's particular title, though, tapped into the
popular interest in all things ante-diluvian, and, by implication, with
the Deluge itself as a critical point of discontinuity. Figuier was plainly
aware of the state of geological thought over the biblical catastrophe.
He thus decided to distinguish two deluges, one confined to northern
Europe, the other to a part of Asia. Both were attributed to natural
causes, and in so doing, he disconnected them firmly from scripture.
The plate (by Riou) that illustrated the European deluge depicted a
cataclysmic flood almost worthy of Turner or Danby. The plate of
the Asiatic deluge (again by Riou) clearly owed its inspiration to
John Martin, complete with tidal vortex, zig-zag lightning and an
ante-diluvian city disappearing beneath raging torrents.[205] In neither
picture, though, was there an ark, once again emphasizing the natural
rather than supernatural form of the event. The English translation of
the book was issued in revised form two years later, based on the sixth
French edition.[206] From his prefatory thesis, Figuier argued for studies
of the history of the earth to be placed in front of the imagination of
youth. Not for them the fantastic worlds of fable and fairytale, of
Aladdin, Jack the Giant Killer and Cinderella.[207] As geology no longer
opposed itself to Christian religion, there was no longer any reason to
keep it from 'Popular Reading'.[208] Paradoxically, of course, the sometimes fearful ante-diluvian creatures that peered out from the pages of
Figuier's book answered to exactly the same imaginative faculties that
fables and fairytales had. The land of goblins and devils, magicians and
demons, was replaced by one of monster reptiles, inhabiting not just
the land, but the sea and the air. Even ancient plant life yielded mon-
strosities, puffed-up versions of forms now found only in ditches and
beneath hedgerows. The cause of these various extinctions became one
of the critical puzzles that preoccupied Charles Darwin as he pondered
the origin of species.

6

Competition, Competition . . .

When on board H.M.S. Beagle, as naturalist, I was much struck
with certain facts in the distribution of the inhabitants of South
America, and in the geological relations of the present to the past
inhabitants of that continent. These facts seemed to me to throw
some light on the origin of species . . .[1]

Prelude

Belief in the Deluge encouraged a belief in the fixity of species.
Ante-diluvian creatures were thus seen as representatives of one phase
of God's successive creations. And after the destruction and death
caused by the Flood, it was entirely logical to claim that God had
then restocked the earth with new creatures. For many early palaeontologists, these new inhabitants were also regarded as being quite unrelated to those of previous creations, a view that seemed to be
reinforced by the startlingly exotic skeletons like those found on the
beaches at Lyme. Even so, ideas about transmutation of species had
been circulating for over half a century. Robert Chambers's *Vestiges* of
1844 had proved a publishing sensation with its speculative account of
'development' in nature.[2] And behind the scenes, well away from public view, Charles Darwin had been steadily seeking to underpin
his parallel evolutionary theory – operating through the medium of
natural selection. If one admitted evolution, the exotic life-forms of the
ante-diluvian era took on an altogether new guise. They became distant
ancestors. There was no longer any need to seek refuge in mystery, or
in calamities like the Flood.

The vital engine of change for Darwin was competition, competition
for food and for life. He was immediately concerned with plants and
with animals, but there was no escaping the fact that the society in
which Darwin lived was also one that was increasingly mediated by
competition. It was apparent in Malthus's idea of human populations
constantly battling against the limitations of subsistence. It was apparent, too, in capitalist political economy, which Marx and Engels were

97 (*facing page*)
Henry De la
Beche's classic
reconstruction of
a Liassic scene in
ancient Dorset-
shire, illustrating
the ferocious
competition for life

soon translating into class war. It was witnessed, in turn, in the various manifestations of 'Social Darwinism', beginning with Herbert Spencer's famous phrase, 'the survival of the fittest'.[3] When Darwin's theory finally broke upon the world, in 1859, it swept like a tidal wave through a society already attuned to its underlying general idea. Even so, the theory's rejection of Omnipotent Design, of Natural Theology, was, for some, a shocking revelation. On Darwin's account, there was no 'Almighty Clockmaker'.[4] Among the artisan and working classes, the book carried with it the implication that man had a monkey ancestry. There was thus nothing really to separate aristocrat from working man. Darwin was signalling hope of social salvation for the masses. For some of their masters, though, the book invited yet more fears of social instability and, worse still, extinction through social revolution.

A Ferment of Money, Ideas and Beliefs

Competition is the completest expression of the battle of all which rules in modern civil society. This battle, a battle for life, for existence, for everything, in case of need a battle of life and death, is fought not between the different classes of society only, but also between the individual members of these classes. Each is in the way of the other, and each seeks to crowd out all who are in his way, and to put himself in their place. The workers are in constant competition among themselves as the members of the bourgeoisie among themselves.[5]

Charles Darwin was the archetypal Victorian man-of-means. Comfortably provided for by both his father and his father-in-law, he was to spend more than half of his years in a secluded former parsonage at Down in north Kent, affecting the life of a country squire.[6] But investments in enterprises like canals and railways, in government stocks, and in land, were inextricably part of the competitive nexus of capital. In between studies as a naturalist, voluminous correspondence with scientific friends and fellow observers, not to mention family life, Darwin was forced to engage with 'odious money matters'.[7] The great railway boom of the 1840s had brought rich profits for some but ruined many more. In October 1852, Darwin had been 'frightened' by information about prospects for his shares in the London and North Western Railway and determined to 'sell all out'.[8] At the time, he had £14,000 invested in railway enterprises, generally.[9] This was a small fortune by the values of the day and a sizeable chunk of his investment portfolio. Earlier that same year, he had faced the necessity to reduce the rental on the Lincolnshire farm that he had purchased in 1845.[10]

The farm was intended to provide insurance against a stock-market collapse, although Darwin also had ideas to be an enlightened land-lord, looking to provide cottagers with allotment gardens. His blood burned with indignation at the stamp laws that made it impossible for the poor man to buy his own quarter acre.[11] Even though a self-confessed free-trader, he wrote in 1852 that he did not like to press too hard on a good tenant.[12] He took a residual pleasure in reviving the paternalist instincts of the eighteenth-century farm squire. It was free trade, nevertheless, that forced him as far away as Lincolnshire to find security in land; the farms around the Down parsonage were by then fetching exorbitant prices.[13] He also had first-hand experience of the intense competition for land and housing in towns and cities. In 1838, in looking for a house to rent in London, in time for his marriage to Emma Wedgwood, he was dismayed by the scarcity of available prop-erties and the way landlords had 'gone mad' in the prices they asked.[14]

The competitive ferment of money was mirrored in the more basic human struggle for shelter, food and fuel. Nowhere was this more apparent than in the growing urban agglomerations, London chief among them. Here were two and a half million human beings, a 'colos-sal centralization' of population, as Engels described it in 1844.[15] An incoming traveller might marvel at the panoramas that the river Thames offered as one navigated inland from the sea estuary to London Bridge. But on the streets Engels found the turmoil of the crowds to have something repulsive about it.[16] During the brief time that Charles Darwin lived in the metropolis, in the late 1830s and early 1840s, he mostly found it a 'vile smoky place'. Misery and vice were everywhere. He felt he was 'stewing in a great den'.[17] For Engels, it was the apparent indifference of the great mass of people to one another that was most striking. Mankind had been reduced to a narrow, self-seeking individualism.[18] The stronger trod the weaker underfoot. Beyond the 'gay world' of Oxford Street, Trafalgar Square and the Strand, down stifling alleys, thick with human and animal effluvia, was where the weak were forced to eke out a living.[19] The houses there were tottering ruins, from cellar to garret. They bred a 'whirlpool of moral ruin', awash with pimps, prostitutes and petty thieves.[20] Nor were the middle orders necessarily immune from the struggle for a liv-ing. Small capitalists, especially, faced perpetual threats from fraud-sters, and bankruptcy nearly always loomed on the horizon. There was no safety net to catch you if you fell.[21] Christian commentators described how those in business floated on oceans of boundless debt, ever prey to being 'shipwrecked' in the perpetual flux that formed the 'capitalist ocean'.[22]

Poverty, misery and worry were inevitable adjuncts of the alternate quickening and slowing of capitalist production. Engels was quite

98   Manchester cotton factory, 1829, archetypal symbol of the industrial revolution

happy to rehearse Adam Smith's dictum that the demand for men necessarily regulated the production of men.[23] When boom gave way to slump, however, the poor working man, in particular, found himself in bitter competition with his fellow workers – just to survive. Invariably the weakest foundered, sinking into the kind of horrific abyss that Elizabeth Gaskell's novel *Mary Barton* so dramatically evoked in the cellar slums of Berry Street in central Manchester: windowpanes mostly 'broken and stuffed with rags', the 'smell so foetid' that it almost knocked men down, children rolling on the wet brick floor through which the 'filthy moisture of the street oozed up', the fireplace 'empty and black'.[24]

From the surplus labour of Adam Smith, it was but a short step to Thomas Malthus's idea of surplus population. The venerable parson's contention that population would always yield the capacity to outstrip food supply made the misery of poverty a perpetual feature of society.[25] He sought reform of the Poor Law and got it – even if thirty years on from the publication of his first essay.[26] Engels, though, saw the social structures of capitalism as the primary enemy of the working man. Without wage labour, he was removed from the market. He could buy neither food nor fuel. It was a passport to destitution, to the pawnbroker, to disease and, too often, to death.

Charles Darwin came upon the new sixth edition of Malthus's classic essay in October 1838, much later recalling how he had read it 'for amusement'.[27] At the time, he was in the throes of collecting facts on variation among plants and animals, gleaning information from skilful breeders and gardeners, from 'domesticated productions'.[28] Darwin soon registered that the struggle for existence, as recounted in Malthusian theory, might have a counterpart in the plant and animal worlds. It then struck him that, under this competitive condition, favourable variations might be preserved and unfavourable ones destroyed.[29] He speculated that in time this could lead to the formation of new species. Here was the theory of evolution by natural selection. It was to inform his scientific investigations for the next twenty years. Variability in organisms was no more a product of divine edict than was the course by which the wind blew.[30] Darwin was reforming the study of nature just as Malthus and fellow political economists were reforming the study of society. And soon after his flight from the 'great smoke' of the metropolis to his country parsonage in the village of Down, Darwin came face to face with the Poor's constant struggle for survival. His wife, Emma, gave away penny bread tickets to try to ease their plight.[31] Darwin himself later took on the role of treasurer for the village's Coal and Clothing Club and for its Sick Club.[32] Abolition of the Corn Laws in 1846 eased bread prices, but competition remained the watchword of society and the key, so it was argued, to social progress. The weak, therefore, still struggled, faltered and then sank to the bottom. The shared experience of poverty and destitution in the close confines of the manufacturing towns led to unrest and lawlessness. In the summer of 1842, some northern and midland towns saw political insurgency begin to take on a 'most formidable and disciplined character'.[33] One observer, touring the manufacturing districts, described how discontent was general, bitter and deep-rooted among a vast proportion of the working population.[34] Chartism, which was to form a potent political force for a decade, was flexing its muscles, prompting the Government to go into emergency session. When the social unrest subsided, an editorial in the *Illustrated London News* described it as having semi-revolutionary aspects,[35] echoing a wider frame of mind that, for several generations, had been 'vitally affected by the idea of revolution'.[36] It was an idea that had as much emotional force as it did rational calculation.[37] There was a widespread dread that some 'wild outbreak of the masses . . . would overthrow the established order and confiscate private property.'[38] The origins of this were to be found in the French Revolution and in the revolutionary tendencies that emerged in British society in the 1790s, helped by the spread of radical propaganda like Tom Paine's *Rights of Man* (1791–2). The struggle among the political classes for the Great Reform Act of 1832 was another critical marker.

The Tory party's reactionary policies during the post-war years had become a powerful driving force for a progressively more radicalized opposition. The Whigs and their social allies launched a crusade to rescue the country from a working-class political revolution. Darwin, along with various members of his family and friends, shared in the *frisson* of excitement that the Reform debate generated. When he watched the coronation procession of William IV in September 1831, he saw so little enthusiasm among the crowds that he thought there would be little chance of a coronation fifty years on.[39] A year later, W. D. Fox, Darwin's old friend (and second cousin) from his Cambridge undergraduate days, wrote how, in Nottingham, they had for some days been 'on the verge of revolution'.[40] The excitement had been extraordinary among the lower orders.[41]

And Thomas Carlyle was soon to add to the fever of dread and expectation among the propertied orders with his vivid account of the French Revolution, published in 1837. Written in the present tense, it had an immediacy that was all too disturbing.[42] He gave readers not a conventional narrative, but a series of graphic accounts that cast history as a perpetual cycle of growth and decay, not unlike the manner of Hutton's Earth-Machine.[43] Thus the revolutionary Terror marked 'that black precipitous Abyss; whither all things have long been tending; where, having now arrived on the giddy verge, they hurl down, in confused ruin; headlong, pell-mell, down, down . . .'.[44] It could have been a description of one of John Martin's apocalyptic paintings. Indeed, Carlyle used the imagery of the Deluge to trace the spread of the Terror throughout France: 'death . . . poured out in great floods'.[45] It offered a horrifying vision.

99   Darwin as a young man in 1840 – portrait by George Richmond

Even when Darwin was circumnavigating the globe, his curiosity about the domestic political scene seemed undiminished. Writing to his sister Caroline from Monte Video, late in 1832, he recorded how all the crew knew the Reform Bill was passed, 'but whether there [was] a King or a republic . . . remain[ed] to be proved'.[46] As the radical lobby in Parliament grew fiercer and more licentious in its attempted supression of establishment abuse and privilege, so interest and excitement intensified. Darwin loved it. 'How famously the Ministers appeared to be going on', he wrote to his sister Catherine from Rio Plata in May–July 1833.[47] Interspersed with the political gossip, though, was news of the struggle for human life. Cholera had been spreading all over England. Alongside the Reform debate, it became the topic of the day. But Fox was able to offer comfort to Darwin by remarking that it was 'not by any means so dreadful a visitor to those of our rank in society'.[48] Not for them the putrid cellar dwellings of central Manchester where 'the fever', as cholera became known, was at its most fatal.[49] Not for them the pauper burial ground, with its wooden mockery of a tombstone, constantly reused, hiding pauper bodies piled one upon the other to within a foot or two of the surface.[50]

The rapid march of liberal opinions had the Anglican establishment central in its sights, especially their Lord Bishops, who helped scupper the initial attempts to pass the Reform Bill. One of Darwin's correspondents, late in 1832, gave the House of Lords only another twenty years of life such was the general cry against those who sat in it.[51] However, there was a yet more serious aspect to this politically charged stream of anti-clericalism. It was the appearance of impiety and unbelief, especially among the lower orders. An edition of the *Illustrated London News* in December 1842, confessed to being horrified by the scale of the distribution and sale of 'low, blasphemous and infamous placards and publications'.[52] They displayed a 'defiling and brutalizing spirit'. The shades of atheism were everywhere, with socialism as one of its commonest bedfellows. Key among the secularist leaders was George Holyoake, whose weekly paper, *The Reasoner*, found a wide readership. Republican and utilitarian in outlook, it reviled the country's 23,000 clergy who, it claimed, crippled the moral energies of men and humiliated their native spirit. It ridiculed the clergy's tracts and bibles which 'chain[ed] the spirit of Progress to musty records and drivelling dogmas.[53] The paper, instead, looked to science to elevate the working man. Genesis and geology did not 'hang together'. Quoting Charles Lyell at the British Association meeting of 1846, the Mississippi River required 100,000 years to form its delta, and yet Moses allowed only 6,000. However much one tried to 'stretch Moses and dock the Mississippi', they would not 'live in the same bed'.[54]

The Church, however, was in no fettle to fight back. Its dominion had been heavily fractured. New sects and new denominations had sprung up in the Christian world.[55] It was the dissenting religions, not the Anglicans, that had provided the primary responses to the poverty and social evils of large-scale industrialism.[56] The new chapels of the industrial towns sprang up to provide a perpetual reproach to the Anglican elites.[57] It was an easy but not altogether inappropriate gibe to claim the Anglican Church to be the Tory party at prayer and the Dissenters to be Radicals who sang hymns.[58] F. D. Maurice wrote in 1833 of being crushed by the Babylonian oppression of 'contradictory opinions, strifes, divisions [and] heresies'.[59] There was competition in the religious world just as there was competition in the economic one. The specialization and division of Adam Smith's political economy could be applied to divinity as well as to nature – except that what the latter ultimately led to was in defiance of all Christian teaching.[60]

Darwin's own religious story was of a slow journey toward unbelief. When he left on the *Beagle* voyage, he admitted to being 'quite orthodox' in religion.[61] But gradually he came to see the Old Testament as providing 'a false history of the world'.[62] 'Disbelief', on his own admission, came over him so slowly that he felt no distress, even though he was continually fearful of the responses that his researches might draw from those of a strong religious persuasion.[63] It was never his intention actually to write atheistically, but he saw too much misery in the world to accept the idea of a designed and beneficent creation. He failed to persuade himself that God had made parasites that fed inside the living bodies of caterpillars, or that cats should play with mice.[64] He became one of a significant group of cultural apostates who later became pillars of Victorian society.[65] In Charles's family, in fact, among the Wedgwoods as much as among the Darwins, dissent had a very substantial pedigree. Unitarianism was their culture: intellectual free-thought within which they sought to 'restore Christianity to its pristine purity and make it a religion of universal happiness'.[66] The mystery and miracle-making of the medieval church was swept aside in favour of a 'wholly material world'.[67] Among the Whig industrialists of northern towns like Manchester, this was the faith of the future. Darwin's grandfather, Erasmus, had cast Unitarianism as 'a feather-bed to catch a falling Christian'.[68] But Darwin, if he ever paused there, soon sank lower and lower, with transmutation (evolution by natural selection) as the guiding star. The material world of Nature, 'clumsy, wasteful, blundering[,] low and horribly cruel', made for the kind of book a 'Devil's chaplain might write'.[69] Here was the *Origin of Species*.

Darwin's book, though, including the tortuous path to its completion, its author a perpetual bag of nerves, plagued by dyspepsia, cannot be understood outside of the cultural world within which it was formed. Darwin was unambiguously an 'authentic creature of his times'.[70] He lived and studied amidst a ferment of money, ideas and beliefs. The key to progress was competition. Free trading, free thinking and the free movement of labour were some of its central ingredients. As Squire Wendover remarked in Mrs Humphry Ward's novel *Robert Elsmere* (1888), 'supply and demand, cause and effect, are enough for me'.[71] In much the same way, the travelling mineral-seller who plied his magnificent specimens of amethystine quartz to Darwin's elder brother, Erasmus, as an undergraduate at Cambridge, was regarded as a 'wretch' of a man for the prices he asked.[72] But geology was then highly fashionable. Everyone had their collecting cabinets. The solid earth was being transformed into a commodity, regulated by Adam Smith's raw calculus of supply and demand.[73]

Some decades later, Karl Marx, after labouring on the study of English political economy in the British Museum reading room for as long as Darwin laboured in Down on the *Origin*, remarked how the classification of goods on British railways demonstrated an even more 'fantastic' range of species and varieties than naturalists had argued existed among plants and animals. Every single good or commodity so identified had its own price rate. In this respect, Darwinism became as much social as natural, the metaphor of variation underpinning the complex and controversial commodity pricing of the railway age.[74]

In politics as in religion, Darwin's world was one of perpetual flux. In the 1830s, whilst the superficial signs were of a country peaceful, powerful and prosperous, underneath there was disturbance and discontent, a whole array of often unconnected sources of agitation for which there seemed no means of checking or subduing.[75] The political and religious establishments closed ranks, taking on siege mentalities, much as the anti-transmutationists recoiled in the face of evolutionary theory. But society, more widely, was becoming exuberant, reckless and crude – not altogether unlike Darwin's Nature.[76] Just as Darwin 'ransacked' the literature of his age for ideas of the way the world worked,[77] so the literate among the masses combed the cheap books and pamphlets of the new steam presses for a world that gave them hope of material salvation. Ideas of transmutation or evolution did exactly that. And the irony of Darwin's work was that it gave a sort of legitimacy to class war. The comfortable Victorian men-of-means thus might be forced to the wall, their tormentors marching to cries not of evolution but revolution. Moreover, with seemingly perpetual political and social turmoil in France, with fears of a French invasion never far from people's minds for much of Darwin's working life, the spectre seemed even more fearful.

Darwin's 'Story'

> My everlasting Species-Book quite overwhelms me with work – It is quite beyond my powers, but I hope to live to finish it –[78]

Charles Darwin was a natural scientist, but he also had a fertile imagination. Although this may seem contradictory to modern eyes, there are senses in which many scientific theories, especially when first advanced, are highly fictional.[79] A truly imaginative mind could be as essential as all the conventional instruments of physical research, so one late-Victorian commentator wrote.[80] Darwin's theory of evolution can readily be cast as a 'form of imaginative history'.[81] In his introduction to the *Origin of Species*, Darwin recalled how philosophers had described such research as concerned with 'that mystery of mysteries'.[82] And so working from a band of 'significant' facts, set alongside a range of unsolved difficulties, he found himself making great leaps of imagination in order to connect them. However, this process began not in the organic world but in geology. In his autobiography, he recalled how, as an undergraduate at Edinburgh University, he found the geology lectures there dull beyond belief and determined never as long as he lived 'to read a book on geology or in any way study the science'.[83] On the *Beagle* voyage, however, clearly prompted by his reading of the first volume of Lyell's *Principles of Geology*, and by the inspiration derived from an enthusiastic and highly successful geological excursion in North Wales with Adam Sedgwick the previous summer,[84] the science became a central preoccupation. On making landfall in new parts of the world, it seemed to Darwin that nothing could appear more hopeless than the chaos of the rocks. But after patient recording, reasoning and prediction, light slowly dawned and the geology became intelligible.[85] In effect, he found himself engaged in a series of fascinating detective stories. These reach their climax in one particular chapter of the *Beagle* narrative, 'Passage of the Cordillera', that is the Andes mountain range of South America. Here Darwin achieves an almost 'cinematic' re-creation of the continent's geological history.[86] Passing through the main valleys, he was struck by the way all of them had, on both sides, 'a fringe or terrace of shingle or sand, rudely stratified, and generally of considerable thickness'.[87] These fringes, he believed, once extended right across the valleys and were thus once united. The shingle and sand were plainly the deposits of mountain torrents where they were checked upon entering a lake or arm of the sea. But now those torrents were steadily wearing away the deposits. The noise of the thousands of stones striking against each other left Darwin 'thinking on time'. The stones became minutes as they made passage seaward. The ocean became their 'eternity'.[88] There was but one explanation that he could see to account for the sequence of change: the rock mass of the Cordillera was gently rising. The sand and shingle terraces were thus remnants of old shorelines from when the sea had penetrated high up the present valley systems. As the land was slowly elevated, so the rivers began down-cutting – through beds sometimes accumulated to thousands of feet in thickness. It was a process impossible for the mind to comprehend, except as an effect that was slow, produced by 'a

100 On the coast of South America: Mount Sarmiento from Warp Bay – from Robert Fitzroy's *Narrative* of the voyage of HMS *Beagle*, 1831–6

cause repeated so often that the multiplier itself conveys an idea, not more definite than the savage implies when he points to the hairs of his head'.[89] Here was Lyell's abyss of time, a vast imaginative realm that also had Darwin wondering of the species of animals that had passed from the face of the earth as the stones of the mountain torrents rattled on their course, night and day, in a giant wasting process that could consume whole continents.

On the return passage to Chile, Darwin saw the plainest evidence yet of his ideas about the submergence and then elevation of continental masses. At 7,000 feet, he found petrified trees. They were embedded in successive thin layers of sandstone. His imagination ran riot: 'I saw the spot where a cluster of fine trees once waved their branches on the shores of the Atlantic when that ocean (now driven back 700 miles) came to the foot of the Andes'.[90] Later, this dry land had sunk to the depths of the ocean and become covered with thick sediments. But later still, 'the subterranean forces exerted themselves' and he now 'beheld the bed of an ocean, forming a chain of mountains more than 7,000 feet in height.'[91] When Lyell heard from Darwin of his conjectures about the South American continent, it was as if a dream had come true: 'what a splendid field you have to write upon', he recorded.[92]

When Darwin returned from the *Beagle*'s five-year circumnavigation, in October 1836, geology still claimed a significant share of his attentions.[93] He read short papers to the Geological Society in London.[94] And writing to his cousin, W. D. Fox, he expressed surprise and pleasure that his geological work had attracted the consideration of men like Lyell.[95] Fox was by then living on the Isle of Wight and Darwin was soon exploring the island: he was so 'very curious to see its geology'.[96] Having seen Wight's remarkable geology, he announced to Fox his intention to 'geologise the parallel roads of Glen Roy' in Scotland.[97] At the time, these formed one of the 'great geological riddles'.[98] These were rock shelves or rock terraces that stretched in parallel along the line of the Glen, having the general appearance of roads. Upon reaching Fort William, at the mouth of the Glen, Darwin enjoyed five days of remarkably fine weather with 'gorgeous sunsets'. He felt as though he had been transported back to the Cordillera. Although he could find no traces of shells, he was convinced that the 'roads' were former sea-beaches.[99] Ten years on, Darwin was corresponding on the matter with Robert Chambers, secret author of the infamous *Vestiges*. Both conceived of the 'roads' as former sea margins, connected to the elevation of the land.[100]

Darwin's penchant for hypotheses in geological interpretation was most acutely observed in his study of coral reefs. He was later to remark how no other work of his was 'begun in so deductive a spirit'.[101] The whole theory was thought out on the west coast of South America before he had ever seen a coral reef.[102] Lyell had claimed that reefs were mostly coral-encrusted rims of barely submerged volcanic craters, coral being known to grow only in shallow water. Darwin, though, linked the reefs to his theory of the subsidence and elevation of land. So as craters sank gradually deeper and deeper new coral was being continuously formed on top of old. These ideas were eventually confirmed when Darwin saw for himself the coral islands and reefs of the Pacific Ocean.[103]

Darwin once described his geological hypotheses as so powerful that, if they had been put into action but for one day, the world would have come to an end.[104] But looking down from the highest crest of the Andean Cordillera, one's mind, undisturbed by minute details, could begin to conceive of and to imagine the stupendous natural forces that had fashioned the world.[105] Travelling also aided this process because observations could be constantly compared, even if the often short time that was spent in different places sometimes encouraged a tendency to draw hasty and superficial conclusions, for want of enough detailed evidence.[106] Transferring such grand theorizing to the organic world, however, was much more difficult. Whilst travelling in South America,

101   Map of part of South America from A. Caldcleugh, *Travels in South America, during the years 1819–20–21*, 1825

for example, Darwin had collected numerous fossils, particularly fossil mammals, but he knew nothing about ways of identifying them. Upon the *Beagle*'s return to England, Darwin thus faced the task of finding someone who could.[107] The most obvious candidate was Richard Owen, the recently appointed Hunterian Professor of Anatomy and already a distinguished comparative anatomist. Crates of Darwin's South American fossils were thus soon carted to the Hunterian Museum in central London where Owen set about the task of studying them. Nearly all of the fossils proved new to zoological science; 'great treasures', as Darwin was to describe them in a letter to his sister Caroline.[108] Rather to Darwin's surprise, they bore some relation to modern South American mammals but they were otherwise members of species that had become extinct. Among them was a giant ground sloth, the size of an elephant, together with a hippopotamus-like animal that had a shell similar to that of a modern armadillo. What fascinated and perplexed Darwin was why these unusual creatures had been rendered extinct and how they came to be succeeded by lesser forms of a similar kind. For Richard Owen, the answer was straight-forward; they were related to successive revolutions in the physical conditions of the earth. Much like the French comparative anatomist Cuvier, Owen attributed the successive life forms apparent in the fossil record as being a function of successive episodes of creation and extinction. In particular, Owen would have no truck with ideas of progressive development in the organic world, or with ideas like trans-mutation. He rejected any notion that the fossil record revealed a movement from lower to higher forms.[109] For Darwin, though, the South American fossils did indeed appear to suggest that species might be mutable, for he had found no evidence from South America's geolo-gy that physical conditions on the continent had altered radically. All he could pick up was evidence of Lyell's 'slow and insensible' lapse of ages.[110]

To extend his speculations, Darwin began to focus on living things rather than their fossilized counterparts. After all, he was now suitably well-versed in observation of nature. He began assiduously to collect information on animal and plant breeding, on nature under domesti-cation. He established an extensive network of correspondents among breeders and plantsmen in an effort to collect 'every sort of fact, which [might] throw light on the origin and variation of species'.[111] He was impatient with authors who saw natural classifications as concerned with the discovery of the laws according to which the Creator had willed the production of organized beings. For Darwin, classification consisted of 'grouping beings according to their actual *relationship*, i.e. their consanguinity, or descent from common stocks'.[112] At the same time, he began a long sequence of nature experiments in the privacy of his country rectory deep in rural Kent.

As work on the *Origin* proceeded, though, Darwin found himself overwhelmed with his 'riches in facts'.[113] At times, he began to despair altogether and wondered whether the project was beyond his powers. He had phases when experiments intended to try to prove his conjec-tures were constantly going wrong: '[a]ll nature is perverse & will not do as I wish it'.[114] He was working on the problem of plant dispersal across seas. He imagined fish swallowing seeds, the fish being swal-lowed by herons, the seeds then being 'voided' hundreds of miles away on some new shore where they would 'germinate splendidly'.[115] In the experiment, however, he described how the fish ejected the seeds 'vehe-mently, and with disgust equal to my own'.[116] Other experiments, also on the face of it failures, did nevertheless help him to 'see a little clearer how the fight [for survival] goes on'. He was performing seed trials in the meadow at Down. In one instance, out of sixteen kinds of seed sown, fifteen germinated, but they subsequently perished at such a rate that he doubted even one would survive to flower.[117] The critical diffi-

102   Down House and garden, *c.* 1860

culty in such experimental work was its infinitely small scale. It
required a mind with an enormous retentive capacity as well as great
agility to be able to relate it to the wider project. The ever-growing
superfluity of facts forced him into an unceasing search for connections, to discover the multiple contingencies and the webs of interde-
pendence. No wonder, as he confessed in July 1857, that he was so
easily muddled.[118] He began to despise himself as a 'poor compiler'.[119]
He did not, though, despise his whole work. He still thought he had
the foundation for a discussion of the origin of species.[120] It was the
ultimate of 'entangled' worlds, an apparently highly disordered nature,
but one that keen observation might, with time, penetrate and unravel.
Geology certainly had had its puzzles, but they were as nothing alongside the apparent unruliness of the organic kingdom.

Darwin's theory, as expounded in the *Origin*, had a number of components. He began with a study of variation. Organisms in a species
vary. Such variations are inherited and associated with sexual repro-

duction. However, he also argued that the nature of living conditions affected variation. The reproductive system, for instance, was 'eminently susceptible' to very slight change in the surrounding conditions.[121] Some cultivated plants will grow with the utmost vigour and yet never set seed – not until some trifling change is made in giving them more or less water at a particular phase of growth. Changed habits in organisms could also produce variation. Wild ducks had heavier wing bones than domestic ducks, for the latter flew less and walked more.[122] Selective breeding of plants and animals by man added to the complexity. Altogether, the organization of plant and animal life appeared 'plastic', offspring always departing in some slight degree from the parental type.[123] On this basis, Darwin claimed that, contrary to much existing thought, there was no fundamental difference between varieties and species. In fact, there was considerable difficulty in distinguishing them at all.[124]

To pursue his study of variation, Darwin took up the study of domestic pigeons. He kept every breed that he could find, perpetually astonished by their diversity. As skeletons, other differences were also apparent. Bones varied in number, length, breadth and in curvature. Had an ornithologist been shown a score of pigeons and been told that they were wild birds, he would have ranked them as 'well-defined species'.[125] However, Darwin was sure, in common with other naturalists, that the different birds had a common ancestor, namely the rock-pigeon.[126] And what he went on to show was how artificial selection by man, over 5,000 years or more, had produced new varieties.[127] The principles followed by the pigeon-fanciers were copied by the horticulturalists. The steady increase in the size of the common gooseberry was one example. Florists' flowers had been astonishingly improved when set alongside drawings of similar plants of thirty years earlier.[128] Much the same applied in livestock-breeding. From artificial selection, it was a logical step to ask why a similar selection process could not also be found in nature: 'species in a state of nature being lineal descendants of other species'.[129] Indeed, the two were less distinct than would initially appear, for man could select only 'variations which are first given to him in some slight degree by nature'.[130] In the final analysis, Darwin was not actually able to demonstrate the assumed lineage of the domestic pigeon, but he did enough to provide a model that could form the focus of a discussion set in the natural world.

The second major component in Darwin's theory was the struggle for existence, inspired by his reading of Malthus.[131] This was critical in understanding how varieties, or 'incipient species', became distinct species.[132] The overwhelming feature about reproduction in nature was its astonishing fecundity, following a geometric rate of increase.[133] But this led rapidly to overpopulation and thus there was an immediate struggle for life. Plants, for example, suffer a vast destruction of their seeds, destroyed by insects and slugs. More vigorous plants gradually wipe out less vigorous ones in the struggle for light, water and soil nutrients.[134] Climate may have a bearing on the process, enhancing mortality, for instance, in conditions of unusual heat or cold. The actions of gazing animals provided a further check, particularly in the growth of trees.[135] More widely, Darwin sought to demonstrate how plants and animals were bound in a complex web of relations. Red clover, for instance, is pollinated only by humblebees, for other bees cannot reach the nectar. Were humblebees to disappear for some reason, red clover would also disappear.[136] Thus the web of ecological relations he described was in continuous flux, in a continuously dynamic state.[137] Darwin also showed how the structure of organic beings was always related to the other organic beings with which they came into competition. The plumed seed of the dandelion, for example, allowed it to be blown over a long distance, even where the land was thickly covered with vegetation. The water-beetle was specially adapted for diving, in order to compete with other aquatic insects.[138] In this way, the 'war of nature' was not necessarily 'incessant'. The 'vigorous, the healthy, and the happy' survived and multiplied.[139]

The pressure of population on resources was critical to Darwin's theory of natural selection for in the struggle for life, 'individuals having any advantages, however slight, over others, would have the best chance of surviving and of procreating their kind'.[140] At the same time, he was sure that variations 'in the least degree injurious' would be destroyed. The concept of natural selection was thus the preservation of favourable individual differences and variations and the destruction of unfavourable ones.[141] As observers, however, Darwin remarked that 'we see nothing of these slow changes in progress, until the hand of time has marked the lapse of ages, and then so imperfect is our view into long past geological ages, that we see only that the forms of life are now different from what they formerly were'.[142] But the clues are apparent when we pause to contemplate the leaf-eating insects that are green, the bark-feeders that are a mottled grey and the red-grouse the colour of heather. The belief must be that such traits are of service to these insects and birds to preserve them from danger and have evolved through natural selection.[143]

The preservation of some variations and not others would not in itself lead to species breaking up and the subsequent formation of new species. For this, Darwin offered separate mechanisms. One of these involved what he called divergence. Within a confined area of nature, all individual beings varying in the right direction, if in different degrees, will tend to be preserved. But in a large area, with districts exhibiting different conditions of life, the individuals of the same species undergo modification, with the newly formed varieties inter-

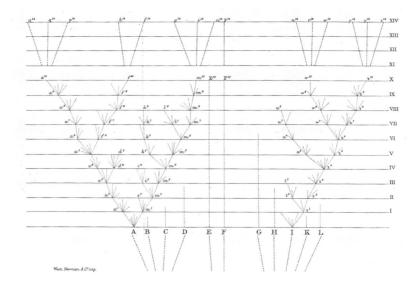

103   The 'tree of evolution' from the *Origin of Species*, 1859

crossing only within the confines of each district.[144] This mechanism could be enhanced by geographical isolation, an idea that first occurred to Darwin when observing the small differences among plants and animals in the various islands of the Galapagos group. The island units themselves were not large enough for the production of new species, but natural selection would, in time, modify species.[145] Darwin had learnt from the island's vice-governor, for instance, that the tortoises differed from one island to the next and that it was always possible to tell which island any captured specimen came from. The same was true, so Darwin eventually concluded, of Galapagos mocking-thrushes and finches, and he was to curse himself for not having systematically labelled his specimens when he first collected them.[146]

Darwin ultimately summarized his ideas in the form of a tree of life:

The green and budding twigs may represent existing species; and those produced during former years may represent the long succession of extinct species. At each period of growth all the growing twigs have tried to branch out on all sides, and to overtop and kill the surrounding twigs and branches, in the same manner as species and groups of species have at all times overmastered other species in the great battle for life. The limbs divided into great branches, and these into lesser and lesser branches, were themselves once, when the tree was young, budding twigs, and this connection of the former

and present buds by ramifying branches may well represent the classification of all extinct and living species in groups subordinate to
groups. Of the many twigs which flourished when the tree was a
mere bush, only two or three, now grown into great branches, yet
survive and bear the other branches; so with the species which lived
during long-past geological periods, very few have left living and
modified descendants. From the first growth of the tree, many a limb
and branch has decayed and dropped off; and these fallen branches
of various sizes may represent those whole orders, families, and genera which have now no living representatives, and which are known
to us only in a fossil state.[147]

Darwin conceived of his tree of life embedded in the crust of the earth,
the crust's sedimentary layers entombing generations of extinct plants
and animals. On the surface, though, the tree was fresh and green,
'ever-branching and beautiful'.[148]

One of the striking features of the style in which Darwin wrote the
Origin is the extent to which it echoes that of Robert Chambers's
Vestiges of 1844.[149] Both authors, for instance, begin by using
the pronoun 'we'. Narrator and reader are thus immediately joined
in enquiry. They become intimates in a journey to the edge of
the unknown. Darwin, in particular, embellishes this feature by his
repeated use of the words 'believe' and 'belief'. In chapter four,
on natural selection, the subsection 'On the Intercrossing of
Individuals' includes the following phrases: 'I am strongly inclined to
believe that'; 'in accordance with the almost universal belief of'; and
'on the belief that'.[150] The theory of evolution is being offered as a kind
of declaration of faith. The tone is 'quiet and unhectoring'.[151] The language is devotional. And the reader is being invited to share in the
devotional act. Both of the books share a desire to anticipate objections, Darwin having one whole chapter that focuses on 'difficulties'.
Such passages assume the manner of confidential discussions. The inti-
macy of author and reader is reinforced.[152] And in the *Origin*, it is
noticeable how Darwin temporarily reverts to more exclusive use of
'we', finding security in its collective resonance in the face of potential
opponents.

Another theme common to the two works is domesticity. In *Vestiges*,
it centres upon images of the family, of mutuality in relations.[153] In the
Origin, it focuses on plant and animal breeding, on pigeon-fanciers
and plantsmen. Indeed, as already noted, Darwin's book actually opens
with a discussion of variation under domestication. The territory is
everywhere familiar, unthreatening in its import. And much the same
reaction flows from the fact that both *Vestiges* and *Origin* were books
of nature, with 'the physical act of turning pages' becoming itself a

'metaphor for scientific exploration'.[154] The intellectual messages they convey may have differed, but both offered accounts of creation that showed affinities with the Bible.[155] At times, in fact, the language reveals shades of the Bible itself, or of Anglican liturgy. Darwin, for example, begins his discussion of sexual selection (in chapter four) with the word 'inasmuch'. He makes frequent use of the invocation 'let us now'. When coupled with the perpetual references to 'belief', one develops a sense of evolutionary theory as a kind of creed. And just as creeds are often constructed around myths, so Darwin effectively presented a whole series of new myths about the history of life on earth. Somewhat paradoxically, he was re-mythologizing the world he had sought to demythologize.[156] In *Robert Elsmere*, one can pick out much the same drift. Reading all the details of Darwin's evidence, and being forced to understand the whole hypothesis, evolutionary theory became a revelation akin to the other (Christian) revelations, even if it did not square with them.[157]

Chambers's *Vestiges* and Darwin's *Origin* differed, though, in the way the former emphasized development, while the latter emphasized variation. Variation was the engine of change. The theory of evolution came out of deviation and divergence.[158] It was also a theory, as indicated at the outset, that relied heavily upon Darwin's (and the reader's, for that matter) imaginative faculty. Repeatedly, he urges the exercise of the imagination. In the struggle for existence, it was necessary 'in our imagination' to give one organic form some advantage over another.[159] Imaginary instances were introduced in his discussion of natural selection.[160] And imagination was involved in comparing the eye and its development to a telescope:

> If we must compare the eye to an optical instrument, we ought in imagination to take a thick layer of transparent tissue, with a nerve sensitive to light beneath, and then suppose every part of this layer to be continually changing slowly in density, so as to separate into layers of different densities and thicknesses, placed at different distances from each other, and with the surfaces of each layer slowly changing in form. Further he must suppose that there is a power always intently watching each slight accidental alteration in the transparent layers; and carefully selecting each alteration which, under varied circumstances, may in any way, or in any degree, produce a distincter image.[161]

If, as Darwin continued, the imagination was allowed to rove over 'millions on millions' of years, one might well arrive at a living telescope 'superior to one of glass'.[162] For the reader, here was the 'insensible made sensible'.[163] Paley's famous contention about the human eye as so complex an organ that it could only have been formed by a divine creator, was being slowly but inexorably discarded.

The problem with Darwin's imagination, though, was that it had a habit of constantly outrunning the evidence.[164] The fact that he gave so much space in the *Origin* to difficulties with his theory is a clear illustration of this. It opened easy lines of attack and ridicule among his detractors. Nowhere were the difficulties more apparent than in the absence of transitional varieties – whether found among living organisms or discovered within the fossil record. The difficulties left the door open for all kinds of theories of deluge and catastrophe. Darwin did, in fact, have a partial theoretical response to the problem, but the chronic gaps and imperfections of the fossil record left him wide open to criticism. For all William Smith's work on distinguishing strata by their characteristic fossils, and for all John Phillips's efforts to refine it, the crust of the earth, as Darwin conceded towards the end of the *Origin*, was 'not a well-filled museum', but 'a poor collection made at hazard and at rare intervals'.[165] As early as 1843, he had written how geology had never revealed and probably would never reveal 'more than one out of a million forms, which have existed'.[166] In parallel, Darwin registered a critical difficulty in scaling time relative to evolutionary theory. Species might remain unmodified over lengthy periods. This may have been especially true at the dawn of life when organic forms were fewer and simpler.[167] Organic change as a measure of time was thus always going to be problematic. Despite this, however, Darwin was still willing to allow his imagination to run the evolutionary story forward to 'a distant futurity'.[168] His 'prophetic glance' foretold that it would be the common and widely spread species, belonging to the larger and dominant groups, that would prevail – in due course procreating new and dominant species in an unending succession.[169] No matter that selection under domestication had failed to produce new species.

One of the special appeals of the *Origin* to the mid-Victorian generation was its accord with passions for natural history. Whether it was collecting butterflies or moths, or exploring the wonders of the shore armed with one of Gosse's little books, Darwin's *Origin* was full of incidents and observations from the natural world. It was not just that it commenced with a study of domestic variations, but chapters like those on Instinct and on Hybrids were full of observations that would have resonated with men who had spent lifetimes raising flowers or keeping bees. Alongside the formal chapters on the idea of evolution, Darwin was effectively offering readers a series of nature stories. He told of how the pear could be far more readily grafted on the quince, which is ranked as a distinct genus.[170] There were plants like the

Hippeastrum which were far more easily fertilized by the pollen of another and distinct species.[171] Then he related a whole sequence of observations and experiments about the honeycombs of bees. Here the tenor of the story becomes very much one of nature's wonders. The bees had evolved a cell shape for their combs that ensured they held the greatest possible amount of honey. It was a feat that neither the mathematician nor the skilful workman could achieve with ease. This was nature at its most fantastic. But Darwin could still ascribe the actions of bees to the progressive refinement of a few simple instincts.[172]

The great range of nature stories in the book as a whole became a kind of metaphor for the diversity of nature itself.[173] In turn, Darwin was echoing the style of much contemporary fiction. Stories within stories were a form of narrative plot that was crucial, for example, to Dickens's writing. Such plots unfolded around the slow uncovering of the connections between the multiple characters.[174] Eccentricity became a route to discovery.[175] Darwin's use of analogies or metaphors was also a form of story-telling, the case of the telescope and the human eye forming one of the most singular. One kind of story was used to invoke another. And what at first seemed sheer fantasy became a clever piece of counter-intuition.[176]

The most powerful story in Darwin's *Origin*, of course, but one that was partly disguised because it was so explosive, concerned the history of the world. It was not merely that it reinforced the abyss of time of Lyell's geology, with all the consequent implications for the account of Creation in the Bible. It added to this by sketching out an altogether new beginning. Rather than an Eden there was merely sea and swamp. There was no man, merely an 'empire of molluscs'.[177] Life came out of a primordial slime. Equally disturbing, though, was what the *Origin* implied about the future. 'Not one living species', he wrote, 'would transmit its unaltered likeness to a distant futurity'.[178] Evolution in Nature required no outside control or direction. Mankind thus appeared to be poised on the edge of the unknown. For a generation increasingly worried about an afterlife, this added to their anxiety. And if life was, as Darwin maintained, a perpetual struggle, what was there to prevent a descent of mankind into barbarism? Such imagery, in turn, had a yet more unsavoury and shocking message. This related to the descent of man. Darwin included no explicit discussion of the origin of man in his book, except for one cryptic sentence at the very end.[179] However, the implications of his ideas were not lost on many readers. He was offering a new genealogy for mankind. So evolution, 'once a mere germ in the mind', as Robert Elsmere recorded, began 'to press, to encroach, to intermeddle with the mind's other furniture'.[180]

Looking into the Unknown

> And no one has a right to say that no water babies exist, till they have seen no water babies existing; which is quite a different thing, mind, from not seeing water babies.

> You must not talk of 'ain't' and 'can't' when you speak of this great wonderful world round you, of which the wisest man knows only the very smallest corner, and is, as the great Isaac Newton said, only a child picking up pebbles on the shore of a boundless ocean.[181]

Charles Kingsley's *The Water Babies* was one of the most popular of Victorian fairytales, even though it appeared first in serial form in the erudite *Macmillan's Magazine* of 1862–3.[182] It was in part inspired by Darwin's theory of evolution. In effect, it adapted Darwin's *Origin* for children, but at the same time touching on a range of problems that Darwin's theory revealed.[183] It also doubled as a satire on the workings of capitalism.[184] And rather like Lewis Carroll's *Alice in Wonderland* of 1865, it became a book that ultimately had almost as much appeal for adult readers as it had for the young.

The hero of *The Water Babies* is Tom, the boy chimneysweep. The early part of the tale sets out the harsh realities of child labour in the early nineteenth century, if suitably sensitized for a young audience. Tom laughs as much as he cries, and accepts his indifferent lot as simply the way of the world. He daydreams of the time when he will be a man, and when he will be a master chimneysweep, like his own master, Grimes. Tom's adventures in fairyland begin when he accidentally descends the wrong chimney whilst sweeping the labyrinth of flues in a great country mansion. Tumbling into the fireplace, blackened from head to toe with soot, he is mistaken for a villain and forced to take flight into the countryside. After many hours, he is exhausted, hot and thirsty and jumps into a clear, cool stream. But somehow he falls asleep in the stream and begins to dream. When he wakes up he has turned into a water baby. He has gills; he is amphibious; and soon he embarks on a series of adventures in an underwater world teeming with life in all shapes and sizes.

As a practised marine zoologist, author of *Glaucus; or, Wonders of the Shore*,[185] Kingsley paints a lively and credible narrative. He takes Tom face to face with a vast array of underwater life. But, in between all the adventures, Kingsley places a spyglass on evolutionary theory. Tom is just one of many hundreds of water babies. Here is the fecundity of life in Darwin's *Origin*. But it also means immense waste. The 'fairy' mother who takes a shine to Tom picks up great armfuls of

104   A cartoon of the contest between Richard Owen and T. H. Huxley as featured in later editions of Charles Kingsley's *The Water Babies*

water babies and then throws them away. Contrary to the pattern of real nature, though, they come 'paddling and wriggling' back.[186] It is a children's story after all. The 'fairy' mother is actually Kingsley's surrogate for natural selection. She picks out good creatures from bad. She made 'old beasts into new all the year round'.[187] At almost every stage, the book mythologizes the various elements of Darwin's theory.[188] Kingsley also parodies the scientific and popular reception of Darwin. If water babies were cast as transitional forms, so the story told, what about catching one, depicting it in one of the illustrated papers, or, better still, cutting it in two halves, one to be sent to Richard Owen, the other to Huxley.[189] The allusion is to the vicious controversy that was by then current over evolutionary theory and over man's ancestry.[190] Owen was convinced that man formed a distinct subclass of mammals and that this was revealed, in particular, in the anatomy of the brain. Huxley, though, with Darwin behind him, argued that whatever organs were studied, 'the structural differences which separate man from the Gorilla and the Chimpanzee are not so great as those which separate the Gorilla from the lowest apes'.[191] Man, in other words, was a descendant of the higher apes, following the manner of Darwinian theory. Descent operated through modification by variation and selection.

Kingsley's allusion to this scientific contest was later to be illustrated in a woodcut that showed cartoon images of Owen and Huxley each peering intently at a water baby alive in a glass flask. The contest was simultaneously being played out in private correspondence, in lectures, in sessions of the British Association for the Advancement of Science, in learned journals, in the gutter-presses, as well as in comic magazines like *Punch*.[192] All testified to the speed with which the ideas in Darwin's *Origin* permeated the different stratas of society. By March 1863, Darwin was writing about his trouble with Owen to his botanist friend Joseph Hooker. He remarked how he longed to be 'in the same boat with all . . . [his] friends, i.e. at open war'.[193] Later, in July, Huxley wrote assuring Darwin that Owen was damning himself 'as fast as is good for us'.[194] Owen's view of nature was a godly one. It did not admit of ideas of transmutation. Darwin's nature at war was thus viewed as a fiction. Owen had previously penned an anonymous attack on the *Origin* for the *Edinburgh Review*, published in the spring of 1860. Darwin had read it with trepidation: it was, he subsequently wrote, 'extremely malignant, clever, & . . . very damaging'.[195]

By the summer of 1860, the sequence of attacks on the book had become a flood. Despite praise from Huxley in the *Westminster Review* of April 1860, in which he claimed it 'superior to any preceding or contemporary hypothesis',[196] and despite a deluge of letters of praise arriving in Down, Darwin had secretly began to harbour doubts about the entire theory.[197] His old geological mentor, Adam Sedgwick, whilst admiring parts of the book, had found others 'utterly false & grievously mischievous'.[198] Above all, he rejected Darwin's foresaking of the 'true method of induction'. There were too many unproven assumptions. There was a dearth of evidence.[199] The geologist William Hopkins had echoed much the same view in an article in *Fraser's Magazine*. He bemoaned the bold manner in which Darwin disposed of 'difficulties'. Just because objections to a theory could not be proven did not mean one could accept it as true.[200] Kingsley picked up on these sorts of objection in *The Water Babies*. Just because evidence was lacking, or lacking in sufficiency, did not mean that there was none to be had: not having seen water babies did not justify claiming they did not exist. Darwin had actually been pursuing a deductive rather than an inductive style of method. Out of his initial hypothesis came a theory, but only when that theory 'explained an ample lot of facts'.[201] The method was much as used by Charles Lyell in his *Principles*. But it seems that geologists were more easily converted than simple naturalists: they were 'more accustomed to reasoning', so Darwin argued.[202]

As Darwin worried over the torrent of attacks, he had learned from his publisher, John Murray, that copies of the *Origin* were still being snapped up. The first edition of around 1,200 copies had been sold out on the day of publication.[203] The second, running to 3,000 copies, was two-thirds gone by June 1860.[204] By November 1860, Murray was calling for a third edition.[205] For those who could not obtain a copy, or could not afford one, the ideas set out in the *Origin* were very rapidly assimilated and appropriated by most of the presses of the day, popular as well as intellectual. The weekly *Chambers's Journal*, for instance, had noted the imminence of the book's publication in its issue of 26 November 1859.[206] And three weeks later, it was offering readers a commentary on it.[207] All living forms of life were 'lineal descendants of those which lived long before', and the reviewer was certain that the 'ordinary succession by generation ha[d] never once been broken . . .

that no cataclysm ha[d] desolated the whole world'.[208] As an inevitable corollary, one could safely infer that 'not one living species will transmit its unaltered likeness to a distant futurity'.[209] For, as already described, Darwin's theory indicated how many species and many genera had become 'utterly extinct' and 'left no descendants'.[210] Five weeks after publication of the *Origin*, the same reviewer had remarked how the book was 'making a sensation'.[211] It was *Vestiges of the Natural History of Creation* of 1844 all over again, with Robert Chambers, its still anonymous author, observing the new sensation from the pages of his popular journal. *The Reasoner*, George Holyoake's atheistical weekly that had been running since 1846, was quick to set Darwin's *Origin* within its secularist project. It offered its readers abstractions from Sunday evening lectures in Tottenham Court Road's Hall of Science on the 'Origin of Man'.[212] It was science against theology writ large. Mechanics and artisans flocked to the cause, whether they came to listen in the debating halls or to read in their institutes, the new temples in the working classes' acquisition of knowledge. It was here, in fact, that Huxley's *Evidences to Man's Place in Nature* of 1863 had started life: as lectures for 'Working Men'.[213] He began to teach 'the great unwashed of their gorilla ancestry'.[214] Hundreds packed in to hear him, to learn how men were 'better sorts of apes', 'a little longer in the leg, more compact in the foot, and bigger in the brain.[215] Nor was there any 'psychical' distinction to dignify or separate man from a chimpanzee or gorilla ancestry.[216] 'Even the highest faculties of feeling and of intellect bega[n] to germinate in lower forms of life'.[217]

Darwin was among the first to receive a copy of Huxley's work. He called it 'the Monkey Book' and was enthralled with the pictures.[218] Joseph Hooker, though, described it to Darwin as 'coarse-looking', and, in the view of some, 'not fit for a gentleman's table'.[219] Even so, Hooker was plainly fascinated by the book's message. In the event, *Man's Place* was a roaring success. Not only did Huxley benefit from his cloth-cap followers, he also was granted a following wind with the publication, earlier in 1863, of Charles Lyell's *Antiquity of Man*.[220] Indeed, in some reviews, the two works were put together, Lyell making man a hundred thousand years old, Huxley giving him a hundred thousand apes for ancestors.[221] The *Saturday Review* of 7 March 1863 described Lyell's book as a 'trilogy', concerned with 'Prehistoric Man, Ice, and Darwin'.[222] For Darwin, the section on ice was his favourite. It examined the chronological relations between human and glacial history.[223] He was disappointed, though, by the book's 'want of originality'.[224] In the parts dealing with species modification, it relied heavily on the work of others.[225] And Darwin was bothered by Lyell's 'excessive caution' in his discussion of such modification.[226] As for prehistoric man, Lyell had been assembling evidence from a whole range of

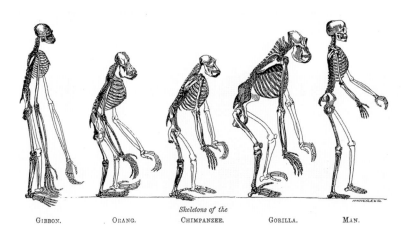

GIBBON.   ORANG.   *Skeletons of the*   GORILLA.   MAN.
                          CHIMPANZEE.

105   One of the illustrations from Huxley's *Evidences to Man's Place in Nature*, 1863, comparing the skeletons of man and apes

archaeological sites. Discoveries of flint weapons in juxtaposition with the bones of extinct animal forms had been steadily pushing back the bounds of human history,[227] and the unearthing of a savage-looking human skull in the Neander valley, near Dusseldorf, in 1857 had added to the brew.[228] Archaeology was putting a spotlight on evolution. Was Neanderthal man the vital link in man's ape ancestry? Opponents of the idea of transmutation had long cited the absence of gradational forms, but negative evidence, so Lyell argued, was not a basis for rejection, and especially in the case of man. 'The pages of the great book of nature' needed much more extensive searching.[229]

Lyell's book was also an immediate bestseller. The first edition of nearly 4,000 copies sold within a couple of months. A second edition of 5,000 lasted only six months, and, by November 1863, there was a third.[230] Lyell was actually out-running Darwin in the sales stakes. One reason for its success was the way it was bitterly attacked by Richard Owen. Writing in a letter to the *Athenaeum*, he accused Lyell of calumny, that is malicious misrepresentation. Lyell had cast doubt on the evidence from which Owen had argued that there were distinct differences between ape and human brains.[231] Darwin remarked what an 'accursed evil' it was that there should be 'all this quarrelling within what ought to be peaceful realms of science'.[232]

What fuelled the fire was that many readers saw Huxley's hand behind the offending parts of Lyell's text.[233] The 'quarrelling' had actually been running for several years. Indeed, it had turned into a kind of soap opera, a perpetual source of copy for magazines like *Punch*, as well as for the output of the penny presses. The question of an ape

THE LION OF THE SEASON.

Alarmed Flunkey. "MR. G-G-G-O-O-O-RILLA!"

106 *Punch* joins in
the debate about
man's descent from
the higher animals

ancestry for man formed a terrible fascination. Punch's writers had
gorillas from London's Zoological Gardens penning enquiries of
Huxley and Owen in the form of poems. The gorillas wanted to know
whether they were 'men' and 'brothers'.[234] The magazine carried car-
toons depicting gorillas in evening dress at fashionable London soirees,
frightening footmen and flunkies all around.[235] Man had been turned
into a monkey deprived of his tale. Soap opera was also a label that did
not sit uncomfortably in the debates that centred around Owen and
Huxley at the annual meetings of the British Association for the
Advancement of Science. One Saturday at the end of June 1860, in the
newly built university museum at Oxford, the Bishop of Oxford,
Samuel Wilberforce, had mounted a frontal assault in which he
ridiculed Darwin's *Origin* and savaged Huxley's championing of it.
Huxley and Owen had already had a vigorous public exchange of
views, and the Bishop was obliging Owen and his supporters with a
spirited finale.[236] Joseph Hooker, who was present throughout, had
written memorably to Darwin after the event: 'the battle waxed hot.
Lady Brewster fainted . . . my blood boiled. . . . He [the Bishop] was
absolutely ignorant of the rudiments of Botanical Science'.[237] Huxley's
response to the Bishop was grave. He judged the speech to have been

bereft of argument and in reply to Wilberforce's question as to whether Huxley found apes on his grandfather's or his grandmother's side, Huxley had retorted that he would rather be an ape than a bishop.[238] It was the stuff of theatre and hugely enjoyed by the hundreds crammed in to hear it. The episode was constantly retold in conversation, in the columns of the press and in caricature. Ultimately, no one version tied with another and the Oxford meeting soon acquired a kind of mythical aura, a war of words that stood proxy for a deep intellectual and social fissure. Publication of Huxley's lectures, *Man's Place*, late in 1863, provided the coda for the soap opera, but not before Huxley had laid the lie to Owen's claims of man's separateness from an ape ancestry. The opportunity came at the Cambridge session of the British Association in October 1862. In full public view, Huxley had an ape brain dissected, exposing its hippocampus, that part of the brain that Owen claimed to make man alone distinct.[239] As *Punch* recorded, apes and monkeys now crowded in dozens to celebrate their kinship.[240]

The contest between Owen and Huxley had a kind of mirror image in a schism that had almost simultaneously been opening among the English clergy. It centred on the publication in February 1860 of *Essays and Reviews*. The volume was the work of a group of liberal Anglicans and took its basic inspiration from the new German biblical criticism. It cast doubt on literal readings of the bible and generated not only powerful intellectual tensions, but tensions of faith and of authority. It seemed to some that all the familiar certainties of belief and of obedience had been turned into a speculative bubble. Creeds reverentially repeated, time out of mind, were shot through with doubt.[241] The outcome was that the authors were confronted by a barrage of criticism and denunciation: they had betrayed the sacred trust accorded to them by God. There were seven separate essays, of which C. W. Goodwin's on the Mosaic Cosmogony dealt with the many attempts that had been made to reconcile Genesis and Geology.[242] These were viewed by Goodwin as universally futile. Men did not require a revelation of those things that they could find out for themselves.[243] The Mosaic narrative of earth history could be only 'the expression of the most vague generalities'.[244] The essay by Baden Powell concentrated on miracles. If one was to be able properly to defend Christianity in an age of science, miracles had to be tied to faith alone. They became incapable of investigation by reason and by physical evidence.[245] In arguing in this manner, Powell was seen by critics as questioning the very nature of the divine.[246] And the same critics recoiled in horror at his reference to 'Mr Darwin's masterly volume on *The Origin of Species*'.[247]

The most controversial of the seven essays, however, was without doubt the final one, by Benjamin Jowett. It dealt with the interpretation of scripture. Educated people were beginning to ask, he wrote, 'not what scripture may be made to mean, but what it does.[248] He claimed that the language of creeds and liturgies often had a disturbing influence on the interpretation of Scripture: words were 'singled out and incorporated in systems like stones taken out of an old building and put into a new one'.[249] The original meaning was lost in the process. Bishop Samuel Wilberforce had cited the authors of the seven essays in his famous address to the British Association meeting in Oxford in June 1860. They were the 'seven against Christ'.[250] Wilberforce and twenty-five of his fellow bishops threatened the authors with indictment for heresy.[251] Ultimately, it was only Powell and Jowett who faced the courts, and Powell died before he could appear: he had supposedly been 'removed to a higher tribunal'.[252]

Not surprisingly, *Essays and Reviews* began to vie with the *Origin* as one of the most popular topics of the day.[253] Indeed, according to the atheistical *Reasoner*, a 'few quiet scholars' had apparently begun to 'displace' Dickens, Thackeray, George Eliot and Wilkie Collins.[254] The publishers could not print copies fast enough.[255] By the end of March 1861, 14,000 had been sold.[256] There was hardly a newspaper, periodical or weekly that had not given it footage. Like Darwin's *Origin*, it also found its way into literature. In *Robert Elsmere*, Robert, as priest, is asked how the 'facts' on which Christian theology is based are known to be facts.[257] The inquiry takes him on an inexorable path towards unbelief. He reads of the new German biblical criticism from the pen of his local squire, who happens to be a distinguished man of letters.[258] The Gospels are revealed as just like other books, 'full of mistakes, and credulous'.[259] The foundations of Christianity are built on passionate acceptance of an 'exquisite fairy tale'.[260] Painfully, Robert realizes that he can no longer believe in an Incarnation and Resurrection.[261] He has no other course but to resign his living.[262] Samuel Butler, in *The Way of All Flesh* (1903), had Ernest Pontifex, another lapsed priest, producing a pastiche of the famous 'Essays'.[263] Butler conceded, though, that 'the fruits of victory were for the most part handed over to those already in possession'.[264] This could equally have applied to *Essays and Reviews*. They resonated to an audience that was already disposed to doubt.[265] And the Darwinians were in one camp, Owen and his followers in the other. Many of the unstated implications of the *Origin* for Christian thought and belief seemed to crystallize in *Essays and Reviews* and in the intellectual ferment that it helped articulate. Darwin's almost 'throw-away' line, at the very end of his book, in which he suggested that it might throw light on the origin of man, became the cue for a vast array of imaginative reflection and straight speculation. As Leonard Jenyns wrote to Darwin in January 1860, after his first reading of the *Origin*, it was one thing to suppose that man was merely a 'modified and no greatly improved orang', it was another to conceive of man's faculties of reason and his moral sense as having been 'obtained from irrational progenitors, by mere natural selection'.[266] This would be 'doing away altogether with the Divine Image which form[ed] the insurmountable distinction between man and the brutes'.[267] There were many others who thought like Jenyns, Charles Lyell for a long time among them. While a few divines might see evolution by natural selection as a continuous chain of Creation, Jenyns had put his finger on the vital stumbling block. But within a decade or so, Darwin had addressed that too. It came in *The Descent of Man* (1871). It was the final assault of science on Christian revelation.[268] And the conservative presses lost no time in seeing the book's potential for disturbing the social order. *The Times* furiously castigated Darwin for publishing at a time when 'the sky of Paris was red with the incendiary flames of the Commune'.[269] What made mat-

ters worse was the fascination that the events in Paris increasingly held for the population at large. The monthly *Fraser's Magazine*, for instance, had a string of articles on the Paris Commune in the second half of 1871. And whilst there was no 'particular sympathy' with the communists, there was a desire for the 'English working man [to] make better terms with capital'.[270]

Descent took its readers on the same kind of comfortable tale of discovery as did the *Origin*. Its underlying argument was that the bodily structure of man, with the exception of the brain, was little different from other animals. But what Darwin did beyond this was to search for 'mental resemblances between man and other animals'.[271] In effect, the culture and society of man could be viewed as a product of evolution as well. Love and hate, sympathy and kindness, defence and aggression, all had a brute ancestry. The human spirit grew from the social instincts of apes, not from some kind of divine intervention.[272] Darwin reminded his readers how 'notorious' it was that man's bodily structure was arranged on the same general model as other mammals.[273] The reproduction of the species, in turn, was strikingly the same, 'from the first act of

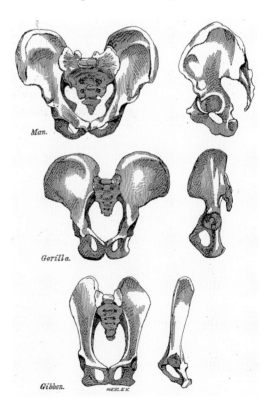

107   Another of the illustrations from Huxley's *Evidences to Man's Place in Nature*, 1863, that so enthralled Darwin. Here it is the pelvic bones of man and apes that are being compared

courtship by the male to the birth and nurturing of the young'.[274] The human embryo, at the earliest period of its development, was likewise indistinguishable from those of other members of the vertebrate kingdom.[275] Darwin told of how he had seen a man who could draw his ears forward, recalling the power of erection in ears common among wild animals.[276] He recalled the fine wool-like hair of the human foetus at six-months gestation. This was, he argued, 'the rudimental representative of the first permanent hair in those mammals which are born hairy'.[277] Study of the character of wisdom teeth among different races of humans revealed quite startling variations. In civilized man, they tended to be more and more rudimentary, the product of feeding largely on soft, cooked food.[278] Darwin concluded by urging his readers frankly to admit their community of descent with the animals. It was arrogant to do otherwise.[279]

The claim was illustrated further with reference to mental powers, where Darwin sought to demonstrate that there were again no fundamental differences between man and the higher mammals.[280] Dogs revealed capacities for love and jealousy, for enjoyment and boredom.[281] The higher animals, generally, demonstrated powers of attention and memory. They could also be observed to show powers of reason. When Eskimo dogs pulling sledges across arctic ice began to diverge and separate, it was often the first sign that travellers had of the ice becoming dangerously thin.[282] Chimpanzees could be observed using stones and sticks as implements.[283] In London's Zoological Gardens, a monkey with weak teeth had been seen breaking nuts with a stone.[284] Sociability was another trait that was clearly seen among animals. Some perform services for each other – removing parasites, for example. Groups of mammals, in particular, often post sentinels to look out for danger, and when attacked, they may defend each other.[285] The upshot of these various kinds of illustration was that, whilst one could concede vast differences between the minds of the 'lowest man' and the 'highest animal', such differences were ones of degree not of kind.[286]

Having addressed directly the question of descent, it was entirely logical for Darwin then to demonstrate variation in the body and mind of man, and, more widely, to trace the progress from barbarism to civilization. Just as Darwin had become fascinated by variation in pigeons, so he sought to remind readers of the astonishing variation that close study of the human anatomy had revealed. Nor was it just bone structures that differed. Arteries could be found running abnormal courses and muscles were eminently variable.[287] As was the case in the animal world, Darwin could relate such variation to variation in conditions of use. But it was also a function of the same general and complex laws that applied to man's progenitors, that is the struggle for existence and natural selection.[288]

The essence of the *Origin* had been contained in its first five chapters, and much the same was true of *The Descent of Man*. One did not have to read very far to gather the gist of the argument. Fundamentally it was about observation, and then connecting that to the idea of transmutation. But the style was determinedly unscientific. It made, instead, for a sedate and comfortable read.[289] It was 'an arm-chair adventure', with the English 'evolving, clambering up from apes, struggling to conquer savagery, multiplying and dispersing around the globe'.[290] It all fitted beautifully with the Victorians' sense of progress.[291] In so doing, it also played to ideas of superior races, including the whole field of eugenics that was to so taint the world in the following century.[292]

The Descent of Man, in two volumes and getting on for 500 pages, was an instant bestseller. The first edition sold out within three weeks. Darwin had worried himself sick at the reactions it would bring, but his worst fears were not in fact realized. It was almost as though people expected something like it.[293] The *Westminster Review* summed up many reactions by remarking that Darwin had been 'anticipated by others', notably Huxley.[294] It considered *Descent* a 'remarkable work', but noted somewhat ominously that 'the instinctive belief in God [was] altogether rejected'.[295] The fiercest criticism came from the anatomy lecturer and catholic St George Mivart, who had once been Huxley's pupil. Not only did he author a devastating review of the book in the *Quarterly Review* of 1871, but shortly before the work's publication, he had launched a series of attacks on natural selection, effectively pre-emptive strikes. These were concerned not merely to refute the theory but to highlight the social dangers that it invited, not least the effects on morals and on religion.[296] Mivart's *Genesis of Species* (1870), for example, had pointed to insuperable difficulties with Darwin's thesis. Natural selection was, in his view, only a secondary and subordinate mechanism in the evolution of the organic world. He refused to accept that the human eye had been formed by natural selection.[297] For the Catholic Church within which Mivart worshipped, Darwinian theory had become an acute problem. Whereas Protestantism had slowly been accommodating to evolutionary theory, Catholicism had been steadily crystallizing its opposition. Darwin appeared to dispense with the possibility of Adam and Eve being possessed of immortal souls.[298] For Darwin, mental phenomena had only a naturalistic explanation. Mivart became the Catholic Church's torch-bearer, heaping abuse on Darwin's ideas. Once again, though, Huxley came to his mentor's rescue. Mivart's stance was shown to be 'as pernicious theologically as it was disastrous scientifically'.[299] Darwin was lifted once more from the 'slough of despond'. It gave him a new burst of energy as he worked on a revised edition of the *Origin*, appending an extra chapter specifically against Mivart's views.[300]

For all Huxley's championing, though, *Descent* did nothing to diminish the imaginative turmoil that the earlier *Origin* had set in motion in men's minds. Not only was there the difficulty of coming to terms with a tadpole-like ancestry, but there was also the opposite difficulty of being left on the edge of a physical and spiritual unknown. If one accommodated evolution from primitive forms, what was the destiny of present life-forms? Man began to seem nothing more than a random phenomenon in the stream of life. As the *Westminster Review* recorded, it was bound to be 'highly distasteful to many readers'.[301] It prompted H. G. Wells, as a student, to speculate how man's descendants might evolve in a manner that humans would be completely unable to recognize. His novel *The Time Machine* (1895) was the eventual outcome.[302] Others began to contemplate the idea of human extinction, perhaps at the hands of a far more barbaric nature, one over which man would be unable to exercise any control.[303] As for the human spirit, the impact of *Descent* was a deepening loss of hope. As the traditional fabric of Christian doctrine disintegrated, with all its beliefs in the ultimate sanctity of man reduced to nothing, a great void opened up in men's minds. The response was a desperate attempt to try to fill it. Spiritualism was one such outcome. People felt a compulsive desire to find ways of looking into 'the beyond'.[304] They shuddered to contemplate the grave as the end of all.

Some might have sought refuge among the classics of imaginative literature. But, even here, there was no escaping Darwin. Charles Dickens was one of Darwin's favourite authors, and *Bleak House* (1852–3) could almost have been a model for the *Origin*. As the novel unfolds, the impression given to the reader is that 'all is confused, random, inchoate'.[305] But gradually it emerges that 'everything is involved with, or connected to, or somehow has a bearing on, everything else'.[306] It is exactly like the variation and diversity of nature that Darwin endeavours to make sense of by means of evolutionary theory. The seemingly never-ending case of Jarndyce and Jarndyce might stand proxy for the continually operating struggle for existence. Scores of people had 'deliriously found themselves made parties in Jarndyce and Jarndyce, without knowing how or why',[307] just as plants and animals were inexorably caught up in the competitive web of life. There seemed no prospect of an end to the lawsuit. It could have been a metaphor for Hutton's or Lyell's ideas of earth history, both of them critical in underpinning Darwin's theory.[308]

Bleak House came before Darwin's *Origin* and George Eliot's *Middlemarch* (1871–2) came after it. But *Middlemarch* reveals many of the same themes. It is a struggle for social existence, between old families and new, from one generation to another. The fictional model is evolutionary, concerned with 'complexity of relations, repetition with variation, progression and continued divergence'.[309] Eliot's prelude to the novel touches upon its Darwinesque themes: 'the varying experiments of Time', 'indefiniteness', 'limits of variation' that are 'wider than anyone would imagine'.[310] A number of Eliot's novels reveal Darwin's presence. It was in Thomas Hardy's writing, though, that Darwin found his major intellectual influence among novelists.[311] Blind interaction as part of Nature's laws became a common level of plot in Hardy.[312] He had already accommodated his mind to the vast deserts of geological time. He now added to this the fateful character of man's existence, with human intelligence and consciousness reinforcing that fate through its realization.[313] When Henry Knight, in *A Pair of Blue Eyes* (1872–3), slipped down part of the cliff face of the Dorset coast, he found himself face to face with a trilobite: 'it was a creature with eyes. The eyes, dead and turned to stone, were even now regarding him ... separated by millions of years in their lives, Knight and this underling seemed to have met in their place of death ... [T]he immense lapses of time had known nothing of the dignity of man'.[314] The wonder, renewal, fecundity and diversity of Nature in Hardy's novels are not just interwoven with characterization, but form a perpetual

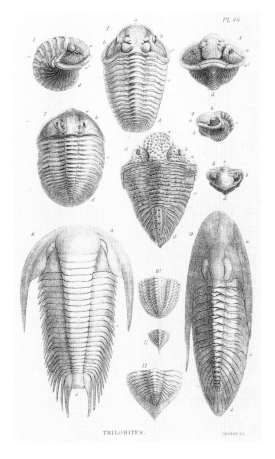

108 Trilobites as illustrated in William Buckland's *Geology and Mineralogy ...*, 1836

reminder of man as within Nature and subject to its laws. The outlook of many of the novels is dour and, at times, tragic. Characters seek happiness, but it constantly evades them. Man seems doomed to maladaptation.[315] Even the heath-dwellers on Egdon were alien to its primitive form.[316] It was a far cry from Kingsley's *Water Babies* where Tom is reborn and restored to life as a man of science.

7

The Prehistoric as Exhibition

Very few people have any idea of the multiplicity of specimens required for the purpose of working out many of the simplest problems concerning the life history of animals or plants. The naturalist has frequently to ransack all the museums, both public and private ... to compose a monograph of a single common genus, or even species, that shall include all questions of its variation. ...[1]

Prelude

There is no standing still, no rest: 'onward, ever onward', is at once the motto of our intellects and our necessities.[2]

Time was the great obsession of the Victorian age. It was revealed in the preoccupation with development, with a life of progress, with perpetual comparisons of 'then' and 'now'. It was seen in the dual themes of cycle and crisis, the rollercoaster of industrial capitalism, with all its attendant social gains and failures. But, above all, it was manifest in the 'ever-accumulating past' – not just the extending history of mankind, but the extended history of the earth and of biological life itself.[3] Understanding this particular obsession, though, was not straightforward. One had to make visible what was largely invisible.[4] Charles Lyell, as already noted, had to make liberal use of his imaginative faculties to resurrect and make more intelligible the cycles of restoration and decay that characterized the Huttonian Earth-Machine.[5] Darwin faced a similar problem: firm evidence of his evolutionary theory was hard to come by.[6] Like Lyell, he was forced into conjecture as an outcome. He constructed a narrative, a story based on a trail of clues that were often imperfect and long-delayed in their decipherment.[7] The new museums of the Victorian age were similarly preoccupied with time. They also performed the role of storytellers. The emergent sciences of geology, palaeontology and archaeology were

109 (*facing page*) Oxford University's Natural History Museum, as depicted in *Building News*, 1859

brought into public view in a manner never seen before.[8] Alongside panoramas, dioramas and 'peep-shows', depicting elements of the ante-diluvian age, museums offered their own narratives on the prehistoric world. They presented the prehistoric world as *exhibition*.[9]

Natural Worlds on Show

> There are two books from which I collect my divinity; besides that written one of god, another of his Servant, Nature, – that universal and public manuscript that lies expanded into the eyes of all.[10]

Regardless of what Robert Chambers had implied in *Vestiges* and regardless of what Darwin signalled in the *Origin*, the new museums were essentially shrines to God's Creation, memorials of divine power and wisdom as expressed in the natural world. Here the public was presented with the full wonder of God's bounty. The tenets of Paley's natural theology were everywhere borne out. And this was reinforced by the ecclesiastical architectural styles in which many of the museum buildings were constructed. Naves and transepts vied with cloisters and cells to give an unmistakeable air of reverence. Here was science in theology writ large. At the same time, the progression of divine creation was easily adapted as a metaphor for the progression of western man, for underpinning the ideology of conquest and empire. Richard Owen, superintendent of natural history at the British Museum, enshrined the principle in an address to the annual meeting of the British Association for the Advancement of Science held at Leeds in 1858. The country's natural history collections had to be worthy of the colonies.[11] For 'no empire in the world had ever so wide a range for the collection of the various forms of animal life as Great Britain'.[12] Nineteenth-century museums thus became veritable showcases of empire, 'where visitors were able to see the trophies of conquest'.[13] The involvement of white settlers in the natural histories of their adopted lands thus became a sort of 'cultural justification for colonization'.[14] This same theme was augmented by the way the fauna of the colonies came to be interpreted. Natural worlds there were presented as 'primitive, unruly and unstable, in need of civilizing dominion and stewardship'.[15]

Owen, of course, was one of the prime defenders of progressive creation, that is flying in the face of Darwin's evolutionary theory.[16] For Owen, just as 'death was balanced by generation', so extinction was 'concomitant with creative power'.[17] And when his central project, London's great Natural History Museum in South Kensington, was finally opened in 1881, *The Times*, in its leader column, was prompted to liken the structure to Noah's Ark.[18] More than twenty years on from Darwin's *Origin*, biblical imagery and biblical perspectives on time still permeated natural history.

The great nineteenth-century museums were, at their most basic level, collections of objects. And 'collecting', as discussed already, was a central pastime for many Victorian men and women.[19] They assembled for their own leisure vast arrays of rock and mineral specimens. They gathered fossils by the score. Here was another manifestation of the prehistoric world as *exhibition*. At quite an early stage, regular dealers emerged who sold custom-made cabinets of such specimens, as an earlier chapter has revealed.[20] By 1859, the year that Darwin published the *Origin*, one dealer in the Strand in London was offering study collections priced from two guineas to one hundred.[21] The natural worlds of the past were being commoditized. One was as likely to see display cabinets in the home as in any museum, library or mechanic's institute. Collections of this sort held no explicit reference to time. But they plainly alluded to time in terms of the variation and diversity of nature, past as well as present, that they documented. And in the organic world of the past, in particular, the world of fossil plants and animals, variation was evidence of progressive creation on the one hand, as well as a vital clue to Darwinian evolution on the other.

The fresh geological excavations that railway construction constantly laid bare afforded another instance of the prehistoric world as *exhibition*. And as passengers were transported through space at speeds of thirty, forty and even fifty miles an hour, they simultaneously became travellers in time.[22] If God-fearing men and women had initially been horrified by the implied threat to Christian doctrine of Lyell's *Principles*,[23] the familiarity with geology that arose from railway travel paradoxically had the potential to ease such anxieties. For the rainbow-like colours of some of the new excavations became an illustration of the wonder of nature. And John Ruskin, whose geological credentials were unimpeachable, saw the variety of colour and hue in natural stone as providing the true colours of architecture.[24] In *The Seven Lamps of Architecture* (1849), he argued that the full range of colour, 'from pale yellow to purple, passing through orange, red and brown', should be entirely at the architect's command.[25] Here, one might claim, was a reassertion of the divine in English rock, especially where church architecture was involved.[26] Even when brick began to be widely substituted for stone, the same variety of hue and colour prevailed, giving rise to forms of decoration that could be viewed as mimicking the sedimentary layers of the earth on the one hand, or as imitative of the medieval brick churches of north Germany on the other.[27] Associations of geology and religion

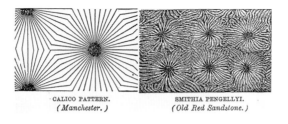

CALICO PATTERN.
(Manchester.)

SMITHIA PENGELLYI.
(Old Red Sandstone.)

110   The pattern of a Manchester calico print compared with the pattern of a coral from the Old Red Sandstone

were still never far away, for all geology's exposure of deep and long-lost pasts.

It was in the decorative arts, though, that the prehistoric world as exhibition appeared in its most subtle forms. When Sir Henry De la Beche wrote his account of the geology of Cornwall,[28] he remarked how the serpentine rock of the Lizard peninsula might be employed for decorative purposes.[29] At the Great Exhibition of 1851, this is exactly what had transpired, for among the exhibition objects were 'specimens of serpentine so beautiful, and made into such elegantly formed obelisks, fonts, chimney-pieces, vases and small ornaments'.[30] A Cornish serpentine industry was by then in full swing, complete with steam-powered saws and cutters.[31] Its products came in an astonishing variety of colours and lustres, cream white to all shades of green and black, translucent to opaque. Few who actually bought such serpentine pieces would have been able to grasp the very remote period of earth history from which the rock derived, or of the 'red-hot haste' of its formation.[32] It was another reminder, though, of the wonders of creation, to be set alongside more familiar examples of rock used for decorative purposes, not to mention the strata of coal and iron ore that were helping to underpin Britain's industrial power.

Hugh Miller took up the theme in his best-selling *Testimony of the Rocks* (1861).[33] One constantly found in the various organisms of the different geological epochs examples of forms that man delighted in reproducing.[34] The delicate diaper-work of some Gothic buildings could be traced to patterns on the Lepidodendra (giant trees) of the Coal Measures, for instance. Indeed, Miller went so far as to claim that one could 'detect among the fossils the germs of numerous designs developed in almost every department of art'.[35] One of the most successful calico patterns ever tried, known as 'Lane's Net', had, so Miller argued, an uncanny similarity to one of the corals of the Old Red Sandstone. So a fabric design that 'so smit the dames of England . . . had been stamped amid the rocks of *eons* of ages before'.[36]

More generally, by the 1820s the use of plant ornament in the decorative arts became much more widespread, with depictions increasingly naturalistic in their form, whether on textiles, on plasterwork or on stone. For some, such blueprints from nature became mere space fillers. For others, including Augustus Pugin and William Morris, natural forms were used in a much more controlled and judicious manner,[37] as the decorative schemes for many leading nineteenth-century museums were soon to bear testament.

The Museum is Born

A Museum is, be it observed, primarily, not at all a place of entertainment, but a place of Education.[38]

The museum movement did not come upon a public entirely unused to display and exhibition. There had long been travelling circuses and menageries, for instance. Fairs were familiar places to see performing bears and monkeys. And as conditions of travel and transport had improved over the late eighteenth century, so the scale and variety of

111   Caricature print of animals on show in London, 1795

112   The Elephant House, Surrey Zoological Gardens

these forms of display multiplied. By the mid-nineteenth century, 'Wombwell's Mammoth Menagerie', for instance, had '20 Monster Carriages, 57 Powerful Cart Horses, 53 Employees, and 700 Animals, Birds and Reptiles'.[39] London was a common starting point for many of these travelling shows, with Piccadilly as the primary venue. The intent was not educative, of course, but sensational. Viewers were fascinated by the exotic and by the unfamiliar, whether lions and tigers, or crocodiles and snakes.

As the population of London grew in the early nineteenth century, it was logical for such 'entertainments' to acquire fixed sites. The outcome was the establishment of zoological gardens, of which Regent's Park and the new Surrey Gardens, at Kennington, south of the river, were leading examples.[40] The Regent's Park 'zoo' was founded by the Zoological Society of London in 1828 and covered some thirty acres. On Sundays it was reserved for Fellows, but on other days open to all. By the time railways and steamboats were in full swing, London's cockneys were also soon found travelling in their thousands to the Kent Zoological Gardens at Rosherville, near Gravesend, where the attractions included a bear-pit, a monkey cavern and a host of tropical birds.[41] Outside the metropolis altogether, Bristol had by then acquired its own zoological gardens at Clifton.[42]

The Surrey Zoological Gardens, as well as those at Regent's Park, were profoundly middle class. Charles Darwin first visited Regent's Park in the company of his elder brother Erasmus while down from

Cambridge as an undergraduate in 1828.[43] In 1850, on one of her rare visits to London, Charlotte Brontë also saw the animals there and sent her father, Patrick Brontë, a detailed account of the ones she observed, including a great Ceylon toad that was almost as big as the family's pet dog, Flossie.[44] Erasmus Darwin, was also familiar with the more 'tawdry' venues where exotic animals were kept.[45] These included the various menageries on display in streets like Piccadilly. And here it was that Charles, in 1828, saw a huge mandrill (a West African baboon) indulging in a pipe and a glass of grog, daily at one o'clock.[46] Charlotte Brontë may have seen a similar sight in Regent's Park, for in 1845 there was brought to the zoological gardens there, from Sierra Leone, a female chimpanzee that, according to the *Illustrated London News*, ate her egg with a spoon, could lock and unlock any door or drawer and could thread any needle.[47] The paper included an engraving of the chimpanzee taking tea, affording a humanized image of animal behaviour that was later to take on a deeply controversial significance, of course, as part of Darwinian theory.

113   The Zoological Gardens at Regent's Park

Display and exhibition in this context was not confined to animals. It also included the plant world. The various zoological gardens were very much gardens in the real sense. Many housed collections of exotic plants. The archetype for the botanical garden, though, was Kew, and in the early 1840s, the Great Palm House was under construction, to the design of Decimus Burton. Readers of the *Illustrated London News* were treated to a detailed account of the process of erecting the giant glasshouse and the plan of its arrangement.[48] With a length of 362 feet and a central height of 62 feet, it was designed to accommodate a whole variety of palms, the so-called 'princes of the vegetable world'. The paper anticipated that once the plants had been arranged 'in natural order', there would be 'no more striking exhibition of the Creator's works'. Here, once more, was the comfortable world of natural theology. The Great Palm House was just one of many hothouses at Kew. What they revealed to visitors was not just a whole new plant world, but a taste of the climate and atmosphere of the tropical regions. Some visitors fainted in the heat. Writers for *Punch* were fascinated by the very strange character of some of the plant forms, particularly the succulents. They had spectators marvelling at trees that 'grew groceries' – like date and cocoa palms, and bread-fruit trees.[49]

114   The Menagerie at Exeter Change, Strand, in 1829, shortly before its removal and demolition

A public acclimatized to and fascinated by the sight of exotic animals and plants of the living world was also a ready audience for hearing and learning about animals and plants from lost worlds. And just as there were travelling menageries, so there were collections of ante-diluvian relics that plied for public entertainment. They included not just organic remains found in Britain, but bones from Europe and North America. 'Koch's Antediluvian Museum' was one of the travelling exhibits at the Egyptian Hall in Piccadilly. It included the skeletal remains of the so-called Missouri Leviathan, a giant extinct reptile.[50] At the Surrey end of Blackfriar's Bridge, meanwhile, was a building that for a time housed the collection of organic remains assembled by Sir Ashton Lever. Here one could see 'an unparalleled display of Nature in Epitome, with Organic Remains, or Fragments of the Ancient World'.[51] Some exhibits and displays of this kind had admission charges that were well beyond the means of the ordinary populace, even including the mechanic and artisan classes. However, a few granted free admission, or made only token charges, particularly where they were set up at fairs.[52]

When debates began in mid-century about the range of public access to the new museums, it was not, therefore, against a background of a population unfamiliar with the practice of 'showing and telling'.[53] But what distinguished the museums of the Victorian generations was that they were concerned less 'to create surprise or provoke wonder',[54] and more to educate as part of a state apparatus that was steadily extending its cultural base and its influence over people's lives. The Public Libraries and Museums Act of 1850 was a critical signpost in this process, enabling municipal boroughs around the land to promote instruction and recreation via these mediums. With statutory backing of this kind, it was certainly likely, as John Ruskin observed, that the museum would be a place of education not entertainment.[55] Charles Kingsley made the educational value of museums, local as well as general, abundantly clear. It was about encouraging new habits of mind. It was about 'the art of seeing, the art of knowing what you see, the art of comparing, of perceiving true likenesses and true differences, and of classifying and arranging what you see'.[56] The opening chapter of Darwin's *Origin* practised exactly such a sequence, and in that respect easily tapped into the domains of the pigeon-breeder and the horticulturalist, as the previous chapter demonstrated. It was unfortunate, if not altogether surprising, though, that the revolutionary narrative of organic evolution that Darwin presented did not initially exercise much effect on the structure and arrangement of museum display. Nowhere was this more clearly seen than in Richard Owen's great Natural History Museum which in both its architectural motifs and the manner of its displays still revealed a false separation of living and extinct species.[57]

The most startlingly modern of the first museums was Bullock's, housed in the Egyptian Hall at No.22 Piccadilly.[58] The first part of the

115 Handbill for Koch's Antediluvian Museum open for exhibition at the Egyptian Hall, Piccadilly in the 1840s. The verses taken from the Dublin Mail re-emphasize the continuing force of biblical allusion

116   The Piccadilly frontage of the Egyptian Hall, 1815, then housing Bullock's Museum

museum, called the London Museum, was accommodated in a large, galleried room, with birds, beasts, fishes, amphibia and insects displayed around its walls.[59] This followed already familiar patterns of display. The second part of the museum, by contrast, conformed to a pattern that today would be described as 'virtual reality'. It occupied a room forty feet high, within which the visitor found a mock tropical forest complete with stuffed examples of most of the quadrupeds that had by then been described by naturalists. The trees were 'correct models from nature, or from the best authorities, of the trees and other vegetable productions of the torrid climes'.[60] The shape and layout of the room (known by the name *Pantherion*) created a 'panoramic effect of distance' which added to the illusion of reality. And the whole effect was enhanced by the particular form of the entrance way. It was a kind of basaltic cavern and served to emphasize the contrast between the crowded metropolis outside and the imaginary tropical world within.[61] It is not clear how far the remains of extinct creatures featured in Bullock's Museum, but it was certainly there that the first Ichthyosaurus found by the Annings at Lyme was displayed in about 1814 or 1815.[62] A final, strikingly prophetic feature of this museum was the way some of the creatures were depicted in combat.[63] Here was nature, as

Tennyson would much later describe it, 'red in tooth and claw'.[64] In general terms, it anticipated Darwin's identification with the struggle for existence, even if Darwinian evolution remained some time away.

A London museum that had entirely different origins was the Hunterian Museum of the Royal College of Surgeons, established in 1813. During the initial period of its existence, the museum was devoted largely to the needs of physiological instruction and training, even though it housed an important collection of natural history specimens, the legacy of the physician John Hunter.[65] However, in 1827 the young comparative anatomist Richard Owen took charge of the museum and set about renovating it. His idea was to extend its focus beyond the specific requirements of medical training so as to embrace the emerging sciences of geology and palaeontology.[66] The architect Charles Barry was brought in to perform the structural alterations to the building in Lincoln's Inn Fields, and it was reopened in grand ceremonial fashion, in the company of Wellington and Peel, in February 1837.[67] Owen had been determined to show examples of extinct as well as living creatures and, on entering the redesigned museum, the eyes first alighted on the fossil shell of a giant extinct armadillo, brought from Buenos Aires. In addition, there was a shell of the Mylodon, an extinct giant sloth, also from South America, and parts of the skeleton of a Megatherium, an extinct giant quadruped that William Buckland had brought to fame in the 1820s.[68] The main room, with its three-storey galleries, was ultimately packed with exhibits. And echoing the museum's origin, it included the eight-foot high skeleton of the giant Irishman Charles O'Brien who died aged twenty-two, reputedly of excess drinking.[69] Even after its makeover, though, the Hunterian remained socially exclusive, open only to Fellows and Members of the Royal College and to visitors introduced by them.[70]

Such social restriction was not true of the Museum of Practical Geology, opened in 1851 in a Renaissance-style building in Jermyn Street, off London's Piccadilly. The museum had its origins not in a professional society but in a series of governmental and other

117   The interior of the Hunterian Museum of the Royal College of Surgeons after its remodelling by Charles Barry, 1837 – from the *Illustrated London News*

118   Interior view of the Museum of Economic Geology, from *The Builder*, 1848

organizations that found interests in common under one roof. These were the Geological Survey, the School of Mines, the Mining Record Office and the Museum of Practical Geology.[71] The last-named institution had been established by the government in 1837, at the suggestion of Sir Henry De la Beche. Also known as the Museum of Economic Geology, it had acquired a government-appointed curator by 1839 whose job it was to manage the collections and, at the same time, keep mining records. It was open to all, free of charge, throughout the year.[72] The museum that replaced it, in 1851, was likewise open to all comers. However, it came to fulfil a much more distinctive public role by virtue of its spacious lecture room which took up the major portion of the new building's ground floor area.[73] Here mechanics and artisans could learn about science. And after 1854, they could learn it from the young T. H. Huxley, by then offering lectures on natural history as a government employee.[74] There Huxley soon began his famous 'lectures for working men', finding such audiences a welcome relief from the 'dilettante' middle class groups that he had faced so often before.[75]

A few years later, *Chambers's Journal* was championing the attractions of the Jermyn Street Museum. But it was at pains to say that it offered little in the way of amusement. It was not a spectacle in the way that some museums were. Its purpose was overwhelmingly educational.[76] And this was demonstrated not only in its lecture series, but in exhibits showing models of mines and pumping engines, and examples of all the tools necessary for mining. Here the artisan saw 'the newest improvements in the tools with which he works' and could gain a knowledge of the 'general plan of mining arrangements in different parts of the country'.[77] As far as the museum's collections of rocks, minerals and fossils went, this, so the journal writer remarked, was where the 'child of the nineteenth century' saw collected treasures from the depths of the ocean, from the summits of the mountain, from the mine, and from the quarry'.[78] There could be no longer any doubting the national utility of scientific education and the importance of 'scientific acquirements'.[79] Here ordinary people could gaze at leisure and gain a very tangible sense of their dominion over nature and the scientific means employed for its achievement.

Many of the earliest public museums were ridiculed by members of the universities. Not only was there the familiar problem of science being equated with unbelief, but, in Oxford and Cambridge, study of physical science had long been opposed.[80] By the 1830s, however, this pattern was beginning to change. At Cambridge, for instance, where Adam Sedgwick held the chair of geology, a competition was launched for the design of a new building to house the University's growing collection of natural history specimens. The winning architect was C. R. Cockerell and the Museum was open by 1842. However, only one wing of the design was actually ever built and the specimens were restricted to the building's lower level. The university's library occupied the rest of the space and so the embryo museum in fact ended up being sidelined by traditional demands.[81] At Oxford, by contrast, a rather different story emerged. The founding of its natural history museum was above all related to the foundation of an Honour School of Natural Science in 1849.[82]

Whereas the fortunes of the Cambridge Museum became tied up with and subordinated to the needs of the University Library, in Oxford it was connected to the needs of scientific study. When built, the Oxford Museum thus contained lecture rooms, professors' offices and, above all, laboratories.[83] The Museum's prime mover was a clergyman, the Reverend Richard Cresswell. A founder of the Ashmolean

119   A very early perspective image of the new Natural History Museum in Oxford, 1861, with semi-detached laboratory off-centre and keeper's house to the right

Society, he saw no conflict between science and religion. He regarded study of natural science, for instance, as 'a kind of religious contemplation'.[84] Just as several decades earlier, William Buckland had sought to anchor parts of the geological record to the biblical deluge,[85] so Cresswell and his supporters saw the collection, display and study of objects from the natural world as a celebration of the wonder of God's creation.[86] Paley's natural theology, in other words, retained a powerful impetus even in the 1850s. As at Cambridge, a competition was held to find a suitable design for the new building, the University granting £30,000 for the cost of the shell of the structure.[87] The competition was eventually won by Thomas Deane and Benjamin Woodward, the Irish partnership responsible for the new museum building at Trinity College, Dublin.[88] Building work commenced in 1855 and the project was largely complete by 1860. The architects' choice of Gothic, conscious or otherwise, reinforced the interweaving of science and religion that the Museum's founders embraced. The contrast was with the Cambridge Museum and the Museum of Practical Geology in London where classical designs had won the day. According to Henry Acland, one of the Museum's key scientific campaigners, as University Reader in Anatomy, Gothic was not only best suited to the general architectural character of medieval Oxford, but it was unrivalled in terms of adaptive capacity, including all the necessary considerations for a scientific building.[89]

By the time Oxford's Natural History Museum opened to the public, there had developed among the population at large, especially among the artisanal and labouring classes, a clear familiarity with, and taste for, the organized display of objects. The Great Exhibition of

1851, including the astonishing extent of its appeal for ordinary folk, had sealed the trend, and in 1857, when the forerunner of the Victoria and Albert Museum opened its doors in South Kensington, it was dedicated to serve an undifferentiated public. Indeed, its admissions policy and hours of opening sought to maximize the numbers of visitors from among the working classes.[90] Whereas in the socially turbulent 1830s and 1840s, government officials would have blanched at any idea of encouraging the crowd in such a manner, with all its predispositions for tumultuous and indisciplined behaviour, by the 1850s, the exhibition and the museum had begun to be perceived not merely as a form of educational enterprise but as an agency of social discipline more widely.[91] It was inevitable, then, that curators and officials would soon begin setting their minds to the provision of a new home for the growing national collection of natural history specimens that was by that time concentrated in London's British Museum. Richard Owen, the curator of natural history there, encouraged some of the country's leading naturalists to petition the government for more space.[92] In parallel, some of the country's leading scientists, T. H. Huxley and Charles Darwin among them, urged the government to separate the Museum's natural history collections into two; one part would be solely for scientific study, the other for popular display.[93] Owen himself dissented from such a division and was determined upon a unified museum, located on a site in South Kensington. Ultimately, this was the government's choice, too, and in 1864 a competition was announced for the design of a suitable building. Out of thirty-three entries, two emerged as front-runners. One was by Captain Francis Fowke, a one-time military engineer and the designer of the 1862 exhibition building that had occupied the site earmarked for the new museum. The other was by the architect Robert Kerr. Fowke's design won the competition, but the Museum's Trustees petitioned their preference for Kerr's.[94] The argument turned on the different genus of building type that each design represented. Fowke's guiding emphasis was the necessity to give the curator 'every possible variety of light, to suit the variety of objects which he ha[d] to display'.[95] This meant a structure not all that dissimilar to his 1862 exhibition building, and included rooms opening out of a longitudinal corridor that had uninterrupted light along one side.[96] Kerr, however, saw such a structure as vulgar, reminiscent of a kind of 'bazaar'. Specimens would appear as commodities.[97] It was too far removed from the exalted view of natural history as a celebration of divine creation. Kerr offered, instead, a much more traditional structure that supposedly reflected the seriousness of scientific endeavour as well as the wider moral imperatives that characterized Victorian attitudes to the natural world.[98] Fowke unfortunately died before any commission to build had been granted, and the gov-

120 The frontage of London's new Natural History Museum in South Kensington, as depicted in *The Graphic*, 1880

ernment decided to engage Alfred Waterhouse to carry out Fowke's plans. The commission, though, was not for the execution of the scheme on the South Kensington site, but on a site alongside the new Thames Embankment.[99] The date was now 1869 and the whole enterprise had been dragging on for over a decade. However, a new museum for the natural history collections was still years away. The new site turned out to be too expensive and so the architect was required to begin again on the South Kensington site.[100] It was thus Easter 1881 before the museum finally opened its doors to the public. The overall style of the building was Romanesque, its dramatic façade on the Cromwell Road measuring 675 feet in length, punctuated by twin central towers, with high pavilions at either end.[101] Inside was a glazed central hall or court, again in Romanesque style, which gave access to the galleries on each side. As Richard Owen had intended, it was a museum building that combined the functions of public display and scientific study.

Owen did not long occupy the directorship of London's Natural History Museum. He was already seventy-seven years old when it opened its doors to the public and within two years he had retired. His professional life, though, had become emblematic of the growth of the nineteenth-century museum movement.[102] Most of Britain's museums were constructed in Owen's working lifetime.[103] All told, there were

240 of them in existence by the time of his death in 1892.[104] In Scotland, for example, the Edinburgh Museum of Science and Art was completed within a year of Oxford's Natural History Museum. Unlike the Oxford museum, though, it was a highly materialist endeavour. It illustrated industrial genius, including the way the store of rocks and minerals of the earth's crust was being transformed into useful objects. It provided examples of the burgeoning decorative arts. In a sense, it became the Scottish capital's very own 'crystal palace'. Above all, it made natural history, in all its various guises, 'co-extensive with art and science'.[105] Outside metropolitan realms, many provincial towns gained museums during Owen's working lifetime. The Ipswich museum, for instance, opened its doors in 1847. One of its founding members was J. S. Henslow, successively Professor of Mineralogy and Professor of Botany at Cambridge from the 1820s.[106] A friend of Darwin, he became one of the pioneers of scientific education in elementary schools.[107] The museum functioned not just as a repository for collections of objects, but as a venue for lectures by notable Victorian scientists. Like so many of its counterparts in county towns up and down the land, it became a classic monument to the aspirations of Victorian popular science.[108] Whilst some of its founders and supporters still saw themselves promoting study of the works of God, others saw the goal in a much more secular light.

Narrating the Prehistoric

> In the arrangement of the many valuable and curious examples of polishable stones which the liberality of our friends has enabled us to bring together, we have always *desired* to employ so much of system as to make these ornamental parts of the fabric *really* and *obviously useful* as part of the exhibition of natural objects.[109]

In 1856, a writer for *Chambers's Journal* described spending a morning at the Museum of Practical Geology in Jermyn Street, Piccadilly.[110] Ascending the steps and then entering first the vestibule and then the great hall, the initial focus of the account is on the variety of rock types of which the structure was composed. The entrance steps were of red granite, the vestibule of Portland stone, while in the hall there were pilasters of Peterhead granite, mixed serpentine and marbles from Devon and Derbyshire. Leaving the hall, though, and in the act of ascending the staircase to the galleries, the emphasis of display altered. A sort of 'geological staircase' was in the making, delineating the progression of the strata. In effect, one read the geological record as one climbed the stairs. And moving on to the galleries, the same form of display was maintained. The lower gallery housed fossils belonging to the most ancient eras of life, the upper gallery fossils belonging to the most recent. The combined vertical and horizontal spaces of the museum structure thus became an analogue for geological time. The writer remarked how there was 'something curiously solemn in thus standing in the visible presence of a world-history'.[111] Assyria, Greece, Rome and Carthage were mere 'yesterdays' by comparison. One caught glimpses of the sheer immensity of time past.

In many of the earliest museums, such historical forms of organization were rare. Instead, the systems of display reflected a mixture of ideas usually focused on the external characteristics of objects. Early nineteenth-century developments in geology and palaeontology, particularly the work of William Smith and Charles Lyell, began the first moves towards historical classification or taxonomy. And, in due course, the system was extended to biology, archaeology and the whole field of human history.[112] However, there were sometimes significant time-lags between scientific advances and their accommodation within museum organization and display. As late as 1888, one museum commentator observed how remarkable it was that, although Darwin's theory had exercised a profound influence on almost all departments of thought, it had still had little effect on museums of natural history.[113] This was above all true of London's Natural History Museum. Not only did it separate living and extinct species by placing them on opposite sides of the central hall, but its architectural decoration echoed the same division between biology and palaeontology.[114]

The interior displays at Oxford's Natural History Museum, first opened in 1860, mirrored some of the features of the Jermyn Street geological museum. However, the emphasis was primarily upon communicating variation in nature rather than historical sequence or progression. The idea was to produce such a display of nature's variety (whether rocks, minerals, plants or animals) as to leave observers with an unrivalled vision of the wonders of God's creation. For Henry Acland, the museum would be nothing less than an 'open book of Nature', revealing all the 'mysteries of life under the guidance of a higher Power'.[115] Moreover, this was to be achieved as much by the form of the building's interior decoration as it was by the objects that were on display. Its Gothic ornamentation, for instance, was to focus on subjects connected with the objects of the building, just as architects of the Middle Ages 'confined their ecclesiastical decorations in sacred edifices'.[116] Like the Jermyn Street museum, the pilasters or columns around the central court illustrated various sorts of British rock. Those

121   The decorated capitals of Oxford's Natural History Museum, from the *Illustrated London News*, 1860

on the lower arcade on the west side, for instance, illustrated the granite series, those on the east side the metamorphic, on the north side the calcareous and on the south side the English marbles.[117] The capitals and bases of the columns were in turn decorated with carvings of various groups of plants and animals, representing different climates and different epochs, and arranged mainly according to their natural orders.[118] This particular work was undertaken by the O'Shea brothers, Irish sculptors of quite remarkable skill. They had been brought over from Dublin by the architects Deane and Woodward. The brothers turned up each day with plants borrowed from the University's botanic garden and proceeded to chisel their likeness in stone. They might have been decorating a medieval cathedral, such was their enthusiasm, knowledge and skill.[119] Such decorative treatment also extended to the delicate wrought iron roof, completed by Skidmore of Coventry, George Gilbert Scott's favourite metalworker.[120] In the spandrels, the decoration took the shape of 'interwoven branches, with leaf and flower, and fruit of lime, chestnut, sycamore, walnut, palm and other trees or shrubs of native or of exotic growth'.[121] In the trefoils of the girders, there were 'leaves of elm, briar, water-lily, passion-flower [and] holly'.[122]

Just how many of the Museum's visitors paused to study the decorative intricacies is, of course, hard to gauge. Ascending to the first-floor galleries, however, no visitor could fail to have been struck by the forest-like appearance of the interior ironwork, an effect emphasized by the diamond glass panes of the roof which gave rise to a dappled lighting effect not unlike that from a real forest canopy. The columns and pilasters that enframed the galleries, forming a cloister, added to this impression, filtering views much as tree stands would do in a woodland glade.

For the various leading members of the Oxford clergy who were so enthused by the project, such natural style of decoration was nothing less than an act of faith. The Museum celebrated divine wisdom. It might be seen by some to represent modern science, detached from all theological concerns, especially given its use of modern materials like cast iron. However, it was the evidence of the almost infinite variety of forms present in the natural world that was the guiding force. Indeed, one might say that manmade materials like iron merely extended this variety. So Oxford science was 'fused with natural theology'.[123] And this was yet more powerfully illustrated in the museum's entrance porch. It was decorated in a manner that demonstrated spiritual and material progression, Adam and Eve at the bases of either side of the arch, the angel of life at its apex.[124] Indeed, the entire edifice was replete with biblical imagery. When coupled with the overall Gothic style of the building, its nave-like interior court, complete with a cloister of the same dimensions as that at Westminster Abbey,[125] not to mention the separate laboratories to the side that took their inspiration from the Abbot's Kitchen at Glastonbury, no visitor could be forgiven for thinking of Oxford's Natural History Museum as a religious shrine, with nature as its central emblem of belief.

For John Ruskin, the many decorative sculptures were vital in telling the purpose of the building, and he was not short of advice for the O'Shea brothers as they carried out their craft.[126] It was originally intended that some of the internal wall faces would be painted in fresco, but this scheme was never executed. However, in some of the rooms leading from the galleries, decoration of the walls took the form of large 'geological' paintings: in one example a picture of the Mer de Glace; in another, one of the lava streams of Vesuvius.[127] There was more subtle geological allusion in the way the building's façade was enlivened by horizontal bands of stone of different colours, mimicking the stratification of some sedimentary rocks. Within a few decades, moreover, there arose on a site opposite the museum an even starker reminder of the sedimentary layers from which so much evidence of earth history had been disinterred. This was Keble College, designed by William Butterfield and named in memory of the English poet and churchman John Keble. Whatever its homage to the polychromy of medieval designs from the Continent,[128] Butterfield used different colour bricks in systems of horizontal banding both as a 're-assertion of the divine in English clays and rock',[129] and as an accommodation to geological discovery. With Keble as the college in Oxford that commemorated the Christian revivalism of the Oxford Movement, such an attempted fusion of religion and science was perhaps particularly appropriate. In the course of time, Oxford's Natural History Museum became a central symbol of the new secularism in the University. And the O'Shea brothers perhaps unwittingly offered portents of this in the monkey carvings that they crafted on the museum's façade, affording signs of man's ape ancestry.[130] In 1860, though, most Victorian instincts were to 'read' the museum much as if it were a bible: constantly alert to signs and signifiers of divine intention.[131]

The Natural History Museum in South Kensington repeated many of the characteristics of its earlier Oxford counterpart. Indeed, the building itself was even more ecclesiastical in style. It was not merely the selection of German Romanesque by the architect, Alfred Waterhouse, that was significant, deviating from Fowke's original plan. Equally critical was the structure's cathedral-like scale and form. Visitors entered by huge wrought-iron gates up a broad flight of steps to face a vast cathedral portal, flanked by twin towers. Once through the entrance doors, they found themselves in a great rectangular cathedral nave, complete with triforium above and arcades below, with chapels to the sides.[132] If one ignored for a moment the roof of iron and glass, it might easily have been a cathedral. What better way to celebrate God's work in the wonder and range of the natural worlds. The building was finished inside and out with buff-coloured terracotta, enlivened with bands of bluish-grey. Waterhouse had a special interest in the facing qualities of the material. Moreover, the inability to maintain colour consistency in manufacture meant that its usage became 'a kind of architectural equivalent of painting in water-colour'.[133] The random variations of hue and shade became a form of ornament in themselves, not in any way inappropriate for a building that cele-

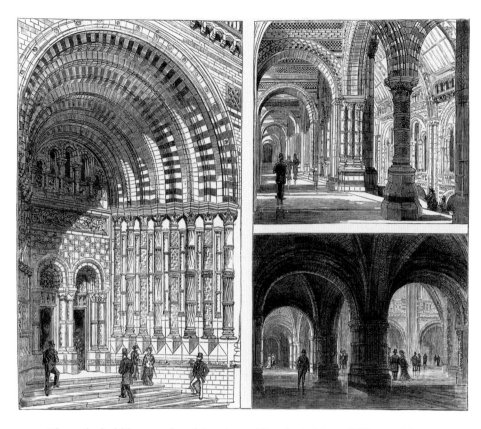

122   The cathedral-like portal and interiors of London's Natural History Museum, as depicted in *The Graphic*, 1880

brated the variety of nature. There was also an extensive range of ornamental sculpture. Waterhouse worked closely with Richard Owen when he was designing the sculptures of extinct species. For living species, though, he needed no guidance producing a range of display that, in some instances, verged on the comical.

Although, in general terms, the medley of ornamental sculptures purported to 'tell a story', it was not one that was framed within any 'evolutionary' sense of time. The Museum divided its exhibits between living and extinct species, the former in the west side of the building, the latter in the east. The ornamental sculptures reflected the same division. Even though it was now 1880, the Museum seemed still to preserve the old ante-diluvian distinctions – of the world as it now is, and the world as it once was. The way the objects on display were arranged also presented a somewhat alien perspective. Passing through the Museum's galleries, one did not follow any progression in the record of the earth's history. The reason lay in the perpendicular form of the galleries, 'arranged like the spine and teeth of a comb'.[134] This necessitated constantly re-tracing one's route. Only in the so-called 'Index Museum', in the entrance area, was there found a display that guided visitors through the ages. Effectively, then, the Museum became not much more than a kind of catalogue, reverting to the earlier genre of spectacle, a sort of Noah's Ark. It recalled the sentiments of Owen's lecture to the British Institution in April 1861. The extent and aims of a National Museum of Natural History had to be about 'imparting and diffusing that knowledge which begets the right spirit in which Nature should be viewed'.[135] Even so, the ornamental creatures gave endless pleasure and amazement to visitors as they walked round the museum. 'Birds, beasts and fishes were on every surface. They peered from a series of splendid panels set in Portland stone gate piers, they swam round the gallery columns, and stood beside the staircase'.[136] A correspondent of the *Daily News* of September 1879 remarked on the long-distant prehistoric time that some of the creatures were reminders of; a world of 'vast steaming swamps and shallow lagoons, bordered by gigantic plants of species now embedded in the strata from which we extract heat and power'.[137] But Waterhouse's decoration was not confined to the animal world. Foliate and clustered patterns on capitals offered allusion to the world of plants. In turn, string courses in the terracotta decoration provided geological reference.[138] There may not have been a 'forest of filigree' in the manner of the Oxford museum,[139] but there could be no mistaking the wider story, the 'Book of Nature', that Waterhouse's creation sought to engage.

'. . . But is this the Prehistoric World at all?'

> For some minutes Alice stood without speaking, looking out in all directions over the country – and a most curious country it was.[140]

> 'A slow sort of country!' said the Queen. 'Now, *here* you see, it takes all the running *you* can do, to keep in the same place. If you want to get somewhere else, you must run at least twice as fast as that!'[141]

One of the most striking features of the great hall of the Natural History Museum in South Kensington was not the skeleton of the sperm-whale placed in the central floor area,[142] but the grand staircase that occupied one end. Beginning as a single broad flight, it breaks into

123 (*right*)   The central court of
Oxford's Natural History Museum, from
the *Illustrated London News*, 1860

124 (*facing page*)   The vast entrance
court of London's Natural History
Museum in South Kensington, showing
part of the main staircase

two at the first landing to give access to the triforium galleries on the
first floor. Then, to reach the second floor galleries, there are further
flights that lead to a bridge spanning the hall, from which a final staircase 'leaps across a gap between the bridge and the end of the hall'.[143]
No cathedral nave ever offered such a spectacle to its worshippers or
visitors. It was an experience unique to the great nineteenth-century
museums and to the great exhibition buildings that had been their progenitors. But even lesser museums adapted the same general design
form. One critical outcome was that museum visitors themselves
became part of the exhibitionary medium. The vast array of elevated
vantage points, whether from galleries or staircases, opened members
of the public to each other's gaze. In Oxford's Natural History
Museum, the entire central court is exposed to constant view from
those passing round the galleries. Edinburgh's Museum of Science and
Art had a similar but elongated central court, with two storeys of galleries on either side from which visitors could 'spectate', looking not
just at the animal skeletons on the floor of the court, but at the groups

of passing people there. The fears that once prevailed in government and in official circles that publicly accessible museums would become prey to beggars, pimps and prostitutes were soon stifled. It was, to quote Jeremy Bentham, a case of 'the democratic aspiration of a society [being] rendered transparent to its own controlling gaze'.[144] Constructed largely of stone and iron, and with such vast open interiors, the sound of footsteps and voices was invariably magnified. Visitors were ever-conscious of each other's presence. Just as laughter and raised voices destroyed the hushed silence of a cathedral's precincts, so it did also in the cathedral-like interiors of museum buildings. The public's heightened sense of self-control was aided by the way passage through a museum was so often carefully prescribed. This was vital to appreciation of the narratives that their exhibits afforded. Not long after its opening in 1881, the Natural History Museum in South Kensington decided to open in the evenings to allow visiting by members of the working classes. Hitherto, this social group had come largely on Sundays and Bank Holidays.

To achieve this, though, artificial lighting had to be introduced. Gas lamps were therefore installed, but confined to the Index Museum,[145] where, it appears, the workmen and their families were to be restricted.[146]

As visitors wandered around halls and galleries, peering at the vast range of exhibits often locked behind glass showcases, they were, of course, being pressed into seeing the world as a picture. Regimented lines of specimens, whether rocks, minerals, insects, crustaceans or plant remains, were seen detached from the reality of which they had once been part. When the British Association for the Advancement of Science undertook a formal survey of the country's provincial museums, it was estimated that these institutions housed some two million geological specimens alone.[147] For many local and provincial museums, the accumulation of specimens appears almost alone to have afforded their primary reason for being.[148] Moreover, according to the British Association, the outcome was too often highly unsystematic.[149] In cases, there was also a tendency to set up organic remains 'in a dramatic or artistic manner merely as sensational ornaments'.[150] In the decorative arts, so the British Association's report remarked, the 'modelling of plant forms' had been carried to 'great perfection',[151] whereas museum exhibits often left much to be desired. Philip Henry Gosse, the incurable naturalist romantic, wrote of them as forming a 'dryasdust' mode of representing the organic world.[152] For the visitors themselves, however, there were comfortable certainties about such vast arrays of objects, all neatly labelled and organized. The imperfections of the fossil record, the intensities of fire and heat that produced such rocks as basalt and serpentine, the contested stories of evolution, all these questions faded from view. In effect, the organic and inorganic worlds of past and present were being sanitized. Moreover, it was too frequently the case, as the British Association's report into provincial museums made plain, that an artificial line was drawn between present and past epochs.[153] The early organization of the Natural History Museum in South Kensington committed this error, as already remarked, and the separation of biology from palaeontology in many museums was another way in which 'natural connection' was obscured.[154] Much the same tendency was apparent in the choice of giant mammal skeletons as the prize exhibits in the metropolitan museums. Despite the life-sized models of dinosaurs in the park at the Sydenham Crystal Palace, London's Natural History Museum initially had a sperm-whale in the great entrance hall when it was opened in the 1880s.[155] Twenty years earlier, the Edinburgh Museum of Science and Art had had a vast whale skeleton suspended in the central court, on a level with the first-floor galleries.[156] Despite the public fascination with the giant extinct lizards (dinosaurs or saurians) in Gideon Mantell's day, well evident in both popular and scientific presses, the skeletons of these fearsome creatures were, in later decades, more often demoted to side or rear corridors. The attraction of whales was that, as examples of living mammals, they evaded the difficult questions that extinct species invited or invoked. Whilst on the one hand, the accumulating number of Victorian museums testified to great historical confidence, the paradox was that their engagement with the historical process was sometimes tenuous, particularly as the museum movement consolidated its place within the emerging cultural edifice.

For young children, visiting museums provided a fertile source of imaginative enquiry, aside from the more strictly didactic objectives that many museum founders initially harboured. At Oxford, the Natural History Museum became (shortly after its opening) a source for some of the stories that Charles Dodgson (Lewis Carroll) created for the young Alice Liddell, later famous as the tales of *Alice in Wonderland*.[157] By the time of Dodgson's death, in 1898, around 159,000 copies had been sold.[158] Alice was the daughter of the Dean of Christ Church and Dodgson a mathematical tutor there. Dodgson developed a close friendship with Alice over the years of her childhood and, together with her sisters, he took her on frequent expeditions in and around Oxford. The newly opened museum became a favourite venue, particularly on rainy afternoons. Here the Liddell children liked to view the stuffed animals and birds, especially if Dodgson was there to make up stories about them.[159] A specially favourite exhibit in the great central court was the remains of the Dodo, a large flightless bird related to the pigeon family, that had been found in Mauritius when the island was first colonized by the Dutch in the early sixteenth century. Within fifty years, though, it had been wiped out. Dodgson identified himself with the Dodo, the name itself mimicking not only his own but his characteristic stammer.[160] The children were also fascinated by the museum's collection of crustaceans and insects.[161] And as one might anticipate of childish enquiry, one of their preoccupations was which insects could sting.[162]

It was not just the creatures, though, that provided Dodgson with inspiration, for both *Wonderland* and *Through the Looking Glass* are shot through with allusions to the strange new worlds that nineteenth-century

125   Sir John Tenniel's illustration of the Dodo in *Alice's Adventures in Wonderland*, 1872

museums were bringing to public view. When Dodgson wrote up the original text of *Wonderland*, it was entitled *Alice's Adventures Under Ground*.[163] He subsequently altered the title to *Wonderland* in case it might be mistaken as a child's introduction to coal-mining.[164] One reason for originally choosing the 'underground' label was plainly connected with the subverted world that Alice's adventures explore and record. The reader experiences a view that involves a near total destruction of the fabric of the 'self-styled, logical, orderly, and coherent' world above ground.[165] But it does not require much of a leap of imagination to see the 'underground' label as simultaneously reflecting the prehistoric worlds that geology, palaeontology and biology had been progressively helping to expose. It was an underground world that appeared equally subverted, with its dragon-like skeletons, its extraordinary plant forms, and its seashore relics buried where it seemed inconceivable that the sea had ever been. At one point, Dodgson thought of having some form of supernatural reference in the book title.[166] Once more, this was not that far removed from the supernatural associations that some of the celebrated geological and palaeontological discoveries early in the century invited.

A related feature of *Alice's Adventures* that resonates with the newly discovered prehistoric worlds is that centred on time. Dodgson's stories relentlessly mock the transformation of time that characterized mid- to late-Victorian society.[167] All kinds of technological apparatus, from railways to the telegraph, were revolutionizing day-to-day conceptions and experiences of time.[168] The passage of time became inexorably linked to the workings of industrial capitalism. When Alice kept the train guard waiting in *Through the Looking Glass*, after failing to have her ticket ready, the admonition of the other passengers was that the guard's time was 'worth a thousand pounds a minute', just as the smoke of the engine was 'worth a thousand pounds a puff'.[169] This extraordinary shift in conceptions of time was thrown into even sharper relief by the progressive realization of the equally extraordinary age of the earth, the unfathomable abyss of time that books like Lyell's *Principles* had exposed. Biblical time, with its certainty of a beginning, became increasingly untenable. What replaced it was something highly inchoate. Much as Alice embarks on 'a frightful journey into meaningless night',[170] so Victorians contemplating geological time found themselves peering into an equally meaningless past. Its scale seemed to defy all contemporary measure. More problematically, it undermined ideas and conventions about the very purpose of life and being itself.

Dissolution of familiar concepts of time in *Alice's Adventures* reached a climax in the 'Mad Tea Party' scene. For the party apparently had 'no temporal beginning' and would probably 'never have an end'.[171] It might easily have been a metaphor for James Hutton's Earth-Machine – 'no vestige of a beginning, – no prospect of an end'.[172] The Wonderland tea-party is frozen in timelessness. It is always teatime. It is always six o'clock.[173] Hutton's successive cycles of decay and restoration in the history of the earth embodied an equivalent timelessness. And thus for Victorian geologists, it was the geological variation across the earth's crust, variation in space (not time), that became their preoccupation, rather like the creatures at the Wonderland tea-party becoming preoccupied with constantly changing their position (in space) at the tea-table.[174]

Dodgson wrote the initial version of *Alice's Adventures* only a few years after the publication of Darwin's *Origin of Species*, and it is not hard to see the chaos of Wonderland as mimicking the puzzling spectrum of variation, randomness, survival, death and extinction that Darwin's book narrated.[175] Alice became distressed at the lack of any familiar, intelligible order in Wonderland. Her predicament mirrored that of many notable Victorians, in their increasingly discomforting crises of faith. Her adventures also echo the mistaken perspective that Darwinian theory so frequently invoked, as when the Duchess's baby devolves into a pig.[176] This mocked the progressive rather than the Darwinian view of evolution, one where there was an ascending order of creatures, with man at its apex.

Virtually all of the animals that inhabit Wonderland are representatives of the living world, even if their strange behaviour might suggest otherwise. The Dodo was plainly one major exception. Another, much stranger one, was the Cheshire Cat. It was rather like a monster from classical mythol-

126 Tenniel's famous tea-party scene from *Alice's Adventures in Wonderland*, 1872

127   The Cheshire Cat as illustrated by Tenniel in *Alice's Adventures in Wonderland*, 1872

"Well, then," the Cat went on, "you see a dog growls when it's angry, and wags its tail when it's pleased. Now *I* growl when I'm pleased, and wag my tail when I'm angry. Therefore I'm mad."

"*I* call it purring, not growling," said Alice.

"Call it what you like," said the Cat. "Do you play croquet with the Queen to-day?"

ogy. The Cheshire Cat in Wonderland 'comes in sections'.[177] It might have had a Chimaera as ancestor. With a lion's head, a goat's body and the tail of a dragon, the Chimaera breathed 'insubstantial flame', just as the Cheshire Cat boasted an 'insubstantial grin', one that remained even after the body of the creature itself had disappeared.[178] For Alice, 'a grin without a cat' was the most curious thing she ever saw in her life.[179] When she had first set eyes on the cat, moreover, she also noticed that it had very long claws and a great many teeth.[180] Dodgson had transformed it into something potentially more threatening. Here he may well have been recalling his student days at Christ Church when he came into contact with the Buckland family household. William Buckland was then still Oxford's leading geologist and his son Frank was at Christ Church at the same time as Dodgson. The Buckland family quarters in the college were legendary for the live animals that were kept there, and also for the many skeletal remains of extinct species that William had assembled over the years.[181] The story went that the Bucklands had tried to eat their way through the entire animal kingdom. Certainly, visitors to their rooms never failed to record the veritable menagerie of creatures that inhabited them, as well as the heads of extinct species that peered down from the walls. If Dodgson had a model for the underworld life of Wonderland, it was here within the walls of his own Oxford college.[182]

Despite William Buckland's persistent view that there had been a universal deluge within recent earth history,[183] he was otherwise instrumental in helping to establish acceptance of a pre-Adamite world of alien animal and plant forms. And what distinguished these forms, apart from their often gruesome features, was their scale. They dwarfed much of the animal and plant life of the present world. In *Alice's Adventures*, there are obvious echoes of this as she repeatedly alters in size. The stalk of a buttercup became something to lean against, one of its leaves making a convenient fan.[184] A mushroom growing nearby was the same height as Alice. She had to stand on tiptoe to see over it.[185] In *Through the Looking Glass*, bees are transformed into elephants and Alice is left speculating about the size of the flowers from which they must extract nectar to make honey.[186] The absurdities in Wonderland are gross, plainly and deliberately so. But there were many who viewed the existence of giant reptiles in a pre-Adamite world as absurd. Wonderland undeniably offers its readers a 'glorious escape from time and place', but its most memorable effects nevertheless derived from specific features of contemporary society.[187]

Alongside their visits to Oxford's new museum, the Liddell children also went with Dodgson to visit the much older University Botanic Garden, to the south of Magdalen College.[188] Here, in the hothouses, they saw all kinds of exotic plants, much like those seen by visitors to Kew. Like so many Victorians, here they came face to face with living forms of an alien scale and appearance. Sir John Tenniel, the distinguished illustrator who contributed to *Alice's Adventures*, included the Botanic Garden's water-lily house in his picture of the Queen's croquet ground. One of Dodgson's own illustrations, relating to Alice opening out like a large telescope, had Alice's head perched on a stalk that might have been a tree trunk from the Carboniferous Coal Forest, or else the trunk of a hot-house palm.

The Botanic Garden, as well as Oxford's new museum, presented their respective natural worlds as exhibition. They organized and codified those worlds for popular consumption. Dodgson's Wonderland was, in turn, another species of such exhibition. But its picture of a chaotic underground world mocked the order and regularity that characterized the behaviour and society of the world above, the world into which museums unambiguously fitted. On the one hand, Wonderland is deeply allusive to that order (necessarily so), but on the other, it attacks it, especially the very basis of identity,[189] much as Darwin's *Origin* raised difficult questions about the identity of species. Both books, in a sense far apart, the one fairytale, the other scientific treatise, also display uncanny similarities. Darwin revealed Nature to be anarchic, just like Wonderland. However, he clothed his message in as clever a piece of narrative as did the creator of Wonderland himself.

128 The Queen's Croquet Ground by Tenniel in *Alice's Adventures in Wonderland*, 1872, with the water-lily house of Oxford's Botanic Garden in the distance

Epilogue

It is midday and the high fern scrubland has started to shimmer in the heat of the sun. The rains have gone, but the ferns and club mosses still form a deep, lush mat over the rolling uplands. From beneath the green fronds comes the din of Triassic insects – lacewings, scorpion-flies and damselflies flitting from plant to plant. Centipedes and millipedes crawl among the roots still damp from the heavy morning's dew.[1]

The epigraph takes one back to an imaginary world of the prehistoric, to the Triassic geological epoch to be specific. One might be forgiven for thinking that it was written by Gideon Mantell. However, it comes from the pen of Tim Haines, writing at the very end of the twentieth century. Over the course of one hundred and fifty years, imaginative responses to the prehistoric, and to the age of reptiles in particular, have come full circle. What fascinated and preoccupied the early Victorians has proved equally absorbing today. Like William Buckland, Gideon Mantell, Robert Chambers and many others before him, Tim Haines, in *Walking with Dinosaurs*, offers readers his own narrative journey back into the past history of the earth:

> Replace grass and open ground with ferns and low, palm-like cycads. Convert any trees you can into conifers – not the familiar plantation-pine type but older species like monkey puzzles – and space then out over the landscape. Pull the birds out of the air, change the noise the insects make and turn up the temperature. Finally, place a herd of enormous, loud reptiles in front of you and watch as groups of alarmingly fast predators stalk them.[2]

Haines offers his readers not an encyclopaedia of the prehistoric, but an experience.[3] And just as early Victorians were mesmerized by the element of Gothic horror that attached to monsters of the prehistoric, so modern audiences have been exposed to a different form of *frisson* in the computer animations of Haines's television documentary, as well as in the ingenuity shown by sound recordists in seeking to reproduce the vocal registers of such lost life forms.

129 (*facing page*) Stegosaurus grazing on the dry highlands of equatorial Laurasia, from Tim Haines's *Walking with Dinosaurs*

Towards the end of his book, Haines addresses the extinction of the dinosaurs some sixty million years ago. The commonly held view is that a comet or meteor collision destroyed most extant life-forms.[4] And, under a chapter headed 'Death from the Skies', Haines constructs a vision that could be straight out of one of the paintings of John Martin of a century and a half before. A group of tyrannosaurus chicks are tearing at their dead mother's flesh when a light appears on the northern horizon:

> It grows in intensity and the tyrannosaurus look up in curiosity, their mottled faces lit by the new light. Their eyes flicker and they return to their mother's carcass. Unknown to them, a comet 10 kilometres across and travelling at 30 kilometres per second has hit a shallow sea 3,000 kilometres to the south. The distant glow they can see represents an explosion of about 100 million megatons, enough to open a hole 200 kilometres wide and 15 kilometres deep in the planet's crust. Life on earth will never be the same again.[5]

A giant shockwave or earth tremor then sweeps across the floodplain and marshes where the young Tyrannosaurus are engaged in their eating orgy. The young reptiles are thrown into a 'maelstrom of flying rocks, wood and vegetation'.[6] In the skies to the south, meanwhile, a dark plume has been forming; temperatures are rising; and soon 'burning hot rocks slam into the ground and spontaneous fires start up among the piles of debris'.[7] The temperature climbs still further. Creatures that survived the initial shockwave are now burned alive. Finally, in an earthquake measuring about 10 on the Richter scale, the mountainsides around explode. A great plume of dust and debris blocks out the sun.[8] The vision is apocalyptic, undisguisedly Martinesque. Despite all the computing ingenuity involved in the making of the television documentary, Haines's companion book utilizes a narrative form tried and tested over generations. Moreover, the case of the comet was one that would have been familiar to Victorian eyes. As already noted, John Martin had a comet in his painting of the Deluge.[9] In the real world, Donati's comet, which appeared in the skies above Victorian Britain in 1858, was widely reported in newspapers and periodicals. On 5 October, it was only fifty-one million miles from the Earth, its trail stretching across thirty degrees of sky.[10] William Dyce's painting *Pegwell Bay* (1860) gave yet further prominence to it.[11] How far the nineteenth-century gaze thought of comets as meteors that might one day collide with the earth is open to doubt. The modern perception is plainly more acute and the fascination with dinosaur extinction, particularly its suddenness, presents a fearful reminder that, one day, another collision might once again wipe out much of current life on earth.

Tim Haines's book is just one publication in a whole series of works that have propelled the dinosaur, as well as the prehistoric more generally, to the forefront of contemporary natural history. There are all manner of encyclopaedias and atlases of dinosaurs,[12] not to mention the host of children's books and pamphlets that has accompanied the revival of interest in these reptiles in national and provincial museum displays. Not only have the skeletons of these creatures been re-positioned or re-suspended in accordance with accumulating scientific understanding of their body movements, but a whole new industry of 'spectacle' has developed alongside them. At Lyme and at Sandown, for example, museums have even been opened devoted specifically to dinosaurs. Yet more widely, the dinosaur has become just another commodity in the true fashion of late capitalist consumption, appearing in toy-form, as beach inflatable and imprinted on all manner of things. Even at the breakfast table, these extinct reptiles stare out at us from the backs of cereal packets, or, somewhat grossly, tumble into cereal bowls, in minute model form. In the world of political cartoonists, the dinosaur has never had it so good. Indeed, natural history and the natural world, more widely, has become a vital inspiration in contemporary political caricature. Meanwhile, at Oxford, set in the lawn outside the Natural History Museum, one can now find a whole trail of concrete casts of dinosaur footprints, along which children delight in jumping. They form a vivid reminder of the early years of scientific nat-

130 *Punch's* cartoon of 1861 showing the antediluvians recognizing a comet, a subtle reminder of the links between man and the prehistoric in the wake of the publication of Darwin's *Origin of Species*

131   The dinosaur museum at Sandown, Isle of Wight, opened in 2001, the building designed in the shape of a Pterosaur

ural history, in the mid-nineteenth century, when the so-called 'lamp of science' was revealing truths about the natural world that were stranger than fiction or fairytale.[13] Commentators then described how the monster reptiles had been 'their own historians'.[14] They had 'described themselves in the gorgeously illuminated volume called the Stone Book, every page of which [was] formed of the solid rock'.[15] This was true, of course, of everything in Nature: 'the planet and the pebble [were] attended by their own shadows'.[16] Parties of schoolchildren discover this same truth outside Oxford's Natural History Museum at the start of the twenty-first century.

Another feature of the Victorians' engagement with the prehistoric that has re-surfaced in recent time is that centred upon past climates. In the mid-nineteenth century, as noted, the Arctic became a source of public fascination – and by virtue not just of the horrors of extreme cold, but the alien sights and sounds that characterized it. Soon there were geologists and popularizers of science who were translating these conditions to Britain. There had been a time, or times, when the country had been enveloped in ice, its climate no different from the Arctic realms that the great nineteenth-century explorers had so often described for spell-bound readers. Today the preoccupation is less with what the country's climate was once like, but with what it might become in the future. The catalyst here has been the realization among scientists that man, by his various activities, is contributing to climate change. And it is not atmospheric cooling that is the focus of concern, but atmospheric warming.[17] One possible outcome is that the British Isles might not only become warmer but wetter, with extremes of rainfall and floods increasingly more frequent. So just as Victorians looked at tales and accounts of Arctic realms (past as well as present) and saw shadows of instability in their otherwise increasingly progressive and comfortable world, so many in the West at the turn of the twenty-first century have developed not dissimilar anxieties. And these focus not just on climate change, but on the stability and survival of life on earth as we know it.

One of the central difficulties faced by scientists who study climate change is specification of the time-scale at which change is measured. What is quick at one scale may be slow at another – in this respect, some of the apparent nonsense of Lewis Carroll's Wonderland is not nearly so nonsensical after all.[18] Similar difficulties over scaling time are apparent in grasping the span of earth history. The modern mind is not faced with dismantling the biblical time-scale as Victorian minds were, but the great abyss of time, in a geological sense, remains perpetually hard to apprehend. The commonest illustrative technique now seems to be to liken the history of the earth, or life upon it, to the twelve hours on the clock-face. As far as life goes, each hour represents 50 million years, giving 600 million years for the entire history of living things. Within such a frame, man appeared at one minute before midnight, mammals at three hours before, and reptiles at six. Taking the entire history of the planet, involving a time-span of some 4,500 million years, man appeared just seconds before midnight, in other words paling into insignificance in time's great abyss.[19] None of this is all that far removed from the efforts of Charles Babbage or writers like Dickens to convey to their nineteenth-century readers the scale of past time. For Dickens, every syllable of his tale of a phantom ship on an ante-diluvian cruise became a thousand years of earth history.[20] For Babbage, humans could not be anything other than very inadequate observers of time.[21] This same view was repeated some 150 years later by the cultural historian Stephen Kern: man seemed a mere parenthesis in the history of the earth, 'infinitesimal' in brevity.[22] And there is a yet further parallel in perceptions of the prehistoric world then and now. As the pasts of the geologists 'rushed away from the present' in the mid-nineteenth century,[23] so greater control was exercised over the way those pasts were presented within society. Much the same is true today. The prehistoric is about much more than dinosaurs, and yet dinosaurs have now become largely synonymous with it. The monsters of fairytale and fable have, in a sense, won back.

Notes

Prologue

1   *Routledge's Guide to the Crystal Palace and Park at Sydenham* (London, 1854), pp. 189–95.

2   A. Desmond, *Huxley: From Devil's Disciple to Evolution's High Priest* (London, 1998), p. 247.

3   *Ibid.*

4   *Routledge's Guide, op. cit.*, p. 185.

5   The date appears to have been 1842 – see Desmond, *Huxley, op. cit.*, pp. 29 and 652.

6   *Routledge's Guide, op. cit.*

7   C. Dickens, *Bleak House* (London, 1853; repr. Oxford, 1996), p. 11.

8   It was described by William Buckland, the famous Oxford geologist, in a paper the following year. See P. J. Bowler, *Fossils and Progress* (New York, 1976), pp. 20–21.

9   Desmond, *Huxley, op. cit.*, p. 356.

10   *Ibid.*, pp. 356–7.

11   T. Hawkins, *Memoirs of Ichthyosauri and Plesiosauri: Extinct Monsters of the Ancient Earth* (London, 1834), p. 13.

12   See H. Jennings, *Pandaemonium: The Coming of the Machine as Seen by Contemporary Observers, 1660–1886* (London, 1985; repr. London, 1995), p. 241.

13   F. S. Williams, *Our Iron Roads: Their History, Construction and Social Influences* (London, 1852), p. 235.

14   See T. Hawkins, *The Book of the Great Sea Dragons, Ichthyosauri and Plesiosauri: Extinct Monsters of the Ancient Earth* (London, 1840).

15   On Henry De la Beche, see P. McCartney, *Henry De la Beche: Observations of an Observer* (Cardiff, 1977).

16   J. H. Brooke, *Science and Religion: Some Historical Perspectives* (Cambridge, 1991), p. 226.

17   W. Paley, *Natural Theology* (London, 1802).

18   See, for example, B. Lightman (ed.), *Victorian Science in Context* (Chicago, 1997); also recent contributions to the *British Journal for the History of Science*.

19   See D. Livingstone, 'Science and Religion: Foreword to the Historical Geography of an Encounter', *Journal of Historical Geography* 20 (1994), pp. 367–83; also J. Scowen, 'A Study in the Historical Geography of an Idea: Darwinism in Edinburgh, 1859–75', *Scottish Geographical Magazine* 114 (1998), pp. 148–56.

20   See especially A. Desmond, *The Politics of Evolution: Morphology, Medicine, and Reform in Radical London* (Chicago, 1989); also A. Desmond and J. Moore, *Darwin* (London, 1992).

21   See L. J. Jordanova and R. Porter (eds.), *Images of the Earth: Essays in the History of the Environmental Sciences* (2nd edn, London, 1997).

22   See D. Harvey, *The Condition of Postmodernity* (Oxford, 1989).

23   Brooke, *op. cit.*, p. 227.

24   A good summary of the perspective is contained in J. A. Secord, *Victorian Sensation: The Extraordinary Publication, Reception and Secret Authorship of Vestiges of the Natural History of Creation* (Chicago, 2000), pp. 518ff.

25   Desmond, *Politics of Evolution, op. cit.*, p. 23.

26 D. Newsome, *The Victorian World Picture* (London, 1997; Fontana ed., London 1998), p. 12.

27 Harvey, *op. cit.*

28 Newsome, *op. cit.*, pp. 6–7, quoting J. E. Baker.

Chapter One

1 G. A. Mantell, *The Medals of Creation: Or First Lessons in Geology and in the Study of Organic Remains*, II (London, 1844), p. 922.

2 The British Geological Survey was established in 1835, initially under the Board of Ordnance. In 1845, it was transferred to the Office of Woods and Forests.

3 See E. Moir, *The Discovery of Britain: The English Tourists, 1540–1840* (London, 1964).

4 Secord has commented how coastal cliffs became a favourite haunt of the Victorian geologist – J. A. Secord, *The Cambrian-Silurian Controversy in Victorian Geology* (Princeton, 1986), p. 26.

5 Alfred, Lord Tennyson to the Rev. F. D. Maurice – see C. Ricks (ed.), *Poems of Tennyson* 2 (2nd edn., London, 1987), p. 498.

6 J. Clarke, *The Delineator: Or, a Picturesque Historical and Topographical Description of the Isle of Wight* (2nd edn., Newport, 1814), p. iii.

7 *Blackwood's Edinburgh Magazine* XXIII (1828), p. 437.

8 H. and A. Noyes, *The Isle of Wight Bedside Anthology* (Bognor, 1951), p. vii.

9 Sir Walter Scott, *The Surgeon's Daughter*, quoted in Noyes, *op. cit.*, p. 126.

10 J. Austen, *Persuasion* (1818; repr. Harmondsworth, 1965), p. 117.

11 See the commentary in the *Gentleman's Magazine* LXXXV (1815), pp. 240–41.

12 'Scratchell's Bay, Isle of Wight', *Penny Magazine* III (1834), p. 135.

13 G. Brannon, *Vectis Scenery, being a series of original and select views exhibiting the picturesque beauties and places of parti-cular interest in the Isle of Wight* (Wootton, 1838), p. 7.

14 *Ibid.*

15 J. Hassell, *Tour of the Isle of Wight*, I (London, 1790) p. 141–2.

16 *Ibid.*, p. 159.

17 See H. C. Englefield, *A Description of the Principal Picturesque Beauties, Antiquities, and Geological Phenomena of the Isle of Wight* (London, 1816), p. 151.

18 'The Isle of Wight – No. 1', *Penny Magazine* V (1836), p. 342.

19 *Ibid.*

20 *Ibid.*, p. 343.

21 *Ibid.*

22 *Ibid.*

23 See G. L. Davies, *The Earth in Decay: A History of British Geomorphology, 1578–1878* (London, 1969), p. 161.

24 L. G. Wilson, *Charles Lyell, the Years to 1841: The Revolution in Geology* (New Haven and London, 1972) pp. 429ff.

25 J. MacCulloch, *A Description of the Western Islands of Scotland, including the Isle of Man*, II (London, 1819) p. 311.

26 For a summary of the series, see J. Prestwich, 'On the Tertiary and Supracretaceous Formations of the Isle of Wight as exhibited in the sections at Alum Bay and White Cliff Bay', *Quarterly Journal of the Geological Society* (hereafter *QJGS*) II (1846), pp. 225–9.

27 G. W. Colenutt, 'An outline of the geology of the Isle of Wight', in F. Morey (ed.), *A Guide to the Natural History of the Isle of Wight* (Newport, 1909), p. 3.

28 J. Hutton, *Theory of the Earth, with Proofs and Illustrations*, I (Edinburgh, 1795), p. 102.

29 A. Insole, et al., *The Isle of Wight*, Geologists' Association Guide No. 60 (London, 1998), p. ii.

30 *Ibid.*, p. 1.

31 See F. Burkhardt and S. Smith (eds.), *The Correspondence of Charles Darwin* (hereafter Corr.), II (Cambridge, 1985–), pp. 29–30, 59.

32 Secord, *op. cit.* p. 28.

33 Hassell, *op. cit.*, I, pp. 141–2.

34 *Ibid.*, p. 159.

35   *Ibid.*, p. 119.

36   *Ibid.*, p. 159.

37   T. Milner, *The Gallery of Nature: A Pictorial and Descriptive Tour through Creation* (London, 1846), p. 742.

38   Englefield, *Principal Picturesque Beauties, op. cit.*, (1816), title page.

39   *Ibid.*, p. v.

40   *Ibid.*

41   *Ibid.*

42   *Ibid.*

43   Wilson, *op. cit.*, p. 96; also Englefield, 'Observations on some remarkable strata of flint in a chalk-pit in the Isle of Wight', *Trans. Linnean Soc.* VI (1802), pp. 103–9.

44   Wilson, *op. cit.*, p. 96; also Englefield, *Principal Picturesque Beauties, op. cit.*, pp. 117ff. Webster became librarian to the Geological Society from 1812 to 1826 and its secretary from 1819 to 1827. In 1841, he was appointed to the first chair of Geology at University College, London – see J. Challinor, 'Thomas Webster's Letters on the Geology of the Isle of Wight, 1811–13', *Proceedings of the Isle of Wight Natural History and Archaeological Society* IV (1949), p. 108.

45   Englefield, *Principal Picturesque Beauties, op. cit.*

46   See S. J. Knell, *The Culture of English Geology, 1815–1851: A Science Revealed through its Collecting* (Aldershot, 2000), p. 12; the French study appeared as G. Cuvier and A. Brongniart, *Essai sur la Géographie Minéralogique des Environs de Paris* (Paris, 1811).

47   W. Phillips, *Selection of Facts* (London, 1818); A. Sedgwick, 'On the Geology of the Isle of Wight', *Annals of Philosophy* n.s. III (1822), pp. 329–55.

48   See J. Wyatt, *Wordsworth and the Geologists* (Cambridge, 1995), pp. 170ff.

49   Englefield, *Principal Picturesque Beauties, op. cit.*, p. 201.

50   *Ibid.*, p. 216.

51   *Ibid.*, pp. 152–3.

52   *Ibid.*, p. 151.

53   *Ibid.*

54   *Ibid.*, p. 201.

55   Wilson, *op. cit.*, p. 101 – of the Appalachian Mountains in the USA.

56   *Ibid.*, pp. 19–20.

57   *Ibid.*, p. 19.

58   See, for example, S. J. Gould, *Time's Arrow, Time's Cycle: Myth and Metaphor in the Discovery of Geological Time* (Cambridge, Mass., 1987; Penguin edn., 1991); J. A. Secord (ed.), *Charles Lyell: Principles of Geology* (Harmondsworth, 1997), introduction; N. A. Rupke, *The Great Chain of History: William Buckland and the English School of Geology, 1814–1849* (Oxford, 1983).

59   Gould, *op. cit.*, p. 177.

60   A. Desmond and J. Moore, *Darwin* (London, 1991; Penguin edn., 1992), p. 116; D. R. Dean, 'Through Science to Despair: Geology and the Victorians', in J. Paradis and T. Postlewait (eds), *Victorian Scence and Victorian Values: Literary Perspectives* (New Brunswick, 1985), p. 114.

61   Gould, *op. cit.*, p. 105.

62   Wilson, *op. cit.*, pp. 111ff; 42–4.

63   *Ibid.*, p. 114.

64   *Ibid.*, p. 115.

65   See W. H. Fitton, 'A stratigraphical account of the section from Atherfield to Rocken-end in the Isle of Wight', *QJGS* II (1846), p. 55.

66   See L. Jenyns, *Memoirs of the Rev. John Stevens Henslow* (London, 1862), p. 13; J. W. Clark and T. M. Hughes, *The Life and Letters of the Reverend Adam Sedgwick*, I (Cambridge, 1890), p. 204; also chapter 6.

67   Clark and Hughes, *op. cit.*, pp. 219, 227.

68   Sedgwick, *op. cit.*

69   *Ibid.*, p. 329.

70   *Ibid.*, pp. 329–30.

71   *Ibid.*, p. 342.

72   Clark and Hughes, *op. cit.*, p. 267.

73   *Ibid.*

74   Knell, *op. cit.*, pp. 144ff.

75   Oxford University Museum (hereafter OUM), Phillips MS., Note Books and Journals, 31.

76   Corr., I, Charles Darwin to W. D. Fox, 6/11/1836 (p. 517).

77   *Ibid.* II, Charles Darwin to J. S. Henslow, 19/11/1837 (p. 59).

78 *Ibid.* VII, Charles Darwin to W. D. Fox, 21/7/1858 (p. 138).

79 See G. Mantell, *The Fossils of the South Downs; or Illustrations of the Geology of Sussex* (London, 1822); idem, *Thoughts on a Pebble* (London, 1842).

80 G. A. Mantell, *Geological Excursions around the Isle of Wight* (London, 1847).

81 *Ibid.*, p. vii.

82 *Ibid.*, p. 418.

83 *Ibid.*, p. vii.

84 E. C. Curwen, *The Journal of Gideon Mantell, Surgeon and Geologist* (London, 1940), p. 141.

85 *Ibid.*, p. 185. The quotation is from Sir Walter Scott's *The Surgeon's Daughter*. Mantell used it in his 1847 guide to the island's geology.

86 Curwen, *op. cit.*, p. 192.

87 Mantell, *Geological Excursions*, *op. cit.*, p. 272 (1847).

88 *Ibid.*, p. 273.

89 *Ibid.*, p. 280.

90 Curwen, *op. cit.* p. 192.

91 *Ibid.*, p. 193.

92 *Ibid.*

93 *Ibid.*, p. 196.

94 *Ibid.*, pp. 196–7.

95 *Ibid.*, p. 202.

96 *Ibid.*, p. 209.

97 Mantell, *Geological Excursions*, *op. cit.*, p. 228.

98 D. R. Dean, *Gideon Mantell and the Discovery of Dinosaurs* (Cambridge, 1999), pp. 218–19.

99 Mantell, *Geological Excursions*, *op. cit.*, p. 403.

100 Curwen, *op. cit.*, p. 210; also *Hampshire Telegraph and Sussex Chronicle*, 12 September 1846.

101 See *Hampshire Telegraph and Sussex Chronicle*, 19 September 1846.

102 G. A. Prestwich, *Life and Letters of Sir Joseph Prestwich* (Edinburgh and London, 1899), pp. 58, 71.

103 *Ibid.*, p. 58.

104 *Ibid.*, p. 309.

105 *Ibid.*, p. 365.

106 G. W. Colenutt, in Morey, *op. cit.*, p. 34.

107 Wilson, *op. cit.*, pp. 97–9.

108 Englefield, *Principal Picturesque Beauties*, *op. cit.*, p. 165.

109 Wilson, *op. cit.*, p. 99.

110 Englefield, *Principal Picturesque Beauties*, *op. cit.*, plate XXXVIII.

111 Mantell, *Geological Excursions*, *op. cit.*, p. 376.

112 *Ibid.*, p. 346.

113 *Ibid.*

114 OUM, Smith MS, Box 15, Folder 2.

115 Clark and Hughes, *op. cit.*, I, p. 219.

116 *Ibid.*, p. 227.

117 G. Reynolds, *Turner* (London, 1969) p. 34 – the identity of the location of the picture has traditionally been given as the Needles, but the coastal outline does not accord with such an attribution.

118 *Ibid.*, pp. 27–8; see also J. S. Dearden, *Turner's Isle of Wight Sketchbook* (Brightstone, Isle of Wight, 1979).

119 See Dearden, *op. cit.*

120 B. Hinton, 'Island celebrities', in AA Ordnance Survey Leisure Guide: Isle of Wight (London, Basingstoke, 1988), p. 26.

121 D. Linnell, *The Life of John Linnell* (Lewes, 1994), p. 37.

122 *Ibid.*

123 *Ibid.*

124 See M. Pointon, *William Dyce (1806–64): A Critical Biography* (Oxford, 1979), pp. 93ff.

125 See *George Fennel Robson (1788–1833)* – abridgement of article from *Lo Studio* – Yale Center for British Art, Department of Prints and Drawings, docket file for C6R.

126 See Yale Center for British Art, Department of Prints and Drawings, B1975.4.1583 – the view is of Steephill Cove.

127 See *ibid.*, B1981.22 – the date of the picture is 1797.

128 Clark and Hughes, *op. cit.*, I, p. 267.

129 Hinton, *op. cit.*, p. 26.

130 J. Phillips, *Memoirs of William Smith* (London, 1844), p. 10.

131 L. R. Cox, 'New light on William Smith and his work', *Proc. Yorks. Geol. Soc.* XXV (1942–5), p. 64; R. A. Reyment, 'William Smith (1769–1839) – Father of English Geology', *Terra Nova* VIII (1996), p. 664. Rupke has argued that Smith's designation

in this manner was essentially a piece of propaganda on the part of the English school of geology. The designation enhanced the School's reputation as empirical and fact-based, in contrast to the theories of the Edinburgh-based Huttonians. In reality, according to Rupke, Smith's influence was ambiguous. Wilson, by contrast, considers that Smith's maps and fossil studies 'exercised a profound influence on geology'. For years, Smith passed on his geological knowledge and use of his hand-drawn maps to leading geologists of the day, William Buckland (the subject of Rupke's study) among them. Moreover, his methods for identifying strata by their fossils echoed the work of Cuvier, Brongniart and Lamarck on the continent. For Wilson, Smith was 'the geological genius of England' – Rupke, *op. cit.*, pp. 191–2; Wilson, *op. cit.*, pp. 76, 42–3, 77–8, 42. An obituary and assessment of Smith's life's work appeared in *the Proceedings of the Geological Society of London* (hereafter *PGSL*) III (1838–42), pp. 248–54; a more recent, popular (and positive) account of Smith is found in S. Winchester, *The Map that Changed the World: The Tale of William Smith and the Birth of a Science* (London, 2001).

132 Phillips, *op. cit.*, pp. 4–5; William Smith obituary address, 21 February 1840, *PGSL* III (1838–42), pp. 248ff.

133 OUM, Smith MS., Box 40, Diary for 1789.

134 Phillips, *op. cit.*, p. 5.

135 See Winchester, *op. cit.*, pp. 66ff.

136 Phillips, *op. cit.*, pp. 6–7; note: all Smith's diaries for this period are missing, although it seems that Phillips had them at his disposal when writing his memoir.

137 Phillips, *op. cit.*, p. 8 ; also Winchester, *op. cit.*, pp. 90ff.

138 Phillips, *op. cit.*, p. 8.

139 *Ibid.*

140 *Ibid.*, p. 9; also Winchester, *op. cit.*, pp. 94ff.

141 Phillips, *op. cit.*, p. 10.

142 *Ibid.*, p. 11.

143 *Ibid.*, pp. 11, 13.

144 *Ibid.*, p. 12.

145 *Ibid.* p. 9.

146 *Ibid.*, p. 14; also Winchester, *op. cit.*, pp. 119ff.

147 Phillips, *op. cit.*, p. 14.

148 *Ibid.*

149 *Ibid.*, p. 15; also Winchester, *op. cit.*, pp. 122–4.

150 Phillips, *op. cit.*, p. 15; the diluvial school is treated in Rupke, *op. cit.*

151 See William Smith obituary address, *op. cit.*, p. 248.

152 W. Smith, *Stratigraphical System of Organized Fossils* (London, 1817), p. vii.

153 *Ibid.*, p. 23.

154 *Ibid.*

155 *Ibid.*; Winchester, *op. cit.*, pp. 78–9.

156 Reyment, *op. cit.*, p. 663.

157 See W. J. Arkell and S. I. Tomkeieff, *English Rock Terms chiefly as used by Miners and Quarrymen* (Oxford, 1953); also Winchester, *op. cit.*, pp. 32ff.

158 E. Meteyard, *The Life of Josiah Wedgwood from his Private Correspondence and Family Papers*, I (London, 1865), p. 500.

159 *Ibid.*

160 *Ibid.*, p. 502.

161 Smith, *op. cit.*, p. 5; see also William Smith obituary address, *op. cit.*; see also Winchester, *op. cit.*, pp. 136–41.

162 Reyment, *op. cit.*, p. 663.

163 *Ibid.*, p. 662.

164 See T. S. Willan, *River Navigation in England, 1600–1750* (London, 1936).

165 See C. Hadfield, *British Canals: An Illustrated History* (London, 1952).

166 OUM, Smith MS, Box 44, Folder 5.

167 *Ibid.*

168 See Cox, *op. cit.*, pp. 24–5; Phillips, *op. cit.*, p. 54; Winchester, *op. cit.*, pp. 209ff.

169 Phillips, *op. cit.*, pp. 57ff.

170 *Ibid.*

171 Reyment, *op. cit.*, p. 662; Cox, *op. cit.*, p. 35.

172 OUM, Smith MS, Box 53, Diary for 1802.

173 *Ibid.*; see also Winchester, *op. cit.*, pp. 158ff.

174 OUM, Smith MS, Box 53, Diary for 1802.

175 *Ibid.*, Diary for 1806.

176 *Ibid.*, Box 54, Diary for 1809.

177 *Ibid.*

178 *Ibid.*, Diary for 1812.

179 *Ibid.*, Diary for 1819–21.

180 Cox, *op. cit.*, p. 25; also Winchester, *op. cit.*, pp. 132–3.

181 Cox, *op. cit.*, pp. 26ff; also Winchester, *op. cit.*, pp. 146–7.

182 Cox, *op. cit.*, pp. 26ff.

183 *Ibid.*; also Winchester, *op. cit.*, pp. 218ff.

184 OUM, Smith MS, Box 40.

185 Smith, *op. cit.*, p. v.

186 *Ibid.*

187 *Ibid.*, p. vi.

188 *Ibid.*, p. v.

189 *Ibid.*, p. vi.

190 Secord, *Charles Lyell*, *op. cit.*, p. xxiv.

191 Reyment, *op. cit.*, p. 663.

192 *Ibid.*; see also Winchester, *op. cit.*, pp. 224ff.

193 See Cox, *op. cit.*, pp. 62ff.

194 *Ibid.*, p. 64.

195 *Ibid.*

196 Reyment, *op. cit.*, p. 664.

197 Cox, *op. cit.*, p. 67.

198 Reyment, *op. cit.*, p. 664. Perusing the large collection of William Smith's surviving papers, though, one wonders just how far this was really the case.

199 Clark and Hughes, *op. cit.*, I, p. 367.

200 E. O. Gordon, *The Life and Correspondence of William Buckland* (London, 1894), pp. 11–12.

201 *Ibid.*

202 See William Smith obituary address, *op. cit.*, p. 249; see also Winchester, *op. cit.*, pp. 280ff.

203 Reyment, *op. cit.*, p. 664.

204 *Ibid.*

205 Cox, *op. cit.*, pp. 65–6.

206 OUM, Smith MS, Box 53, Diaries for 1789–1807.

207 *Ibid.*

208 T. Roscoe, *A Handbook for Travellers along the London & Birmingham Railway* (London, 1839), p. 54.

209 R. Fortey, *The Hidden Landscape: A Journey into the Geological Past* (Pimlico edn., London, 1994) p. 3.

210 Mantell, *Geological Excursions*, *op. cit.*, pp. 66ff.

211 *Ibid.*, p. 66.

212 Curwen, *op. cit.*, p. 144.

213 *Ibid.*, p. 145.

214 *Ibid.*, p. 146; see also Dean, *Gideon Mantell and the Discovery*, *op. cit.*, p. 11.

215 S. Spokes, *Gideon Algernon Mantell* (London, 1927), p. 176.

216 *Ibid.*

217 Curwen, *op. cit.*, p. 242.

218 *Ibid.*

219 See Roscoe, *op. cit.*, p. 31.

220 W. C. Williamson, *Coals and Coal Plants: A Lecture* (London and Glasgow, 1876), p. 11.

221 See Rupke, *op. cit.*, Part I and pp. 194–9.

222 Williamson, *op. cit.*, p. 11.

223 *Ibid.*, p. 13; Rupke refers to the plants as Stigmaria – Rupke, *op. cit.*, p. 199.

224 Williamson, *op. cit.*, p. 13.

225 J. Hawkshaw, 'Description of Fossil Trees found in the Excavation of the Manchester and Bolton Railway', *PGSL* III (1838–42), pp. 139–40 and idem., 'Further Observations on the Fossil Trees found on the Manchester and Bolton Railway', *PGSL* III (1838–1842), pp. 269–70; there was a further paper by J. E. Bowman, 'On the Character of the Fossil Trees discovered near Manchester on the line of the Manchester and Bolton Railway', *PGSL* III (1838–42), pp. 270–75 in which the fossil trees were connected to the formation of coal by gradual subsidence.

226 F. Burr, 'Notes on the Geology of the line of the proposed Birmingham and Gloucester Railway', *PGSL* II (1833–8), pp. 593–5.

227 H. E. Strickland, 'Series of Coloured Sections of the Cuttings on the Birmingham and Gloucester Railway', *PGSL* III (1838–42), pp. 313–4; also idem., 'Description of Cuttings across the Ridge of Bromsgrove Lickey, on the line of the Birmingham and Gloucester Railway', *PGSL* III (1838–42), pp. 446–8.

228 'Railway Sections', *PGSL* III (1838–42), p. 473.

229 Report of the Tenth Meeting of the British Association for the Advancement of Science (hereafter BAAS) (London, 1841), pp. xxviii and xxxiii.

230   Report of the Twelfth Meeting of the BAAS (London, 1843), p. 38.

231   Report of the Eleventh Meeting of the BAAS (London, 1842), p. 67; Report of the Twelfth Meeting of the BAAS, *op. cit.*, pp. 39–40.

232   Report of the Twelfth Meeting of the BAAS, *op. cit.*, p. 39.

233   *Ibid.*

234   *Ibid.*

235   Report of the Thirteenth Meeting of the BAAS (London, 1844), p. 295.

236   The first issue of the *QJGS* appeared in 1845. It was preceded by the various series under the title *Proceedings*, or *Transactions*, of the Geological Society of London.

237   F. W. Simms, 'Account of the section of the strata between the Chalk and the Wealden clay in the vicinity of Hythe, Kent', *PGSL* IV (1843–5), p. 206.

238   See note 65.

239   See Dean, *Gideon Mantell and the Discovery*, *op. cit.*, pp. 48–9; on dynamic geology, see M. J. S. Rudwick, *The Great Devonian Controversy: The Shaping of Scientific Knowledge Among Gentlemanly Specialists* (Chicago and London, 1985), p. 45.

240   See, for example, F. W. Simms, 'Account of the strata observed in the excavation of the Bletchingley tunnel', *QJGS* I (1845), pp. 90–91; Idem, 'On the Junction between the Lower Greensand and the Wealden at the Teston Cutting', *QJGS* I (1845), pp. 189–90; J. Prestwich and J. Morris, 'On the Wealden Strata exposed by the Tunbridge Wells Railways', *QJGS* I (1845), pp. 397–405.

241   Simms, 'Account of the strata', *op. cit.*, p. 357.

242   R. I. Murchison, 'On the Distribution of the Flint Drift of the South-East of England on the Flanks of the Weald and over the Surface of the South and North Downs', *QJGS* VII (1851), p. 383.

243   J. Prestwich, 'On the probable age of the London Clay and its relations to the Hampshire and Paris Tertiary Systems', *QJGS* III (1847), p. 363.

244   J. Prestwich, 'On the Structure of the Strata between the London Clay and the Chalk in the London and Hampshire Tertiary Systems', *QJGS* VI (1850), p. 269.

245   Ibid., x (1854), p. 88 – this particular paper extends to 170 pages and incorporates details of railway sections from nine separate sites.

246   E. Hull, 'On the Physical Geography and Pleistocene Phenomena of the Cotteswold Hills', *QJGS* XI (1855), p. 491.

247   T. F. Jamieson, 'On an Outlier of Lias in Aberdeenshire', *QJGS* XV (1859), p. 132.

248   R. I. Murchison, 'On the Silurian Rocks of the South of Scotland', *QJGS* VII (1851), pp. 159–60.

249   J. Phillips, 'On some sections of the strata near Oxford', *QJGS* XVI (1860), p. 115.

250   See W. Jardine, *Memoirs of Hugh Edwin Strickland* (London, 1858), p. cclix; also *QJGS* x (1854), p. xxv.

251   Davies, *op. cit.*, p. 201.

252   *Fraser's Magazine* XXVI (1842), p. 363.

253   *Ibid.*, p. 365.

254   *Ibid.*, p. 369; also *Gentleman's Magazine* n.s. XVIII (1842), p. 187.

255   K. Lyell (ed.), *Life, Letters and Journals of Sir Charles Lyell* II (London, 1881), p. 129.

256   Curwen, *op. cit.*, p. 210.

257   See also Knell, *op. cit.*, p. 262.

258   *Ibid.*, p. 235.

259   Gordon, *op. cit.*, p. 30; botanical study was extended in a similar way – see D. E. Allen, *The Naturalist in Britain: A Social History* (2nd edn., Princeton, 1994), p. 110.

260   The prime example is Chat Moss west of Manchester – see M. Drayton, *Poly-olbion* (London, 1622; tercentenary edn., Oxford, 1969) IV, p. 536.

261   W. Paley, *Natural Theology* (London, 1802).

Chapter Two

1   See the useful commentary in 'The Anatomy of Time', *Chambers's Journal* XII (July–Dec., 1859), part 1, pp. 35–7.

2   S. Smiles, *The Life of George Stephenson* (1857; repr. London, 1903), p. vii.

3   M. Freeman, *Railways and the Victorian Imagination* (London and New Haven, 1999), pp. 78ff.

4   C. Dickens, *Dombey and Son* (1848; repr. Oxford, 1982), p. 236.

5   R. Fortey, *Life: An Unauthorised Biography* (London, 1997), p. 10.

6   *Ibid.*, p. 28.

7   S. Butler, *The Way of All Flesh* (London, 1903; Oxford, 1993), pp. 54–5.

8   The title is taken from part one of R. J. Chorley, A. J. Dunn and R. P. Beckinsale, *A History of the Study of Landforms*, 1 (London, 1964).

9   J. Hutton, 'The Theory of the Earth', *Trans. Royal Soc, Edinb.*, 1 (1788), p. 7.

10 J. Playfair, *Illustrations of the Huttonian Theory of the Earth* (Edinburgh, 1802), p. 119.

11 J. Hutton, *Theory of the Earth, with Proofs and Illustrations*, 11 (Edinburgh, 1795), p. 561.

12 *Ibid.*, 1, pp. 16–17.

13 Hutton, 'The Theory of the Earth', *op. cit.*, pp. 7ff.

14 Hutton, *Theory of the Earth, with Proofs*, 11, *op. cit.*, p. 562.

15 G. L. Davies, *The Earth in Decay: A History of British Geomorphology, 1578–1878* (London, 1969), p. 154; also Hutton, 'The Theory of the Earth', *op. cit.*, pp. 3ff.

16 P. Rossi, *The Dark Abyss of Time: The History of the Earth and the History of Nations from Hooke to Vico* (Chicago and London, 1984), p. 115.

17 Davies, *op. cit.*, p. 174.

18 Gould, *Time's Arrow, Time's Cycle: Myth and Metaphor in the Study of Geological Time* (1987; repr. London, 1991), pp. 78–9; also Hutton, 'The Theory of the Earth', *op. cit.*, p. 96.

19 Davies, *op. cit.*, pp. 155, 174.

20 *Ibid.*, pp. 175–6.

21 Hutton, 'The Theory of the Earth', *op. cit.*, p. 8.

22 *Ibid.*, p. 8.

23 See Rossi, *op. cit.*, p. 117.

24 D. R. Dean, *James Hutton and the History of Geology* (Ithaca, 1992), p. 264.

25 *Ibid.*

26 Hutton, 'The Theory of the Earth', *op. cit.*, p. 96.

27 Davies, *op. cit.*, p. 10.

28 *Ibid.*

29 See J. H. Brooke, *Science and Religion: Some Historical Perspectives* (Cambridge, 1991), p. 272.

30 *Ibid.*, p. 14.

31 *Ibid.*, p. 13.

32 *Ibid.*

33 *Ibid.*, p. 14.

34 Hutton, 'The Theory of the Earth', *op. cit.*, p. 9.

35 *Ibid.*, p. 89.

36 Rossi, *op. cit.*, p. 116.

37 Hutton, 'The Theory of the Earth', *op. cit.*, p. 89.

38 R. Porter, *The Making of Geology: Earth Science in Britain, 1660–1815* (Cambridge, 1977), pp. 186–7.

39 C. Lyell, *Principles of Geology* (1830–33; repr., London, 1997), p. 16.

40 Hutton, 'The Theory of the Earth', *op. cit.*, p. 7.

41 Lyell, *op. cit.*, p. 16.

42 See R. Mudie, *The Modern Athens: A Dissection and Demonstration of Men and Things in the Scottish Capital* (Edinburgh, 1825).

43 *Ibid.*, p. 53.

44 E. de Selincourt (ed.), *Journals of Dorothy Wordsworth*, 1 (London, 1941), p. 385.

45 *Ibid.*

46 *Ibid.*, 11, p. 344.

47 Mudie, *op. cit.*, pp. 4–6.

48 *Ibid.*, p. 187.

49 See A. C. Chitnis, *The Scottish Enlightenment: A Social History* (London, 1976).

50 A. Desmond and J. Moore, *Darwin* (1991; repr., London, 1992), pp. 21–2.

51 Mudie, *op. cit.*, p. 302.

52 *Ibid.*, pp. 276–8, 287.

53 *Ibid.*, p. 285.

54 G. Y. Craig, *James Hutton's Theory of the Earth: The Lost Drawings* (Edinburgh, 1978), p. 2; see also V. A. Eyles, 'Introduc-

tion', in *James Hutton's System of the Earth, 1785; etc. . . .* (Darien, Conn., 1970), pp. xiff.

55 *Ibid.*

56 J. Playfair, 'Biographical Account of the Late Dr. James Hutton, F. R. S. Edin', *Trans. Royal Soc. Edin.* v (1805), p. 98.

57 J. Kay, *A Series of Original Portraits and Caricature Etchings* i, i (1842), p. 57.

58 Playfair, 'Biographical Account', *op. cit.*, p. 98.

59 Porter, *op. cit.*, p. 189.

60 It has been argued that one likely influence on Hutton's ideas was G. H. Toulmin's *Antiquity and Duration of the World* (London, 1780) – see D. B. McIntyre, 'James Hutton and the Philosophy of Geology', in C. C. Albritton, *The Fabric of Geology* (Reading, Mass., 1963), pp. 8ff.

61 See Craig, *op. cit.*, p. 1.

62 Porter, *op. cit.*, p. 190; see also B. Hilton, *The Age of Atonement: The Influence of Evangelicalism on Social and Economic Thought, 1785–1865* (Oxford, 1991), pp. 149–50.

63 *Ibid.*

64 Craig, *op. cit.*, p. 2.

65 Davies, *op. cit.*, p. 157.

66 *Ibid.*, p. 158; see Eyles, *op. cit.*, pp. xvii–xix.

67 Chorley, Dunn and Beckinsale, *op. cit.*, p. 48.

68 *Ibid.*, p. 48; Davies, *op. cit.*, p. 158.

69 Chorley, Dunn and Beckinsale, *op. cit.*, p. 48.

70 See *Analytical Review* i (1788), pp. 424–5, quoted in D. R. Dean, 'James Hutton and his Public, 1785–1802', *Annals of Science* xxx (1973), pp. 89–105.

71 Quoted in Dean, *James Hutton and his Public, op. cit.*, p. 92.

72 Quoted in *ibid.*, p. 95.

73 Quoted in *ibid.*, p. 98.

74 Chorley, Dunn and Beckinsale, *op. cit.*, pp. 48ff; Davies, *op. cit.*, pp. 188ff.

75 Quoted in Chorley, Dunn and Beckinsale, *op. cit.*, p. 49.

76 *Ibid.*; see also chapter 5, p. 179–80.

77 Lyell, *op. cit.*, p. 21.

78 Chorley, Dunn and Beckinsale, *op. cit.*, pp. 49–50; see also chapter 5, p. 179.

79 Lyell, *op. cit.*, p. 20.

80 Davies, *op. cit.*, pp. 186ff.

81 Porter, *op. cit.*, p. 198.

82 H. Cockburn, *Memorials of His Times* (Edinburgh, 1856), p. 80; also J. Walker, 'Edinburgh in Spatial, Social and Symbolic Conflict: The King's Birthday Riot of 1792', B. A. Dissertation, School of Geography, University of Oxford, 1995, p. 40.

83 See Desmond and Moore, *op. cit.*, for later manifestations.

84 Mudie, *op. cit.*, pp. 182–3.

85 See E. P. Thompson, *The Making of the English Working Class* (London, 1963).

86 Davies, *op. cit.*, p. 187.

87 Porter, *op. cit.*, p. 198.

88 *Ibid.*, p. 159.

89 *Ibid.*, p. 177.

90 *Ibid.*, p. 158.

91 *Ibid.*, pp. 178–9; Gould, *op. cit.*, pp. 66ff; see also the discussion in Porter, *op. cit.*, pp. 184–5.

92 Playfair, 'Biographical Account', *op. cit.*, pp. 44ff; see also E. B. Bailey, *James Hutton – the Founder of Modern Geology* (London, 1976), pp. 4, 6–7, 23–4; also Dean, *James Hutton and the History of Geology*, pp. 6–7.

93 Playfair, *Illustrations of the Huttonian Theory*, p. 61.

94 *Ibid.*, p. 88.

95 Dean, *James Hutton and the History of Geology*, p. 13.

96 Porter, *op. cit.*, p. 191.

97 Davies, *op. cit.*, p. 190.

98 Gould, *op. cit.*, p. 95.

99 *Ibid.*

100 Playfair, *Illustrations of the Huttonian Theory, op. cit.*, pp. 102ff.

101 *Ibid.*, p. 119.

102 *Ibid.*, p. 117.

103 *Ibid.*

104 See Gould, *op. cit.*, pp. 1ff.

105 Playfair, *Illustrations of the Huttonian Theory, op. cit.*, p. 120.

106 *Ibid.*, p. 126.

107 See A. Desmond, *Huxley: From Devil's*

Disciple to Evolution's High Priest (London, 1998).

108 The line is taken from Tennyson – see note 160.

109 See J. F. C. Harrison, *Early Victorian Britain, 1832–51* (London, 1979), p. 150; D. Newsome, *The Victorian World Picture* (London, 1997), pp. 195–6.

110 B. M. G. Reardon, *Religious Thought in the Victorian Age: A Survey from Coleridge to Gore* (2nd edn. London, 1995), p. 212.

111 See *Hymns Ancient and Modern: Historical Edition* (London, 1900), pp. 403–4, 829.

112 I. Watts, *The Knowledge of the Heavens and Earth Made Easy: or, The First Principles of Astronomy and Geography Explain'd . . . etc.* (London, 1726).

113 *Ibid.*, pp. 443, 16–17.

114 *Ibid.*, pp. 556, 807.

115 See I. Bradley, *Abide with Me: The World of Victorian Hymns* (London, 1997), p. 135.

116 Lyell, *op. cit.*, title page.

117 *Ibid.*, p. 3.

118 Gould, *op. cit.*, p. 105.

119 *Ibid.*, pp. 104–5.

120 *Ibid.*, p. 105.

121 Chorley, Dunn and Beckinsale, *op. cit.*, p. 144.

122 *Ibid.*, p. 145.

123 Lyell, *op. cit.*, p. 30.

124 See N. A. Rupke, ' "The End of History" in the Early Picturing of Geological Time', *History of Science* XXXVI (1998), pp. 61–90.

125 Lyell, *op. cit.*, p. 30.

126 *Ibid.*

127 Davies, *op. cit.*, p. 13.

128 Lyell, *op. cit.*, p. 31.

129 *Ibid.*, p. 32.

130 *Ibid.*, p. 36.

131 *Ibid.*, p. 37.

132 *Ibid.*

133 *Ibid.*, p. 35.

134 *Ibid.*, p. 38.

135 *Ibid.*

136 For a fuller account of Lyell's earlier life and work, see L. G. Wilson, *Charles Lyell, the Years to 1841: The Revolution in Geology* (New Haven and London, 1972).

137 *Ibid.*, p. 152.

138 *Ibid.*, p. 153.

139 *Ibid.*, p. 152.

140 *Ibid.*

141 Lyell, *op. cit.*, p. 39.

142 *Gentleman's Magazine* CII, i (1832), p. 46.

143 *Ibid.*, I, n.s. (1834), p. 77.

144 *Quarterly Review* XLIII (1830), p. 413.

145 *Ibid.*

146 G. Mantell, *The Geology of the South-East of England* (London, 1833), p. 24.

147 *Ibid.*, p. 32.

148 See D. R. Dean, 'Through Science to Despair: Geology and the Victorians', in J. Paradis and T. Postlewait (eds.), *Victorian Science and Victorian Values: Literary Perpectives* (New Brunswick, 1985), p. 114.

149 Chorley, Dunn and Beckinsale, *op. cit.*, p. 147.

150 *Ibid.*, p. 159.

151 Lyell, *op. cit.*, p. 110.

152 *Ibid.*, p. 230.

153 *Ibid.*, p. 115.

154 *Ibid.*

155 *Chambers's Journal* VI (July–Dec., 1846), p. 182.

156 *Household Words* III (1851), p. 492.

157 There were only eight officially sanctioned volumes in the series. Babbage's volume was thus a kind of snub. C. Babbage, *The Ninth Bridgewater Treatise: A Fragment* (1837; repr. London, 1967).

158 *Ibid.*, pp. 87–8.

159 *Penny Cyclopaedia of the Society for the Diffusion of Useful Knowledge* XI (1838), p. 147.

160 See D. R. Dean, *Tennyson and Geology* (Lincoln, 1985), p. 6; this and the following extracts from Tennyson are taken from C. Ricks (ed.), *The Poems of Tennyson*, 3 vols (2nd edn., London, 1987).

161 *Ibid.*, II, p. 19.

162 *Ibid.*, p. 30.

163 *Ibid.*, p. 8.

164 *Household Words*, VI (1857), p. 357.

165 *Ibid.*

166 See *Hymns Ancient and Modern, op. cit.*, p. 663. This is the first line of a hymn written in 1819 by Bishop Reginald Heber.

167 Playfair, *Illustrations of the Huttonian Theory*, op. cit., p. 388.

168 *The Penny Illustrated News* I, no. 20 (1849), p. 154.

169 See Lyell, op. cit., p. 69.

170 *Ibid.*, p. 48.

171 *Ibid.*

172 *Ibid.*, p. 64.

173 *Ibid.*

174 Chorley, Dunn and Beckinsale, op. cit., p. 211.

175 Davies, op. cit., p. 283.

176 See Gillispie, *Genesis and Geology: A Study of the Relations of Scientific Thought, Natural Theology, and Social Opinion in Great Britain* (Cambridge, Mass., 1951) p. 151 – Agassiz himself remained committed to catastrophism and in this sense his studies on glaciers gave it a new lease of life.

177 Chorley, Dunn and Beckinsale, op. cit., pp. 333ff; Davies, op. cit., pp. 290–91 – Ramsay proposed a Permian glaciation as early as 1852.

178 C. C. Loomis, 'The Arctic Sublime', in U. C. Knoepflmacher and G. B. Tennyson (eds.), *Nature and the Victorian Imagination* (Los Angeles, 1977), p. 97; on this aspect of Alpine glaciation, see K. Flint, *The Victorians and the Visual Imagination* (Cambridge, 2000), pp. 119ff.

179 *Ibid.*

180 *Ibid.*, p. 98; see also J. L. Lowes, *Road to Xanadu: A Study in the Ways of the Imagination* (Boston, new edn., 1930), pp. 135ff.

181 See Lowes, op. cit., p. 140.

182 R. Holmes (ed.), *Coleridge: Selected Poems* (London, 1996), p. 83.

183 Loomis, op. cit., p. 99.

184 *Ibid.*; also M. Shelley, *Frankenstein, or the Modern Prometheus* (1818; repr. London, 1993), p. 24.

185 Shelley, op. cit., pp. 158–9.

186 *Ibid.*, p. 159.

187 *Ibid.*, p. 20.

188 Loomis, op. cit., p. 99.

189 C. Brontë, *Jane Eyre* (1847; repr., London, 1969) pp. 2–3.

190 *The Penny Illustrated News* I, xx (1849), pp. 153–4.

191 See M. H. Cohen (ed.), *The Letters of Lewis Carroll*, I (Cambridge, 1979), p. 39.

192 *Household Words*, XII (1856), pp. 479ff.

193 Loomis, op. cit., pp. 106–8.

194 *Ibid.*, p. 109.

195 *Punch* XVIII (Jan.–June 1850), p. 87.

196 *Ibid.*, VIII (Jan.–June 1845), p. 149.

197 See the discussion in Flint, op. cit., p. 124.

198 *Ibid.*, p. 119.

199 D. Wordsworth, 'Journal of Tour in the Continent 1820', in E. de Selincourt (ed.), *Journals of Dorothy Wordsworth* (London, 1941), p. 286; the *Mer de Glace* was sometimes also known as the Arveiron glacier, referring to the name of the river discharging from its snout.

200 See N. Rogers (ed.), *The Complete Poetical Works of Percy Bysshe Shelley*, II (Oxford, 1975) – the poem 'Mont Blanc', pp. 78–9.

201 Se J. D. Forbes, *Travels through the Alps of Savoy and other parts of the Pennine Chain, with Observations of the Phenomena of Glaciers* (2nd edn., Edinburgh, 1845), p. 22.

202 Davies, op. cit., p. 289.

203 *Ibid.*

204 *Ibid.*

205 F. Burkhardt and S. Smith (eds.), *The Correspondence of Charles Darwin* (hereafter Corr.), II (Cambridge, 1986), Charles Darwin to Charles Lyell, 12/3/1841 (p. 286).

206 N. Barlow (ed.), *The Autobiography of Charles Darwin, 1809–1882* (London, 1958), p. 70.

207 Corr., II, Charles Darwin to W. H. Fitton, 28/6/1842 (pp. 321–2).

208 See Gillispie, op. cit., p. 151.

209 *Chambers's Journal* XIV (July–Dec., 1850), pp. 147–50.

210 *Ibid.*, p. 148.

211 *Ibid.*, IV, n.s. (July–Dec., 1855), p. 48.

212 Davies, op. cit., p. 287.

213 *Ibid.*, pp. 291–2.

214 Lyell, op. cit., p. 181 (vol. 2 – title page).

215 Lady Constance Rawleigh, speaking about Robert Chambers's *Vestiges* – in B. Disraeli, *Tancred or The New Crusade* (1847; repr. Edinburgh, 1927), p. 113.

216 See Ricks, *op. cit.*, II, p. 372 (notes). The lines were probably inspired by Lyell.

217 S. Gliserman, 'Early Victorian Science Writers and Tennyson's "In Memoriam": A Study in Cultural Exchange', *Victorian Studies* XVIII (1975) part II, p. 451.

218 Ricks, *op. cit.*, 2, p. 373 (notes).

219 G. de Beer, 'Introduction', in R. Chambers, *Vestiges of the Natural History of Creation* (1844; repr. Leicester, 1969), p. 12.

220 *Ibid.*, p. 15.

221 *Ibid.*, p. 16.

222 *Ibid.*, p. 18.

223 M. Millhauser, *Just Before Darwin: Robert Chambers and Vestiges* (Middletown, Conn., 1959), p. 5.

224 Beer, *op. cit.*, p. 24.

225 *Ibid.*, pp. 31–2.

226 See J. A. Secord, *Victorian Sensation: The Extraordinary Publication, Reception and Secret Authorship of Vestiges of the Natural History of Creation* (Chicago, 2000), p. 3, also pp. 34ff; Darwin's *Origin* did not overtake *Vestiges* until the 1890s – *ibid.*, p. 526.

227 K. Tillotson, *Novels of the Eighteen-Forties* (Oxford, 1954) p. 22–3.

228 Secord, *op. cit.*, p. 348.

229 *Ibid.*, p. 41.

230 See Desmond, *op. cit.*, p. 209.

231 *Ibid.*, p. 233.

232 *Ibid.*

233 Millhauser, *op. cit.*, p. 4.

234 Secord, *op. cit.*, p. 14.

235 *Ibid.*, p. 7.

236 *Ibid.*, p. 109.

237 *Ibid.*, p. 90.

238 A. N. Wilson, *The Victorians* (London, 2002), p. 100.

239 Chambers, *op. cit.*, p. 388.

240 *Ibid.*, pp. 152–3.

241 *Ibid.*, p. 153.

242 *Ibid.*, p. 146.

243 *Ibid.*, p. 153.

244 *Ibid.*, p. 192.

245 *Ibid.*, p. 195.

246 *Ibid.*

247 *Ibid.*, p. 202–3.

248 *Ibid.*, p. 4.

249 *Ibid.*, p. 272.

250 Quoted in Millhauser, *op. cit.*, p. 5.

251 Beer, *op. cit.*, p. 23.

252 Quoted in *ibid.*, p. 31.

253 Disraeli, *op. cit.*, p. 113; see also the commentary in Secord, *op. cit.*, pp. 188–9.

254 A. Sedgwick, 'Natural History of Creation', *Edinburgh Review* LXXXII (1845), p. 3.

255 For a full list, see Millhauser, *op. cit.*, pp. 229–32.

256 See, for example, *Chambers's Journal* I and II (1844).

257 *Ibid.* IV (July–Dec., 1845), pp. 1–3.

258 *Ibid.* VII (Jan.–June, 1847), p. 24.

259 *Ibid.*

260 Gillispie, *op. cit.*, p. 162.

261 Beer, *op. cit.*, p. 32.

262 *Ibid.*, p. 33.

263 *Ibid.*, p. 34.

Chapter Three

1 J. Phillips, part of his presidential address to the Geological Society, 1859 – see *Quarterly Journal of the Geological Society* (hereafter *QJGS*) XV (1859), p. lxi.

2 H. Miller, *Sketch-book of Popular Geology* (Edinburgh, 1859), p. 87.

3 I have been unable to establish whether the Society was specifically a Bradford one.

4 The book originated as a series of lectures. For further commentary on Miller, see M. Shortland (ed.), *Hugh Miller's Memoir: From Stonemason to Geologist* (Edinburgh, 1995); also idem., *Hugh Miller and the Controversies of Victorian Science* (Oxford, 1996).

5 C. C. Gillispie, *Genesis and Geology: A Study in the Relations of Scientific Thought, Natural Theology, and Social Opinion in Great Britain, 1790–1850* (Cambridge, Mass., 1959), p. 172.

6 *Ibid.*, p. 181.

7 See the advertisement in the back leaf of the edition.

8 *Ibid.*

9 H. Miller, *The Testimony of the Rocks; or Geology in its Bearings on the Two*

Theologies, Natural and Revealed (Edinburgh, 1871), pp. 172–3.

10 *Ibid.*, p. 174

11 *Ibid.*; see also the commentary in J. H. Brooke, *Science and Religion: Some Historical Perspectives* (Cambridge, 1991), p. 272.

12 H. Miller, *The Old Red Sandstone* (Edinburgh, 1841); see also Gillispie, *op. cit.*, p. 172.

13 *Ibid.*, p. 9.

14 *Ibid.*

15 *Ibid.*, pp. 1, 17.

16 See Shortland (ed.), *Hugh Miller's Memoir*, *op. cit.*, pp. 64–5.

17 H. Miller, *The Cruise of the Betsey; or A Summer Ramble among the Fossiliferous Deposits of the Hebrides* (Edinburgh, 1858); the account was written originally for the religious paper *Witness* in connection with a bitter dispute between the Free and Established Churches of Scotland.

18 *Ibid.*, p. 104.

19 *Ibid.*, p. 79.

20 *Ibid.*, pp. 79–80.

21 P. Gosse, *Omphalos: An Attempt to Untie the Geological Knot* (London, 1857); Philip Gosse was made FRS in 1856 – see E. Gosse, *Father and Son* (London 1907; Penguin edn., 1983), p. 256; for a re-evaluation of Gosse's life and work, see A. Thwaite, *Glimpses of the Wonderful: The Life of Philip Henry Gosse, 1810–1888* (London, 2002).

22 E. Gosse, *op. cit.*, p. 125.

23 See chapter 4.

24 E. Gosse, *op. cit.*, p. 105.

25 *Ibid.*, p. 113.

26 Gillispie, *op. cit.*, p. ix.

27 J. H. Brooke, *Science and Religion: some historical perspectives* (Cambridge, 1991) p. 5.

28 *Ibid.*, p. 4.

29 J. A. Secord, 'Introduction' to C. Lyell, *Principles of Geology* (London, 1830–33; Penguin edn., 1997), p. xi.

30 *Ibid.*, p. xii.

31 Quoted in Secord, *op. cit.*, p. xxiv.

32 *Ibid.*, p. xxix.

33 R. Gilmour, *The Victorian Period: The Intellectual and Cultural Context of English Literature, 1830–1890* (London, 1993), p. 132.

34 A. N. Wilson, *The Victorians* (London, 2002), p. 169.

35 W. Brande, *Outlines of Geology* (London, 1817), p. 20.

36 *Ibid.*, p. 26.

37 For fuller discussion, see R. Porter, *The Making of Geology: Earth Science in Britain, 1660–1815* (Cambridge, 1977), pp. 170ff; Gillispie, *op. cit.*, chapters two and three.

38 See T. Roscoe, *The Book of the Grand Junction Railway* (London, 1839), p. 53.

39 See W. Feaver, *The Art of John Martin* (Oxford, 1975), pp. 1ff.

40 *Ibid.*, p. 70.

41 See B. Hilton, *The Age of Atonememt: The Influence of Evangelicalism on Social and Economic Thought, 1785–1865* (Oxford, 1991), p. 147.

42 See chapter five.

43 Feaver, *op. cit.*, pp. 6, 196.

44 See M. D. Paley, *The Apocalyptic Sublime* (New Haven and London, 1986), pp. 2, 5–6; also T. Balston, *John Martin, 1789–1854: His Life and Works* (London, 1947), p. 18.

45 J. Laurence, *Geology in 1835: A Popular Sketch of the Progress, Leading Features and Latest Discoveries of this Rising Science* (London, 1835), p. 13.

46 See chapter two.

47 J. Morton, *The Natural History of Northamptonshire* (London, 1712), p. 97.

48 *Ibid.*, p. 134.

49 J. Whitehurst, *An Inquiry into the Original State and Formation of the Earth* (2nd edn., London, 1786) pp. 181ff.

50 *Ibid.*, p. 181.

51 See R. Laudan, *From Mineralogy to Geology: The Foundations of a Science, 1650–1830* (Chicago, 1987), pp. 87ff.

52 G. F. Richardson, *Geology for Beginners, Comprising a Familiar Explanation of Geology and its Associated Sciences* (London, 1842), p. 65.

53 Laurence, *op. cit.*, p. 16.

54   W. D. Conybeare and W. Phillips, *Outlines of the Geology of England and Wales* (London, 1822), pp. ii–iii.

55   *Ibid.*

56   *Ibid.*, p. iii.

57   *Ibid.*

58   See chapter one.

59   *Ibid.*

60   J. Farey, *General View of the Agriculture and Minerals of Derbyshire; With Observations on the Means of their Improvement*, 3 vols (London, 1811–17); the other early Board of Agriculture reports that included geological surveys and maps were those on the North Riding of Yorkshire, Nottinghamshire and Devonshire.

61   R. Bakewell, *An Introduction to Geology* (2nd edn., London, 1815), p. 30.

62   *Ibid.*, p. 46.

63   Miller, *Sketch-book, op. cit.*, p. 114.

64   J. Scafe, *A Geological Primer in Verse* (London, 1820), pp. 7, 13.

65   *Punch* I (July–Dec., 1841), p. 157.

66   See Porter, *op. cit.*, p. 182.

67   G. B. Greenough, *A Critical Examination of the First Principles of Geology* (London, 1819), p. 1.

68   *Ibid.*, pp. 2, 25.

69   Mantell, *The Geology of the South-East of England* (London, 1833), p. 11.

70   K. Lyell, *Life and Letters and Journals of Sir Charles Lyell, Bart*, I (London, 1881), p. 121.

71   *Westminster Review* 1838, quoted in S. F. Cannon, *Science in Culture: The Early Victorian Period* (London, 1978), pp. 4–5.

72   Mantell, *op. cit.*, p. 11.

73   Porter, *op. cit.*, p. 180.

74   See pp. 36ff.

75   M. J. S. Rudwick, *The Great Devonian Controversy: The Shaping of Scientific Knowledge by Gentlemanly Specialists* (Chicago, 1985).

76   Scafe, *op. cit.*, p. 17.

77   Conybeare and Phillips, *op. cit.*, p. xv.

78   *Ibid.*, p. xiii.

79   *Ibid.*, p. lviii.

80   From J. M. W. Turner's Devonshire Coast No. 1 sketchbook, quoted in J. Hamilton, *Turner and the Scientists* (London, 1998), p. 118.

81   See M. Pointon, 'Geology and Landscape Painting in Nineteenth-Century England', in L. J. Jordanova and R. Porter (eds.), *Images of the Earth: Essays in the History of the Environmental Sciences* (2nd edn., London, 1997), p. 96.

82   *Ibid.*, p. 95; see also M. Shortland, 'Darkness Visible: Underground Culture in the Golden Age of Geology', *History of Science* XXXII (1994), pp. 1–61.

83   See chapter one.

84   See Shortland, *op. cit.*, pp. 11ff.

85   M. W. Thompson (ed.), *The Journeys of Sir Richard Colt Hoare through England and Wales, 1793–1810* (Gloucester, 1983), p. 156.

86   *Ibid.*

87   *Ibid.*, p. 157.

88   'Peak Cavern', *Penny Magazine* III (1834), p. 148.

89   See J. Egerton, *Wright of Derby* (London, 1990).

90   See L. Hawes, *Presences of Nature: British Landscape, 1780–1830* (New Haven and London, 1982), pp. 118–19.

91   *Ibid.*

92   Quoted in Egerton, *op. cit.*, p. 191.

93   *Ibid.*, p. 225.

94   *Gentleman's Magazine* XCIX (1829), pp. 434–5.

95   The painting is attributed to Vincent – see M. Cormack, *A Concise Catalogue of Paintings in the Yale Center for British Art* (New Haven and London, 1985), p. 232.

96   Quoted in Hawes, *op. cit.*, p. 175.

97   See F. Greenacre, *Francis Danby, 1793–1861* (London, 1988), pp. 127–8.

98   Thompson, *op. cit.*, p. 248.

99   J. MacCulloch, 'A Sketch of the Mineralogy of Sky', *Transactions of the Geological Society of London* (hereafter *TGSL*) III (1816), p. 12.

100 M. Pointon, *William Dyce, 1806–64: A Critical Biography* (Oxford, 1979), p. 173.

101 *Ibid.*

102 *Ibid.*, p. 174.

103   See A. Wilton, *Turner and the Sublime* (London, 1980), p. 44.

104   *Ibid.*, pp. 35, 40.

105   Pointon, *op. cit.*, p. 99.

106   R. Ayton and W. Daniell, *A Voyage Round Great Britain . . . etc*, 8 vols. (London, 1814–25; facsimile edn., London, 1978).

107   C. Klonk, *Science and the Perception of Nature: British Landscape Art in the Late Eighteenth and Early Nineteenth Centuries* (New Haven and London, 1996), p. 67.

108   *Ibid.*, pp. 70ff; also Shortland, *op. cit.*, pp. 5–8.

109   *Ibid.*, p. 75.

110   *Ibid.*, p. 68.

111   J. MacCulloch, 'On Staffa', *TGSL* II, (1814), pp. 501–9.

112   Ayton and Daniell, *op. cit.*, III, p. 37.

113   *Ibid.*, I, opposite pp. 48, 52.

114   *Ibid.*, IV, opposite pp. 54, 55.

115   *Ibid.*, IV, opposite pp. 84, 88.

116   *Penny Magazine* I (1832), pp. 236, 238.

117   See Klonk, *op. cit.*, p. 89.

118   *Ibid.*, p. 87.

119   See Wilton, *op. cit.*, p. 127.

120   T. D. Whitaker, *History of Richmondshire*, I (London, 1823), p. 413.

121   Wilton, *op. cit.*, p. 129.

122   E. Shanes, *Turner's Picturesque Views of England and Wales, 1825–1838* (London, 1979) p. 25.

123   Thompson, *op. cit.*, p. 141.

124   See chapter one.

125   C. Ray, *A Picturesque Tour through the Isle of Wight* (London, 1825), plate IX.

126   See chapter one.

127   See Hamilton, *op. cit.*, p. 117; also opening quote.

128   See M. House, *Geology of the Dorset Coast* (Geologists' Association Guide, London, 1969), p. 105.

129   See A. Goudie and D. Brunsden, *Classic Landforms of the East Dorset Coast* (London, 1997), p. 29.

130   Shanes, *op. cit.*, p. 20.

131   See House, *op. cit.*, p. 59.

132   See D. Brunsden and A. Goudie, *Classic Landforms of the West Dorset Coast* (London, 1997), p. 34.

133   Shanes, *op. cit.*, p. 20.

134   *Ibid.*

135   Brande, *op. cit.*, pp. 1–2.

136   Rudwick, *Great Devonian*, *op. cit.*, p. 37.

137   *Ibid.*, pp. 17ff.

138   H. De la Beche, *How to Observe Geology* (London, 1835), p. v.

139   Miller, *Old Red Sandstone*, *op. cit.*, pp. 1–2, 15.

140   Secord, *op. cit.*, p. xi.

141   See J. Topham, 'Science and Popular Education in the 1830s: The role of the Bridgewater Treatises', *British Journal for the History of Science* XXV (1992), pp. 397–430.

142   Miller, *Cruise of the Betsey*, *op. cit.*, p. 88.

143   J. Wyatt, *Wordsworth and the Geologists* (Cambridge, 1995) pp. 174–5.

144   G. Mantell, *The Wonders of Geology*, II (London, 1838) pp. 315–6.

145   *Illustrated London News* XXI (July–Dec., 1852), p. 50.

146   Mantell, *The Wonders*, *op. cit.*, II, p. 317.

147   G. A. Mantell, *The Medals of Creation; or First Lessons in Geology and in the Study of Organic Remains*, II (London, 1844) p. 908.

148   *Ibid.*, pp. 893ff.

149   *Ibid.*, p. 936.

150   *Ibid.*, p. 885.

151   *Ibid.*, p. 891.

152   Rudwick, *Great Devonian*, *op. cit.*, p. 41.

153   See J. A. Secord, *Controversy in Victorian Geology: The Cambrian-Silurian Dispute* (Princeton, 1986).

154   See Wyatt, *op. cit.*, pp. 7ff.

155   *Chambers's Journal* VI, (July–Dec., 1846), p. 256.

156   A. Desmond and J. Moore, *Darwin* (London, 1991; Penguin edn., 1992) pp. 254–5.

157   See p. 202.

158   *Quarterly Review* XLIII (1830), p. 412.

159   See J. G. Paradis, 'The Natural Historian as Antiquary of the World: Hugh Miller and the Rise of Literary Natural History', in Shortland (ed.), *Hugh Miller and the Controversies*, *op. cit.*, pp. 122–5.

160   *Blackwood's Edinburgh Magazine* XLIV (1838), pp. 386ff.

161   *Ibid.*, XXVII (1830), pp. 263–4.

162   See Rudwick, *Great Devonian*, *op. cit.*,

p. 38; also P. McCartney, *Henry De la Beche, Observations on an Observer* (Cardiff, 1977).

163 H. De la Beche, *Sections and Views Illustrative of Geological Phenomena* (London, 1830) p. iii.

164 See S. J. Knell, *The Culture of English Geology, 1815–1851: A Science Revealed Through its Collecting* (Aldershot, 2000), pp. 281–3.

165 McCartney, *op. cit.*, p. 28.

166 E. O. Gordon, *The Life and Correspondence of William Buckland* (London, 1894), pp. 28–30.

167 *Punch* XII (Jan.–June, 1847) – chronology for May.

168 M. J. S. Rudwick, 'The Emergence of a Visual Language for Geological Science, 1760–1840', *History of Science* XIV (1976), pp. 149–95.

169 See W. M. Ivins, *Prints and Visual Communication* (Cambridge, Mass., 1969).

170 T. Webster, 'Observations on the Purbeck and portland Beds', *TGSL*, 2nd ser., II (1829), p. 37.

171 cross-ref ch 1.

172 Ayton and Daniell, I, *op. cit.*, p. iii.

173 *Ibid.*, p. iv.

174 *Ibid.*

175 *Ibid.*, IV (London, 1820), p. 5.

176 *Ibid.*, II (London, 1815), p. 25.

177 *Ibid.*, I, p. iii.

178 *Ibid.* I, p. 52.

179 *Ibid.*, I, p. 94.

180 *Ibid.*, III (London, 1818), p. 36.

181 *Ibid.*, p. 38.

182 *Ibid.*, I, p. 34.

183 *Ibid.*, p. 52.

184 *Ibid.*

185 See Rudwick, 'Emergence of a Visual Language', *op. cit.*, p. 174.

186 *Ibid.*, p. 175.

187 *Ibid.*, pp. 175 and 194. The *camera obscura* involved the fitting of a lens to a movable box. Objects were reflected through the lens aperture on to the wall of a darkened room. The result was of great value to the draughtsman and, later, the photographer.

188 *Ibid*, p. 175.

189 E. C. Curwen, *The Journal of Gideon Mantell, Surgeon and Geologist* (Oxford, 1940), p. 11.

190 See B. Hall, *Description of the Camera Lucida, an Instrument for Drawing in True Perspective, and for Copying, Reducing, or Enlarging other Drawings* (London, 1830).

191 *Ibid.*

192 De la Beche, *Sections and Views*, *op. cit.*, p. viii.

193 *Ibid.*

194 *TGSL* II, 2nd ser. (1829).

195 W. D. Conybeare and M. Buckland, *Ten Plates comprising a Plan, Sections, and Views, Representing the Change Produced on the Coast of East Devon, between Axmouth and Lyme Regis . . .* (London, 1840).

196 See R. Hamblyn, 'Private Cabinets and Popular Geology; The British Audiences for Volcanoes in the Eighteenth Century', in C. Chard and H. Langdon (eds.), *Transports: Travel, Pleasure and Imaginative Geography, 1600–1830* (New Haven and London, 1996), p. 200.

197 De la Beche, *Sections and Views*, *op. cit.*, p. viii.

198 See chapter one; see also J. Morell and A. Thackray, *Gentlemen of Science: Early Years of the British Association for the Advancement of Science* (Oxford, 1981), chapter two.

199 *Chambers's Journal* IV (July–Dec., 1845) p. 65.

200 *Ibid.*

201 *Ibid.*

202 Gillispie, *op. cit.*, p. 188, quoting Adam Sedgwick.

203 *Gentleman's Magazine* n.s. XVIII (1842), p. 188.

204 *Chambers's Journal* XII (July–Dec., 1859), pp. 225–6.

205 *Ibid.*, p. 225.

206 *Ibid.*

207 See pp. 23–4

208 Gordon, *op. cit.*, pp. 28, 30.

209 Lyell, *op. cit.*, II (London, 1881) pp. 42–3.

210 *Gentleman's Magazine* n.s. XVIII (1842), p. 187.

211 *Punch* III (July–Dec., 1842) p. 20.

212 *Ibid.*

213   *Ibid.*
214   *Ibid.*
215   *Ibid.*
216   *Household Words* V (1852), p. 352.
217   *Ibid.*
218   C. Dickens, *Sketches by Boz*, II (London, 1836), p. 338.
219   *Ibid.*, p. 340.
220   *Ibid.*, p. 341.
221   *Ibid.*, pp. 348ff.
222   See Knell, *op. cit.*, pp. 49ff.
223   Porter, *op. cit.*, p. 133.
224   *Ibid.*, p. 134.
225   See also A. C. Todd, 'Origins of the Royal Geological Society of Cornwall', *Transactions of the Royal Geological Society of Cornwall* XIX (1957–64), pp. 179–184.
226   *Punch* I (July–Dec., 1841), p. 141.
227   *Ibid.*
228   See Porter, *op. cit.*, pp. 146–7.
229   See Secord, *Controversy in Victorian Geology*, *op. cit.*, pp. 24ff.
230   *The Penny Cyclopaedia of the Society for the Diffusion of Useful Knowledge* XI (London, 1838), p. 131.
231   *Chambers's Journal* XVI (July–Dec., 1861) p. 356.
232   *Ibid.*
233   *Ibid.*
234   *Ibid.*
235   *Ibid.*
236   Rudwick, 'Emergence of a Visual Language', *op. cit.*, p. 159; see also J. M. Harrison, 'Nature and Significance of Geological Maps', in C. C. Albritton (ed.), *The Fabric of Geology* (Reading, Mass., 1963), pp. 225–6.
237   See Knell, *op. cit.*, p. 282.
238   J. B. Harley, 'Maps, knowledge, and power', in D. Cosgrove and S. Daniels (eds.), *The Iconography of Landscape* (Cambridge, 1988), pp. 277–311.
239   See J. C. Stone, 'Imperialism, Colonialism and Cartography', *Transactions of the Institute of British Geographers* XIII, n.s. (1988), pp. 57–64.
240   L. R. Cox, 'New light on William Smith and his work', *Proc. Yorks. Geol. Soc.* XXV (1942–5), p. 26.

241   *Ibid.*, p. 25.
242   Oxford University Museum (hereafter OUM), Smith MS, Box 13, Folder 1.
243   Secord, *Controversy in Victorian Geology*, *op. cit.*, pp. 30–32.
244   *Ibid.*, p. 30.
245   See J. A. Secord, 'King of Siluria; Roderick Murchison and the Imperial Theme in Nineteenth-Century British Geology', *Victorian Studies* XXV (1982), pp. 413–42.
246   Secord, *Controversy in Victorian Geology*, *op. cit.*, p. 30.
247   Quoted in *ibid.*, p. 29; see also N. A. Rupke, '"The End of History" in the Early Picturing of Geological Time', *History of Science* XXXVI (1998), pp. 61–90.
248   Cox, *op. cit.*, p. 26; see also S. Winchester, *The Map that Changed the World: the Tale of William Smith and the Birth of a Science* (London, 2001), p. 147.
249   Cox, *op. cit.*, p. 26.
250   See p. 30.
251   Cox, *op. cit.*, p. 25.
252   Rudwick, 'Emergence of a Visual Language', *op. cit.*, p. 162.
253   *Ibid.*, pp. 161–3.
254   See *TGSL* II, 2nd ser. (1829).
255   Farey, *op. cit.*, I, plate II.
256   *TGSL* II, 2nd ser. (1829), plates XVI and XXX.
257   See *Report of the First and Second Meetings of the British Association for the Advancement of Science* (London, 1833), p. 584.
258   See Rupke, *op. cit.*
259   Rudwick, 'Emergence of a Visual Language', *op. cit.*, p. 167.
260   OUM, Smith MS, Box 15, Folder 4.
261   Rudwick, 'Emergence of a Visual Language', p. 167; also Rupke, *op. cit.*, pp. 61–8.
262   Rudwick, 'Emergence of a Visual Language', p. 168.
263   De la Beche, *Sections and Views*, *op. cit.*, pp. 41–2.
264   Cox, *op. cit.*, p. 67.
265   Rudwick, *Great Devonian Controversy*, *op. cit.*, p. 46; also Rupke, *op. cit.*, pp. 68–9.

266 Rudwick, 'Emergence of a Visual Language', *op. cit.*, p. 169.

267 Cox, *op. cit.*, p. 61.

268 See *Transactions of the Royal Geological Society of Cornwall* I (1818), plate 4.

269 Rudwick, 'Emergence of a Visual Language', *op. cit.*, p. 170.

270 *Ibid.*

271 De la Beche, *Sections and Views, op. cit.*, p. 7.

272 See H. C. Englefield, *A Description of the Principal Picturesque Beauties, Antiquities, and Geological Phenomena of the Isle of Wight* (London, 1816).

273 OUM, Buckland MS, C42.

274 Rudwick, 'Emergence of a Visual Language', *op. cit.*, p. 171.

275 *Ibid.*, pp. 171–2.

276 T. Sheppard, 'William Smith; his Maps and Memoirs', *Proc. Yorks. Geol. Soc.* XIX (1914–22), facing p. 152.

277 Curwen, *op. cit.*, p. 201.

278 G. A. Mantell, *Geological Excursions round the Isle of Wight and along the adjacent coast of Dorsetshire* (London, 1847), p. viii.

279 Curwen, *op. cit.*, p. 202.

280 Mantell, *Geological Excursions*, p. 418.

281 Mantell, *The Medals of Creation; or First Lessons in Geology and in the study of Organic Remains*, II (London, 1844), pp. 987–8.

282 cross-ref. chapter one.

283 De la Beche, *Sections and Views, op. cit.*, p. vii.

Chapter Four

1 Revelation 12: 7–9.

2 G. Mantell, *The Geology of the South-East of England* (London, 1833), p. 315.

3 J. Carey and A. Fowler (eds.), *The Poems of John Milton* (London, 1968), p. 608.

4 See, for example, the entry under 'Dragons' in *The Harmsworth Encyclopaedia* (London, 1905).

5 *Penny Magazine* XII (1843), p. 432.

6 *Ibid.*

7 J. Fowles, *The French Lieutenant's Woman* (London, 1969; Granada edn., 1977), p. 7.

8 J. Austen, *Persuasion* (London, 1818; Penguin edn., Harmondsworth, 1965), p. 117.

9 *Ibid.*

10 Fowles, *op. cit.*, p. 62.

11 See H. Woodward, *The Jurassic Rocks of Great Britain: III – The Lias of England and Wales* (London, 1893), pp. 60–61.

12 D. Cadbury, *The Dinosaur Hunters: A Story of Scientific Rivalry and the Discovery of the Prehistoric World* (London, 2000), p. 8.

13 *Ibid.*, pp. 7–8.

14 See R. Fortey, *The Hidden Landscape: A Journey into the Geological Past* (London, 1993; Pimlico edn., London, 1994), p. 179.

15 See chapter three.

16 R. Bakewell, *An Introduction to Geology: Illustrative of the General Structure of the Earth* (London, 2nd edn., 1815), p. 470.

17 See H. Torrens, 'Mary Anning (1799–1847) of Lyme; "The Greatest Fossilist the World ever Knew"', *British Journal for the History of Science* XXVIII (1995), p. 259; see also 'The Fossil-finder of Lyme Regis', *Chambers's Journal* VIII (July–Dec., 1857), p. 383.

18 'The Fossil-finder', *op. cit.*, p. 383.

19 *Ibid.*

20 Torrens, *op. cit.*, p. 259.

21 W. D. Lang, 'Mary Anning (1799–1847), and the Pioneer Geologists of Lyme', *Proc. Dorset Nat. Hist. and Arch. Soc* LX (1938), p. 145.

22 See M. Batey, *Jane Austen and the English Landscape* (London, 1996), p. 126.

23 See G. Roberts, *The History and Antiquities of the Borough of Lyme Regis and Charmouth* (London, 1834), p. 288.

24 Torrens, *op. cit.*, pp. 258–9.

25 *Ibid.*, p. 259.

26 *Ibid.*, pp. 259–60; also 'The Fossil-finder', *op. cit.*, p. 383.

27 Torrens, *op. cit.*, p. 260.

28 *Gentleman's Magazine* LXXXIII (1813), p. 618.

29 E. Home, 'Some account of the fossil remains of an animal. . . .', *Philosophical Transactions of the Royal Society* CIV (1814), pp. 571–7.

30   Torrens, *op. cit.*, p. 260.
31   T. Milner, *The Gallery of Nature: A Pictorial and Descriptive Tour through Creation* (London, 1846), p. 721.
32   *Ibid.*
33   Book I of *Paradise Lost*, quoted in Milner, *op. cit.*, p. 722; see also Carey and Fowler, *op. cit.*, p. 473.
34   'The Natural History of the Dragon', *Chambers's Journal* IX (Jan.–June, 1848), p. 263.
35   Milner, *op. cit.*, p. 722.
36   *Ibid.*
37   *Ibid.*
38   Torrens, *op. cit.*, pp. 263–4.
39   See Milner, *op. cit.*, p. 722; also W. Buckland, *Geology and Mineralogy Considered with Reference to Natural Theology*, I (London, 1836), p. 202.
40   Milner, *op. cit.*, p. 722.
41   Torrens, *op. cit.*, p. 264.
42   See *Lyme Regis: Our Town and Neighbourhood* (Lyme Regis, 1894), p. 38.
43   See 'The Fossil Finder', *op. cit.*, p. 383.
44   The diary of Lady Harrier Silvester, quoted in Torrens, *op. cit.*, p. 265.
45   *Ibid.*
46   W. D. Lang, 'Three Letters by Mary Anning, "fossilist", of Lyme', *Proc. Dorset Nat. Hist. and Arch. Soc.* LXVI (1944), p. 169.
47   *Ibid.*, p. 171.
48   W. D. Lang, 'More about Mary Anning, including a newly found letter', *Proc. Dorset. Nat. Hist and Arch. Soc.* LXXI (1949), p. 185.
49   W. Buckland, 'On the Discovery of a New Species of Pterodactyl in the Lias at Lyme Regis', *Transactions of the Geological Society of London* (hereafter *TGSL*) 2nd ser., III (1835), pp. 217–22.
50   *Ibid.*
51   'The Natural History of the Dragon', *op. cit.*, p. 264.
52   Torrens, *op. cit.*, p. 266.
53   See S. J. Knell, *The Culture of English Geology, 1815–1851: A Science Revealed Through its Collecting* (Aldershot, 2000), pp. 197ff.
54   'The Fossil-finder', *op. cit.*, p. 383.
55   *Ibid.*; see also Torrens, *op. cit.*, p. 269.
56   Lang, 'Mary Anning (1799–1847) and the Pioneer Geologists', *op. cit.*, p. 159.
57   *Ibid.*, p. 151.
58   Lang, 'Three Letters by Mary Anning', *op. cit.*, p. 171.
59   *Ibid.*, p. 170.
60   See D. R. Dean, *Gideon Mantell and the Discovery of Dinosaurs* (Cambridge, 1999); also D. Cadbury, *The Dinosaur Hunters: A True Story of Scientific Rivalry and the Discovery of the Prehistoric World* (London, 2000); also E. C Curwen, (ed.), *The Journal of Gideon Mantell, Surgeon and Geologist* (Oxford, 1940); also S. Spokes, *Gideon Algernon Mantell* (London, 1927).
61   See Curwen, *op. cit.*, pp. 9, 20, 27, 39, 48, 52, 58.
62   See Dean, *op. cit.*, pp. 73ff.
63   *Ibid.*, pp. 84–3; see also Curwen, *op. cit.*, p. 52.
64   Mantell, *op. cit.*, p. 315.
65   This, it later turned out, was a flawed exercise. Animal biology could not support such conjectures. However, this was not the contemporary perception.
66   See opening quotation.
67   Dean, *op. cit.*, pp. 155–6.
68   G. F. Richardson, *Sketches in Prose and Verse (second series), Containing Visits to the Mantellian Museum* (London, 1838), pp. 6–7.
69   *Ibid.*, pp. 24–5.
70   Quoted in *Ibid.*, p. 16.
71   *Ibid.*, pp. 16–17.
72   See Curwen, *op. cit.*, p. 72.
73   Dean, *op. cit.*, p. 76; Curwen, *op. cit.*, p. 39.
74   Dean, *op. cit.*, p. 69.
75   *Ibid.*, p. 68, n. 14.
76   *Ibid.*, p. 77.
77   See W. Buckland, 'Notice on the Megalosaurus or great Fossil Lizard of Stonesfield', *TGSL* n.s. I (1824), pp. 390–96.
78   Dean, *op. cit.*, p. 77.
79   Cadbury, *op. cit.*, p. 109.
80   *Ibid.*
81   Dean, *op. cit.*, p. 79.

82 Ibid.

83 See Curwen, op. cit., p. 89.

84 Dean, op. cit., p. 111; also Mantell, op. cit., p. 316.

85 Curwen, op. cit., p. 110.

86 Dean, op. cit., p. 112.

87 Ibid., p. 113.

88 Ibid., p. 114; see also Curwen, op. cit., p. 110.

89 Dean, op. cit., pp. 132–5; Curwen, op. cit., pp. 120–23.

90 Dean, op. cit., p. 135.

91 Dean, op. cit., p. 136.

92 Curwen, op. cit., p. 125.

93 See Dean, op. cit., pp. 166–7; W. Feaver, The Art of John Martin (Oxford, 1975), p. 146.

94 G. Mantell, The Wonders of Geology, I (London, 1838), p. 368.

95 Ibid.

96 Ibid.

97 See the commentary 'Nature at War', in Chambers's Journal VII (Jan.–June, 1847), pp. 24ff.

98 See p. 4.

99 B. Disraeli, Lothair (London, 1870), p. 83.

100 L. Figuier, The World before the Deluge (2nd edn., London, 1867), p. 5.

101 J. Parkinson, Organic Remains of a Former World: An Examination of the Mineralized Remains of the Vegetables and Animals of the Ante-Diluvian World, I (London, 1804), p. 61.

102 Ibid., pp. 62–3 – quoting the work of Robert Plot on the natural history of Oxfordshire and of Staffordshire.

103 See J. Morton, The Natural History of Northamptonshire (London, 1712), p. 253.

104 See J. Lindley and W. Hutton, The Fossil Flora of Great Britain, I (London, 1831), p. 151.

105 Parkinson, op. cit., p. 61.

106 William Stukeley, for example, quoted in Parkinson, op. cit., p. 63.

107 Parkinson, op. cit., p. 278.

108 Ibid.

109 Ibid., p. 283.

110 Morton, op. cit., pp. 188ff.

111 Ibid., p. 243.

112 E. Meteyard, The Life of Josiah Wedgwood from his Private Correspondences and Family Papers, I (London, 1865), p. 500.

113 J. Whitehurst, An Inquiry into the Original State and Formation of the Earth (2nd edn., London, 1786), p. 48.

114 Ibid., p. 50.

115 Ibid.

116 Morton, op. cit., p. 133.

117 Mantell, The Wonders, op. cit., II p, 328.

118 See Milner, op. cit., p. 714; also 'Tracks of Ancient Animals in Sandstone', Chambers's Journal XVIII (July–Dec., 1852), p. 251.

119 'Tracks of Ancient Animals', op. cit., p. 251.

120 See J. Playfair, Illustrations of the Huttonian Theory of the Earth (Edinburgh, 1802), p. 5.

121 See Woodward, op. cit.

122 H. Miller, Sketch-book of Popular Geology (Edinburgh, 1859), p. 115.

123 Ibid., p. 114.

124 Mantell, Geology of the South-East of England, op. cit., p. 343.

125 See R. Bakewell, An Introduction to Geology (2nd edn., London, 1815), p. 344.

126 Milner, op. cit., p. 741.

127 Ibid., p. 739.

128 W. D. Conybeare and W. Phillips, Outlines of the Geology of England and Wales (London, 1822), p. xii.

129 G. B. Greenough, A Critical Examination of the First Principles of Geology (London, 1819), p. 300.

130 Ibid., p. 302.

131 Mantell, Geology of the South-East of England, op. cit., pp. 345–6; see also chapter one.

132 W. T. Brande, Outlines of Geology (London, 1822), p. 300.

133 Mantell, The Wonders, op. cit., pp. 317ff.

134 Ibid.; also Dean, op. cit., pp. 47–9.

135 See Cadbury, op. cit., pp. 80–81.

136 Mantell, Geology of the South-East of England, op. cit., p. 232.

137 Ibid., pp. 285–6.

138 Ibid., pp. 340–41 – quoting W. H. Fitton.

139 See 'Old Bones', Household Words VIII (1854), p. 83.

140 Conybeare and Phillips, op. cit., p. xi.

141   Milner, *op. cit.*, p. 742; G. Mantell, *The Medals of Creation; or First Lessons in Geology and in the Study of Organic Remains*, II (London, 1844), p. 893.

142   See Knell, *op. cit.*, pp. 3–4.

143   See 'A Leaf from the Oldest Books', *Household Words* XIII (1856), pp. 500–01.

144   See Conybeare and Phillips, *op. cit.*, p. xi.

145   See C. Clay, *Geological Sketches and Observations on Vegetable Fossil Remains etc Collected in the Parish of Ashton-under-Lyne, from the Great South Lancashire Coal Field* (London, 1839), pp. 36–7.

146   See Williamson, *Coals and Coal Plants: A Lecture* (London, 1876), p. 20.

147   Lindley and Hutton, *op. cit.*, p. 151.

148   See chapter one.

149   Lindley and Hutton, *op. cit.*, I, p. 9.

150   Williamson, *op. cit.*, p. 24.

151   Miller, *Sketch-book, op. cit.*, p. 175.

152   See Fortey, *op. cit.*, p. 149; also R. Garner, *The Natural History of the County of Stafford* (London, 1844), p. 198.

153   'Mineral Kingdom', *Penny Magazine* II (1833), p. 10.

154   Miller, *Sketch-Book, op. cit.*, p. 79.

155   Some of the features discussed in this section are explored further in G. Beer, 'Travelling the other way', in N. Jardine, J. A. Secord and E. C. Spary (eds.), *Cultures of Natural History* (Cambridge, 1996), pp. 322–37.

156   See 'The late Dr. Mantell', *Illustrated London News* XXI (July–Dec., 1852), p. 501.

157   W. Buckland, *Geology and Minerology Considered with Reference to Natural Theology*, 2 vols (London, , 1836).

158   Parkinson, *op. cit.*, I, p. vi.

159   *Ibid.*

160   *Ibid.*, p. 4.

161   *Ibid.*

162   *Ibid.*, p. vii.

163   For example, *The Wonders of Geology, op. cit.*; *Medals of Creation, op. cit.*; earlier works like *The Fossils of the South Downs* (London, 1822) and *Illustrations of the Geology of Sussex* (London, 1827) were much more in the nature of scientific accounts.

164   *Punch* XIV (Jan.–June, 1848), p. 76.

165   *Ibid.*, XVI (Jan.–June, 1849), p. 111.

166   G. F. Richardson, *Geology for Beginners, Comprising a Familiar Explanation of Geology and its Associated Science. . . .* (London, 1842), p. 21; see also *The Arabian Nights Entertainment or, The Thousand and One Nights*, 4 vols (London, 1811).

167   Mantell, *The Wonders, op. cit.*, I, pp. 373–4.

168   See Milner, *op. cit.*, p. 729.

169   'Our Phantom Ship on an Antediluvian Cruise', *Household Words* III (1851), p. 492.

170   See some of the references in *Punch* for indications of the widespread popularity of these attractions: for example, XVII (July–Dec, 1848), p. 181; XXIII (July–Dec., 1852), p. 85; see also A. N. Wilson, *The Victorians* (London, 2002), pp. 93–4.

171   'Our Phantom Ship', *op. cit.*, p. 492.

172   *Ibid.*

173   *Ibid.*

174   Miller, *Sketch-book, op. cit.*, p. 76.

175   *Ibid.*, p. 77.

176   *Ibid.*, p. 79.

177   *Ibid.*

178   *Ibid.*, p. 152.

179   *Ibid.*

180   *Ibid.*, p. 159.

181   *Ibid.*

182   *Ibid.*, p. 162.

183   H. Miller, *The Cruise of the Betsey; or, A Summer Ramble among the Fossiliferous Deposits of the Hebrides* (Edinburgh, 1858), p. 79.

184   *Ibid.*, p. 15.

185   *Ibid.*

186   See W. Butcher, *Verne's Journey to the Centre of the Self* (London, 1990), p. 61.

187   J. Verne, *Journey to the Centre of the Earth* (Penguin edn., 1994), p. 104.

188   *Ibid.*, p. 163.

189   *Ibid.*, p. 164.

190   *Ibid.*, p. 166.

191   *Ibid.*, p. 185.

192   *Ibid.*, p. 186.

193   *Ibid.*, p. 187.

194   P. Costello, *Jules Verne: Inventor of Science Fiction* (London, 1978), p. 81.

195   *Ibid.*, p. 83.

196   *Ibid.*, p. 82.

197   Butcher, *op. cit.*, p. 62.

198   *Ibid.*

199   *Ibid.*, p. 61.

200   *Ibid.*, p. 68.

201   *Ibid.*, p. 64.

202   See L. Lynch, *Jules Verne* (New York, 1992), pp. 37–8.

203   Translated by Riou, with fifty-two illustrations. Another translation appeared by Malleson in 1876; in the same year, Routledge published a children's version, as part of 'Every Boy's Library'.

204   Richardson, *Geology for Beginners*, *op. cit.*, p. 481.

205   'Pargate-super-Mare', *Chambers's Journal* IV (July–Dec., 1855), p. 242.

206   See D. Allen, 'Tastes and Crazes', in Jardine, Secord and Spary, *op. cit.*, pp. 394–407.

207   'Pargate-super-Mare', *op. cit.*, p. 242.

208   See C. Kingsley, *Glaucus; or, Wonders of the Shore* (4th edn., London, 1859), pp. 1–3.

209   *Ibid.*, p. 4.

210   *Ibid.*, p. 7; see also D. E. Allen, *The Naturalist in Britain: A Social History* (2nd edn., Princeton, 1994), pp. 121ff.

211   J. A. Secord, *Victorian Sensation: The Extraordinary Publication, Reception and Secret Authorship of Vestiges of the Natural History of Creation* (Chicago, 2000), p. 172.

212   *Ibid.*, p. 463.

213   Kingsley, *op. cit.*, pp. 4–5; see also D. Allen, 'Tastes and Crazes'.

214   Kingsley, *op. cit.*, p. 5.

215   'Mineral Kingdom', *Penny Magazine* II (1833), p. 178.

216   *Ibid.*, pp. 347–9.

217   *Ibid.*, pp. 450–52.

218   'The Fossil Elk of Ireland', *Penny Magazine* III (1834), pp. 299–300.

219   'Science in its Position and Prospects', *Chambers's Journal* V (Jan.–June, 1856), p. 297.

220   'The Ipswich Museum of Natural History for the Working-Classes', *Chambers's Journal* IX (Jan.–June, 1848), pp. 103–4.

221   See P. H. M. Cooper, *Fossils, Faith and Farming* (Sudbury, 1999), p. 118.

222   H. Ward, *Robert Elsmere* (London, 1888), p. 179.

223   See A. Secord, 'Corresponding Interests: Artisans, Gentlemen and Nineteenth-Century Natural History', *British Journal for the History of Science* XXVII (1994), p. 385.

224   C. Dickens, *Hard Times* (London, 1854; Penguin edn., 1989), pp. 53–5.

225   *Blackwood's Edinburgh Magazine* XLII (1837), p. 226.

226   *Ibid.*, LXXX (1856), p. 184ff.

227   Richardson, *Geology for Beginners*, pp. 74ff.

228   *Ibid.*, p. 481.

229   *Ibid.*, p. 504.

230   Kingsley, *op. cit.*, p. 38.

231   *Ibid.*, pp. 31ff.

232   *Ibid.*, p. 37.

233   E. Gosse, *Father and Son* (London, 1907; Penguin edn., 1983), p. 125.

234   *Ibid.*

235   See W. M. Ivins, *Prints and Visual Communication* (Cambridge, Mass., 1969).

236   Richardson, *Sketches in Prose and Verse*, *op. cit.*, pp. 8–9.

237   *Ibid.*, p. 9.

238   'The Gallery of Natural Magic' in the Regent's Park Colosseum may also have taken cue from Mantell's mural – see Secord, *op. cit.*, p. 440.

239   See Cadbury, *op. cit.*, pp. 289–90; Dean, *op. cit.*, pp. 260–61; Curwen, *op. cit.*, p. 289.

240   Cadbury, *op. cit.*, p. 293.

241   'Progress of the Crystal Palace at Sydenham', *Illustrated London News* XXIII (July–Dec., 1853), p. 599.

242   *Ibid.*

243   *Ibid.*

244   *Routledge's Guide to the Crystal Palace and Park at Sydenham* (London, 1854), p. 189.

245   *Geological Wonders of London and its Vicinity* (London, 1862), p. 14.

246 *Ibid.*, pp. 14–15.
247 *Ibid.*, p. 12; see also prologue.
248 'Progress of the Crystal Palace', *op. cit.*, p. 600.
249 'The Crystal Palace at Sydenham', *op. cit.*, p. 322.
250 *Ibid.*, p. 342.
251 *Ibid.*, p. 343.
252 *Punch* XXVIII (Jan.–June, 1855), Almanack.
253 *Ibid.*, p. 50.
254 R. Owen, *Geology and Inhabitants of the Ancient World* (London, 1854).
255 See Cadbury, *op. cit.*, pp. 295, 299.
256 *Blackwood's Edinburgh Magazine* LXXVIII (1855), p. 226.
257 'Progress of the Crystal Palace', *op. cit.*, p. 600; also ' The Crystal Palace at Sydenham', *op. cit.*, p. 322.
258 See chapter two.

Chapter Five

1 Genesis 7: 11–12.
2 From Byron's *Heaven and Earth* – as quoted in John Martin's catalogue entry for his Deluge painting of 1826 – see W. Feaver, *The Art of John Martin* (Oxford, 1975), p. 13.
3 W. Garret, *An Account of the Great Floods in the Rivers Tyne, Tees, Wear, Eden etc in 1771 and 1815* (Newcastle, 1818), pp. 5, 9, 13, 14, 41 and 43.
4 'The Great Moray Floods', in *Blackwood's Edinburgh Magazine* XXVIII (1830), p. 151 – a review of T. D. Lauder, *An Account of the Great Floods of August 1829 in the Province of Moray and Adjoining Districts* (Edinburgh, 1830); see also the commentary in J. G. Paradis, 'The Natural Historian as Antiquary of the World; Hugh Miller and the Rise of Natural History', in M. Shortland (ed.), *Hugh Miller and the Controversies of Victorian Science* (Oxford, 1996), pp. 132–3.
5 *Ibid.*, p. 162.
6 See the commentary in R. Porter, *Enlightenment: Britain and the Creation of the Modern World* (London, 2000), p. 226 – Dick's work was re-issued in a third edition in 1873.
7 Garret, *op. cit.*, p. 9.
8 *Gentleman's Magazine* n.s. IX (1838), p. 201.
9 F. Burkhardt and S. Smith (eds.), *The Correspondence of Charles Darwin* (hereafter Corr.), 1 (Cambridge, 1985), Catherine Darwin to Charles Darwin, 27–30/1/1834 (p. 365).
10 Lauder, *op. cit.*, pp. 405–6.
11 G. Eliot, *The Mill on the Floss* (London, 1860; Heron edn., London, 1969), p. 481; an interesting analysis of the way Eliot's novel may be construed as embracing geological debates of the day is contained in J. Smith, *Fact and Feeling: Baconian Science and the Nineteenth-Century Literary Imagination* (Madison, 1994), pp. 121–51.
12 See opening epigraph.
13 Porter, *op. cit.*, p. 297.
14 *Ibid.*
15 *Blackwood's Edinburgh Magazine* XIII (1823), p. 63.
16 *Ibid.* p. 77 – the article also formed a review of Thomas Moore's poem 'The Loves of the Angels' (1823).
17 See N. Cohn, *Noah's Flood: The Genesis Story in Western Thought* (New Haven and London, 1996).
18 W. Bassett, *Molech; or, The Approach of the Deluge; A Sacred Drama* (London, 1826) – see preface.
19 J. E. Reade, *The Deluge; A Drama in Twelve Scenes* (London, 1839), p. 79
20 *Ibid.*
21 J. McHenry, *The Antediluvians; or, The World Destroyed* (London, 1839)
22 *Blackwood's Edinburgh Magazine* XLVI (1839), p. 126
23 Porter, *op. cit.*, p. 85
24 A. Desmond and J. Moore, *Darwin* (Penguin edn., Harmondsworth, 1982), p. 129
25 S. Gessner, *The Death of Abel* (London, 1818), p. viii.
26 Porter, *op. cit.*, pp. 226–7.
27 See L. R. Matteson, 'John Martin's "Deluge": A Study in Romantic Catastrophe', *Pantheon* XXXIX (1981), p. 220.
28 T. Balston, *John Martin, 1789–1854: His Life and Works* (London, 1947), p. 16.

29 Porter, *op. cit.*, p. 227.

30 Balston, *op. cit.*, p. 96.

31 *Ibid.*, p. 98.

32 *Ibid.*, p. 99.

33 French royalty awarded prizes to them – *ibid.*, p. 90.

34 *Ibid.*, p. 88.

35 M. D. Paley, *The Apocalyptic Sublime* (New Haven and London, 1986) pp. 7–8.

36 *Ibid.*, p. 8.

37 Cited in R. Verdi, 'Poussin's "Deluge"; the Aftermath', *Burlington Magazine* CXXIII (1981), p. 393.

38 *Ibid.*

39 H. Milton, *Letters on the Fine Arts, Written from Paris in the Year 1815* (London, 1816), p. 102.

40 *Ibid.*

41 *Ibid.*

42 See Verdi, *op. cit.*, p. 393.

43 *Ibid.*, p. 397.

44 *Ibid.*

45 *Ibid.*, p. 398.

46 *Ibid.*, p. 394.

47 *Ibid.*, p. 397.

48 Paley, *op. cit.*, pp. 106–7; Verdi, *op. cit.*, p. 397.

49 Paley, *op. cit.*, p. 106.

50 Book XI, lines 734–41.

51 Paley, *op. cit.*, p. 107.

52 *Ibid.*, p. 138.

53 Feaver, *op. cit.*, pp. 26–7.

54 M. J. Campbell, *John Martin: Visionary Printmaker* (York, 1992), p. 2.

55 *John Martin: Master of Mezzotint, 1789–1854*, exh. cat., Alexander Postan Fine Art (London, 1974).

56 Balston, *op. cit.*, pp. 88–9.

57 *Ibid.*, p. 89.

58 *Edinburgh Review* XLIX (1829), p. 469.

59 Sir Wyke Bayliss, quoted in Balston, *op. cit.*, p. 89.

60 Campbell, *op. cit.*, p. 96.

61 Balston, *op. cit.*, pp. 103–4.

62 Campbell, *op. cit.*, p. 97.

63 *Ibid.*, p. 96.

64 The theme is from Thomas Grey's poem of the same name.

65 See *Tomlinson's Comprehensive Guide to the County of Northumberland* (London, n.d., but *c*. 1918), p. 157.

66 *Ibid.*, p. 156.

67 *Ibid.*

68 Balston, *op. cit.*, p. 105.

69 *Ibid.*

70 *Ibid.*, p. 103.

71 *Ibid.*

72 Campbell, *op. cit.*, p. 97.

73 *Ibid.*, p. 96.

74 See Paley, *op. cit.*, p. 140; Campbell, *op. cit.*, p. 96.

75 Quoted in Paley, *op. cit.*, p. 140.

76 Matteson, *op. cit.*, p. 223.

77 *Ibid.*

78 W. Buckland, *Reliquiae Diluvianae, or Observations on the Organic Remains Contained in Caves, Fissures, and Diluvial Gravel and other Geological Phenomena, Attesting to the Action of a Universal Deluge* (London, 1823); see also later in this chapter.

79 Matteson, *op. cit.*, p. 223.

80 M. L. Pendered, *John Martin, Painter: His Life and Times* (London, 1923), p. 136.

81 Paley, *op. cit.*, p.143.

82 *John Martin*, Exh. cat., Fine Arts Museum of San Francisco (San Francisco, 1997); C. Johnstone, *John Martin* (London, 1974), p. 75; Paley, *op. cit.*, p. 143.

83 Johnstone, *op. cit.*, p. 75.

84 'Royal Academy Exhibition', *Blackwood's Edinburgh Magazine* XLVIII (1840), p. 383.

85 Matteson, *op. cit.*, p. 227.

86 'Royal Academy Exhibition', *op. cit.*

87 Paley, *op. cit.*, p. 113.

88 *Ibid.*, p. 116.

89 *Ibid.*, p. 113.

90 *Ibid.*, p. 116.

91 See chapter two; also G. E. Finley, 'The Deluge Pictures: Reflections on Goethe, J. M. W. Turner and Early Nineteenth-Century Science', *Zeitschrift fur Kunstgeschichte* LX (1997), p. 543.

92 J. Gage, *J. M. W. Turner: 'A Wonderful Range of Mind'* (New Haven and London, 1987), p. 222.

93 Paley, *op. cit.*, p. 116.

94 *Blackwood's Edinburgh Magazine* LIV (1843), pp. 192–3.

95 F. Greenacre, *Francis Danby, 1793–1861* (London, 1988), p. 114.

96 See Paley, *op. cit.*, pp. 177–8; Greenacre,

op. cit., pp. 26–8; E. Adams, *Francis Danby: Varieties of Poetic Landscape* (New Haven and London, 1973), pp. 174–5.

97   Greenacre, *op. cit.*

98   'Royal Academy Exhibition', *op. cit.*, p. 386.

99   *Ibid.*, p. 114.

100   *Ibid.*

101   Adams, *op. cit.*, p. 103.

102   *Ibid.*, p. 103.

103   'A Pictorial Rhapsody', *Fraser's Magazine* XXII (1840), p. 122.

104   *Blackwood's Edinburgh Magazine* XXX (1836), p. 83.

105   *Ibid.*, XXIV (1828), pp. 36–7.

106   J. Barker, *The Brontës* (London, 1994), p. 197.

107   See *Punch* XXIX (1855), p. 130.

108   S. Butler, *The Way of All Flesh* (London, 1903; Oxford, 1993), p. 43.

109   A. Catcott, *A Treatise on the Deluge* (2nd edn., 1768), pp. 166ff.

110   *Ibid.*, p. 168.

111   *Ibid.*, p. 166.

112   *Ibid.*, p. 256.

113   *Ibid.*, pp. 263–4.

114   See the discussion in G. L. Davies, *The Earth in Decay: A History of British Geomorphology, 1578-1878* (London, 1969), pp. 108–110.

115   See chapter two.

116   Cohn, *op. cit.*, p. 104.

117   *Ibid.*, p. 104.

118   C. C. Gillispie, *Genesis and Geology: A Study in the Relations of Scientific Thought, Natural Theology, and Social Opinion in Great Britain, 1790-1850* (Cambridge, Mass., 1951) p. 55.

119   Cohn, *op. cit.*, p. 105; also R. Kirwan, *Geological Essays* (London, 1799).

120   Cohn, *op. cit.*, p. 106; also Davies, *op. cit.*, p. 144.

121   Cohn, *op. cit.*, p. 107.

122   Davies, *op. cit.*, p. 142.

123   See Gillispie, *op. cit.*, p. 59; Davies, *op. cit.*, p. 136.

124   See Gillispie, *op. cit.*, pp. 56ff – they appeared as a series of letters in the *Monthly Review*.

125   J. A. de Luc, *Letters on the Physical History of the Earth* (London, 1831).

126   Cohn, *op. cit.*, p. 111.

127   G. Cuvier, *Essay on the Theory of the Earth* (Edinburgh, 1813), pp. 171–2.

128   *Ibid.*, p. 176.

129   Cuvier, *op. cit.*, p. vi.

130   Matteson, *op. cit.*, p. 222.

131   Cohn, *op. cit.*, p. 113.

132   See prefatory comment to 'Cain; A Mystery' (1821) – see J. J. McGann and B. Weller, *Lord Byron: The Complete Poetical Works*, VI (Oxford, 1991), p. 229.

133   P. Wilton, *Geology and other Poems* (London, 1818), p. 22.

134   See D. R. Dean, *James Hutton and the History of Geology* (Ithaca, 1992), pp. 184ff.

135   W. Buckland, *Vindiciae Geologicae; or the Connexion of Geology with Religion Explained* (Oxford, 1820), p. 24.

136   *Ibid.*

137   *Ibid.*, pp. 23–4.

138   See S. J. Knell, *The Culture of English Geology, 1815-1851: A Science Revealed Through its Collecting* (Aldershot, 2000), pp. 171ff; see also C. McGowan, *The Dragon Seekers: The Discovery of Dinosaurs during the Prelude to Darwin* (London, 2002), pp. 57ff.

139   See *Gentleman's Magazine* XCII (1822), pp. 160, 352–3.

140   Buckland, *Reliquiae Diluvianae*, *op. cit.*

141   *Gentleman's Magazine* n.s. X (1838), p. 604; see also V. A. Eyles, 'Scientific Activity in the Bristol Region in the Past', in C. M. MacInnes and W. F. Whittard (eds.), *Bristol and its Adjoining Counties* (Bristol, 1955), pp. 132–3.

142   W. Buckland, *Reliquiae Diluvianae*, *op. cit.*, p. 226.

143   *Ibid.*

144   *Ibid.*, p. 227.

145   *Ibid.*

146   *Ibid.*, especially plate 25.

147   See Gillispie, *op. cit.*, p. 108.

148   *Gentleman's Magazine* XCIII (1823), pp. 440–41, 528–30.

149   Buckland, *Vindiciae Geologicae*, *op. cit.*, p, 14.

150   *Ibid.*, p. 19.

151   See A. Sedgwick, 'On diluvial formations', in *Annals of Philosophy* n.s. X (1825), pp. 18–37.

152   *Gentleman's Magazine* CI (1831), p. 20.

153   *Ibid.*, CII (1832), p. 538.

154   See, for example, *ibid.*, XCIV (1824), pp. 32–4, 548; XCV (1825), pp. 391–2, 628.

155   C. Lyell, *Principles of Geology*, III (London, 1830–33), p. 274.

156   See Cohn, *op. cit.*, pp. 118–19.

157   W. Buckland, *Geology and Mineralogy Considered with Reference to Natural Theology*, 2 vols. (London, 1836).

158   N. A. Rupke, *The Great Chain of History: William Buckland and the English School of Geology* (Oxford, 1983), p. 19; see also J. R. Topham, 'Beyond the "Common Context": The Production and Reading of the Bridgewater Treatises', *Isis* LXXXIX (1998), pp. 233–62.

159   *Gentleman's Magazine* n.s. VII (1837), p. 115.

160   *Ibid.*, p. 126.

161   See Gage, *op. cit.*, pp. 218ff; also J. Hamilton, *Turner and the Scientists* (London, 1998), pp. 115ff.

162   Hamilton, *op. cit.*, p. 118; Gage, *op. cit.*, p. 220.

163   See Hamilton, *op. cit.*, p. 118.

164   See Finley, *op. cit.*, pp. 547–8.

165   Gage, *op. cit.*, p. 222.

166   *Ibid.*, p. 220.

167   *Ibid.*

168   Hamilton, *op. cit.*, p. 125.

169   See chapter four.

170   See Balston, *op. cit.* p. 90; Pendered, *op. cit.*, pp. 132–3.

171   Balston, *op. cit.*, p. 90.

172   See Matteson, *op. cit.*, p. 223.

173   Paley, *op. cit.*, p. 139.

174   *Ibid.*, pp. 142–3.

175   Balston, *op. cit.*, p. 91.

176   See Paley, *op. cit.*, p. 142; Todd, *op. cit.*, p. 95.

177   Todd, *op. cit.*, p. 117.

178   *Ibid.*, p. 104.

179   Gillispie, *op. cit.*, p. 152.

180   See *Gentleman's Magazine* n.s. XII (1839), p. 162.

181   *Ibid.*, CI (1831), p. 22.

182   *Ibid.*, n.s. I (1834), p. 77.

183   G. Fairholme, *A General View of the Geology of Scripture . . . etc* (London, 1833), p. 14.

184   *Ibid.*, p. 229.

185   *Ibid.*, p. 431.

186   *Ibid.*, p. 23.

187   *Gentleman's Magazine* n.s. VIII (1837), p. 377.

188   G. Penn, *A Comparative Estimate of the Mineral and Mosaical Geologies* (London, 1822).

189   Idem., *Conversations on Geology . . . etc* (London, 1828); see also Cohn, *op. cit.*; also see chapter seven.

190   Cohn, *op. cit.*, p. 123.

191   See Todd, *op. cit.*, pp. 106–7.

192   *Gentleman's Magazine* CI (1831), p. 345.

193   *Ibid.*, n.s. XXII (1844), p. 521.

194   See chapter three.

195   H. Miller, *The Testimony of the Rocks* (Edinburgh, 1861), p. 317.

196   *Ibid.*

197   *Ibid.*, p. 318.

198   J. Duns, *Biblical Natural Science; being the Explanation of all References in Holy Scripture to Geology, Botany, Zoology, and Physical Geography* (Glasgow, n.d.).

199   D. Page, *Geology and Modern Thought* (Edinburgh, 1866), p. 5.

200   Cohn, *op. cit.*, p. 124.

201   *Gentleman's Magazine* n.s. XIII (1840), p. 392.

202   Cohn, *op. cit.*, p. 124.

203   See D. Linnell, *The Life of John Linnell* (Lewes, 1994).

204   M. J. S. Rudwick, *Scenes from Deep Time: Early Pictorial Representations of the Prehistoric World* (Chicago, 1992), p. 212.

205   *Ibid.*, p. 211.

206   *Ibid.*, p. 212.

207   L. Figuier, *The World Before the Deluge* (London, 1866 edn.), p. 1.

208   *Ibid.*, p. 7.

Chapter Six

1   C. Darwin, *The Origin of Species by Means of Natural Selection* (London, 1859;

Penguin edn., Harmondsworth, 1985), p. 65.

2   See chapter two.

3   See H. Spencer, *An Autobiography*, 2 vols. (London, 1904).

4   J. W. Burrows, 'Introduction' in Darwin, *Origin*, *op. cit.*, p. 40.

5   F. Engels, *The Condition of the Working Class in England* (original German edn., Leipzig, 1845; first English edn., London, 1892; Granada edn., London, 1969), p. 108.

6   In 1851, the joint annual income of Charles and his wife had passed £3,000. In all, they had over £80,000 worth of investments at that time. See A. Desmond and J. Moore, *Darwin* (London, 1991; Penguin edn., Harmondsworth, 1992), p. 396.

7   F. Burkhardt and S. Smith (eds.), *The Correspondence of Charles Darwin* (hereafter *Corr.*), V (Cambridge, 1985–) – Charles Darwin to Josiah Wedgwood III, 21/10/1852 (p. 99).

8   *Ibid.*

9   *Ibid.*, Charles Darwin to John Higgins, 9/4/1854 (p. 191).

10 *Ibid.*, Charles Darwin to John Higgins, 19/6/1852 (p. 94).

11 *Ibid.*, IV, Charles Darwin to J. S. Henslow, 25/7/1845 (p. 228).

12 *Ibid.*, V, Charles Darwin to John Higgins, 19/6/1852 (p. 94).

13 A. Desmond and J. Moore, *op. cit.*, p. 327.

14 *Corr.*, *op. cit.*, II, Charles Darwin to Emma Wedgwood, 27/11/1838 (pp. 128–9).

15 F. Engels, *op. cit.*, p. 57.

16 *Ibid.*

17 *Corr.*, II, Charles Darwin to Leonard Jenyns, 10/4/1837 (p. 16); also Charles Darwin to W. D. Fox, 28/8/1837 (p. 39).

18 Engels, *op. cit.*, p. 60.

19 *Ibid.*, pp. 60–61.

20 *Ibid.*

21 See D. Newsome, *The Victorian World Picture* (London, 1998; Fontana edn., London, 1999), pp. 76ff.

22 See B. Hilton, *The Age of Atonement: The Influence of Evangelicalism on Social and Economic Thought, 1785–1865* (Oxford, 1991), pp. 147–8.

23 *Ibid.*, p. 113.

24 E. Gaskell, *Mary Barton: A Tale of Manchester Life* (London, 1848; Penguin edn., Harmondsworth, 1996), p. 60.

25 T. R. Malthus, *An Essay on the Principle of Population* (London, 1798).

26 This was the controversial Poor Law Amendment Act of 1834.

27 N. Barlow (ed.), *The Autobiography of Charles Darwin, 1809–1882* (London, 1958), p. 120.

28 *Ibid.*, p. 119.

29 *Ibid.*, p. 120.

30 *Ibid.*, p. 87.

31 A. Desmond and J. Moore, *op. cit.*, p. 333.

32 *Corr.*, V, Charles Darwin to G. H. Turnbull, 28/10/1854 (p. 223).

33 *Illustrated London News* 1 (May–Dec., 1842), p. 225.

34 W. C. Taylor, 'Notes of a Tour in the Manufacturing Districts … etc.', *Westminster Review* XXXVIII (1842), p. 391.

35 *Illustrated London News* 1 (May–Dec., 1842), p. 242.

36 W. L. Burn, *The Age of Equipoise: a study of the mid-Victorian Generation* (London, 1964), p. 66.

37 *Ibid.*

38 W. E. Houghton, *The Victorian Frame of Mind, 1830–1870* (New Haven and London, 1957), p. 55.

39 *Corr.*, I, Charles Darwin to Susan Darwin, 9/9/1831 (p. 147).

40 *Ibid.*, W. D. Fox to Charles Darwin, 30/6/1832 (p. 245).

41 *Ibid.*

42 T. Carlyle, *The French Revolution: A History*, 3 vols. (London, 1837; reprinted London, 1888).

43 See chapter two.

44 Carlyle, *op. cit.*, p. 172.

45 *Ibid.*, p. 188.

46 *Corr.*, I, Charles Darwin to Caroline Darwin, 24/11/1832 (pp. 276–7).

47 *Ibid.*, Charles Darwin to Catherine Darwin, 22/5 to 28/9/1832 (p. 312).

48 *Ibid.*, W. D. Fox to Charles Darwin, 29/9/1832 (p. 267).

49 Gaskell, *op. cit.*, p. 61.

50 *Ibid.*

51 *Corr.*, I, J. M. Herbert to Charles Darwin, 1–4/12/1832 (p. 288).

52 *Illustrated London News* I (May–Dec., 1842), p. 497.

53 *The Reasoner*, 3 June 1846.

54 *Ibid.*, 30 September 1846.

55 See *Illustrated London News* II (Jan.–June, 1843), p. 379.

56 B. Colloms, *Victorian Country Parsons* (London, 1977), p. 19.

57 *Ibid.*

58 *Ibid.*, p. 20.

59 Quoted in W. E. Houghton, *op. cit.*, p. 72.

60 See A. Desmond, *Huxley: From Devil's Disciple to Evolution's High Priest* (Penguin edn., Harmondsworth, 1997), p. 191.

61 Barlow, *op. cit.*, p. 85.

62 *Ibid.*

63 *Ibid.*, p. 87.

64 *Corr.*, VIII, Charles Darwin to Asa Gray, 22/5/1860 (p. 224).

65 See F. M. Turner, *Contesting Cultural Authority: Essays in Victorian Intellectual Life* (Cambridge, 1993), p. 44.

66 Desmond and Moore, *op. cit.*, pp. 8–9.

67 *Ibid.*, p. 9.

68 *Corr.*, VI, Charles Darwin to T. V. Wollaston, 6/6/1856 (p. 134).

69 *Ibid.*, Charles Darwin to J. D. Hooker, 13/7/1856 (p. 178).

70 J. Moore, 'Socializing Darwinism: Historiography and the Fortunes of a Phrase', in L. Levidow (ed.), *Science in Politics* (London, 1986), p. 39.

71 H. Ward, *Robert Elsmere* (London, 1888), p. 259.

72 *Corr.*, I, E. A. Darwin to Charles Darwin, 10/1/1825 (p. 11).

73 For further illustration of fossils as commodities, see S. J. Knell, *The Culture of English Geology, 1815–1851: A Science Revealed Through its Collecting* (Aldershot, 2000), pp. 122ff.

74 K. Marx, *Capital*, II (first English edn., London, 1887; Penguin edn., Harmondsworth, 1978), p. 228.

75 See Burn, *op. cit.*, p. 59.

76 *Ibid.*, p. 60.

77 See Moore, *op. cit.*, p. 46.

78 *Corr.*, VI, Charles Darwin to Charles Lyell, 13/4/1857 (p. 377).

79 G. Beer, *Darwin's Plots: Evolutionary Narrative in Darwin, George Eliot and Nineteenth-Century Fiction* (London, 1983), p. 3.

80 See K. Flint, *The Victorians and the Visual Imagination* (Cambridge, 2000), p. 117.

81 *Ibid.*, p. 8.

82 Darwin, *Ibid.*, p. 65.

83 Barlow, *op. cit.*, p. 52.

84 *Ibid.*, pp. 69–70; this field excursion, as well as Darwin's geological training more generally, is examined in J. A. Secord, 'The Discovery of a Vocation: Darwin's Early Geology', *British Journal for the History of Science* XXIV (1991), pp. 133–57.

85 *Ibid.*, p. 77.

86 See J. Smith, *Fact and Feeling: Baconian Science and the Nineteenth-Century Literary Imagination* (Madison, 1994), p. 105.

87 C. Darwin, *The Voyage of the Beagle* (London, 1845; Wordsworth edn., London, 1997), p. 300.

88 *Ibid.*, p. 301.

89 *Ibid.*

90 *Ibid.*, p. 316.

91 *Ibid.*, some of these ideas are explored more fully in F. H. T. Rhodes, 'Darwin's Search for a Theory of the Earth: Symmetry, Simplicity and Speculation', *British Journal for the History of Science* XXIV (1991), pp. 193–229.

92 *Corr.*, I, Charles Lyell to Charles Darwin, 26/12/1836 (p. 532).

93 See the extended analysis in S. Herbert, 'Charles Darwin as a Prospective Geological Author', *British Journal for the History of Science* XXIV (1991), pp. 159–92.

94 See Rhodes, *op. cit.*, pp. 196–7.

95 *Corr.*, II, Charles Darwin to W. D. Fox, 7/7/1837 (p. 29).

96 *Ibid.*, pp. 29–30; see also p. 21.

97 *Ibid.*, Charles Darwin to W. D. Fox, 15/6/1838 (p. 91).

98 Desmond and Moore, *op. cit.*, p. 91.

99 *Corr.*, II, Charles Darwin to Charles Lyell, 9/8/1838 (p. 96).

100 Ibid., iv, Robert Chambers to Charles Darwin, 5/10/1847 (pp. 82–4); also Charles Darwin to Robert Chambers, 6 1848 (p. 148).

101 Barlow, op. cit., p. 98.

102 Ibid.

103 Ibid., p. 98; also N. Barlow (ed.), Charles Darwin and the Voyage of the Beagle (London, 1945), pp. 243–4.

104 Corr., i, Charles Darwin to J. S. Henslow, 11/7/1831 (p. 125).

105 Darwin, Voyage, op. cit., p. 477.

106 Ibid., p. 479.

107 For extended discussion, see Desmond and Moore, op. cit., pp. 204ff.

108 Corr., i, Charles Darwin to Caroline Darwin, 9/11/1836 (p. 518).

109 On Owen, see N. A. Rupke, Richard Owen: Victorian Naturalist (New Haven and London, 1994).

110 Cross-ref chapter 2.

111 Corr., ii, Charles Darwin to J. S. Henslow, November 1839 (p. 238).

112 Ibid., Charles Darwin to G. R. Waterhouse, 26/7/1843 (p. 375).

113 Ibid., vi, Charles Darwin to W. D. Fox, 8/2/1857 (p. 335).

114 Ibid., v, Charles Darwin to W. D. Fox, 7/5/1855 (p. 326).

115 Ibid., Charles Darwin to J. D. Hooker, 15/5/1855 (p. 330).

116 Ibid.

117 Ibid., vi, Charles Darwin to J. D. Hooker, 3/6/1857 (p. 407).

118 Ibid., Charles Darwin to John Lubbock, 14/7/1857 (p. 430).

119 Ibid., Charles Darwin to J. D. Hooker, 2/5/1857 (p. 389).

120 Ibid.

121 Darwin, Origin, op. cit., p. 72.

122 Ibid., p. 74.

123 Ibid., p. 75.

124 Ibid., p. 111.

125 Ibid.

126 Ibid.

127 Ibid., p. 87.

128 Ibid., p. 91.

129 Ibid., p. 89.

130 Ibid., p. 96.

131 See the useful discussion in P. J. Bowler, Charles Darwin: The Man and his Influence (Oxford, 1990), pp. 82ff.

132 Darwin, Origin, op. cit., p. 115.

133 Ibid., p. 117.

134 Ibid., p. 121.

135 Ibid., p. 123.

136 Ibid., p. 125; humblebees are bumblebees in contemporary parlance.

137 Ibid., p. 126.

138 Ibid., 128.

139 Ibid., p. 129.

140 Ibid., pp. 130–31.

141 Ibid., p. 131.

142 Ibid., p. 133.

143 Ibid.

144 Ibid., p. 148.

145 Ibid., p. 151.

146 Darwin, Voyage, op. cit., pp. 374–5.

147 Darwin, Origin, op. cit., pp. 129–30.

148 Ibid., p. 130.

149 See chapter 2, pp. 81.

150 Darwin, Origin, op. cit., pp. 142–3.

151 A. N. Wilson, The Victorians (London, 2002), p. 231.

152 J. A. Secord, Victorian Sensation: The Extraordinary Publication, Reception, and Secret Authorship of Vestiges of the Natural History of Creation (Chicago, 2000), p. 105.

153 Ibid., p. 101.

154 Ibid.

155 Ibid.

156 See Beer, op. cit., p. 114; also G. Levine, Darwin and the Novelists: Patterns of Science in Victorian Fiction (Cambridge, Mass., 1988), p. 117.

157 Ward, op. cit., pp. 170–71.

158 Beer, op. cit., p. 80.

159 Darwin, Origin, op. cit., p. 129.

160 Ibid., pp. 138–42.

161 Ibid., p. 219.

162 Ibid.

163 Levine, op. cit., p. 99.

164 Ibid., p. 99.

165 Darwin, Origin, op. cit., p. 457; on John Phillips, see the discussion in Knell, op. cit., pp. 240ff.

166 Corr., ii, Charles Darwin to G. R. Waterhouse, 26/7/1843 (p. 376).

167 Darwin, The Origin, op. cit., p. 458.

168 *Ibid.*, p. 459.
169 *Ibid.*
170 *Ibid.*, p. 277.
171 *Ibid.*, p. 268.
172 *Ibid.*
173 See Beer, *op. cit.*, p. 114; also Levine, *op. cit.*, p. 95.
174 Beer, *op. cit.*, p. 47.
175 Levine, *op. cit.*, p. 111.
176 *Ibid.*, p. 113.
177 Beer, *op. cit.*, p. 127.
178 Darwin, *Origin*, *op. cit.*, p. 459.
179 *Ibid.*, p. 458 – 'Light will be thrown on the origin of man and his history'.
180 Ward, *op. cit.*, p. 278.
181 C. Kingsley, *The Water Babies* (London, 1863; Penguin edn. Harmondsworth, 1995), pp. 68–9.
182 See Wilson, *op. cit.*, p. 295.
183 Beer, *op. cit.*, p. 123.
184 Wilson, *op. cit.*, p. 299.
185 cross-ref. Ch 4.
186 Kingsley, *op. cit.*, p. 205.
187 *Ibid.*, p. 270.
188 Beer, *op. cit.*, pp. 137–8.
189 Kingsley, *op. cit.*, pp. 68–9.
190 For extended commentaries, see Rupke, *op. cit.*; also Desmond and Moore, *op. cit.*
191 See T. H. Huxley, *Man's Place in Nature and other anthropological essays* (London, 1894), p. 144. The first three essays of this volume were originally published as 'Man's Place in Nature' in 1863. They began life as a series of lectures for working men, the first of which were given in 1860.
192 Kingsley had a 'friendly' correspondence with Darwin – see Wilson, *op. cit.*, p. 299.
193 *Corr.*, XI, Charles Darwin to J. D. Hooker, 17/3/1863 (p. 239).
194 *Ibid.*, T. H. Huxley to Charles Darwin, 2/7/1863 (p. 517).
195 *Ibid.*, VIII, Charles Darwin to Charles Lyell, 10/4/1860 (p. 154).
196 'Science', *Westminster Review* XVII (1860), p. 569.
197 *Corr.*, VIII, Charles Darwin to J. D. Hooker, 2/7/1860 (p. 272).
198 *Ibid.*, VII, Adam Sedgwick to Charles Darwin, 24/11/1859 (p. 396).

199 *Ibid.*
200 *Fraser's Magazine* LXII (1860), p. 84.
201 *Corr.*, VIII, Charles Darwin to Asa Gray, 18/2/1860 (p. 91).
202 *Ibid.*, Charles Darwin to A. R. Wallace, 18/5/1860 (p. 220).
203 *Ibid.*, VII, Charles Darwin to T. H. Huxley, 24/11/1859 (p. 393).
204 *Ibid.*, VIII, note 2 (p. 246).
205 *Ibid.*, Charles Darwin to T. H. Huxley, 22/11/1860 (p. 487).
206 *Chambers's Journal* XII (July–Dec., 1859), p. 351.
207 *Ibid.*, pp. 388ff.
208 *Ibid.*, p. 390.
209 *Ibid.*
210 *Ibid.*
211 *Ibid.*, p. 431.
212 *The Reasoner* XXIV (1861), p. 110.
213 This is clear from an advertisement within the book – see Huxley, *op. cit.*
214 Desmond, *op. cit.*, p. 292.
215 Huxley, *op. cit.*, p. 152.
216 *Ibid.*
217 *Ibid.*
218 *Corr.*, XI, Charles Darwin to T. H. Huxley, 18/2/1863 (p. 148).
219 *Ibid.*, J. D. Hooker to Charles Darwin, 26/2/1863 (p. 179).
220 C. Lyell, *The Geological Evidences of the Antiquity of Man, with Remarks on the Theory of the Origin of Species by Variation* (London, 1863).
221 Desmond, *op. cit.*, p. 315.
222 *Corr.*, XI, note 1 (p. 218).
223 *Ibid.*, Charles Darwin to J. D. Hooker, 24–5/2/1863 (p. 172).
224 *Ibid.*, Charles Darwin to J. D. Hooker, 17/3/1863 (p. 239).
225 *Ibid.*, note 4 (p. 219).
226 *Ibid.*, Charles Darwin to Asa Gray, 23/2/1863 (p. 166).
227 Lyell, *op. cit.*, 3rd edn. (1863), p. 1.
228 *Corr.*, XI, note 19 (p. 229).
229 Lyell, *op. cit.*, 3rd edn. (1863), p. 498.
230 *Corr.*, XI, note 6 (p. 232).
231 *Ibid.*, Charles Darwin to J. D. Hooker, 24/2/1863 (p. 174).
232 *Ibid.*

233 *Ibid.*, J. D. Hooker to Charles Darwin, 26/2/1863 (p. 179).
234 *Punch* XL (Jan.–June, 1861), p. 206.
235 *Ibid.*, p. 213.
236 *Corr.*, VIII, J. D. Hooker to Charles Darwin, 2/7/1860 (p. 270).
237 *Ibid.*
238 The episode is fully treated in Desmond, *op. cit.*, pp. 277–81.
239 *Ibid.*, p. 307; see also *Corr.*, VIII, appendix VI (p. 592).
240 *Punch* XLII (July–Dec., 1862), p. 164.
241 There is an extended commentary in B. M. G. Reardon, *Religious Thought in the Victorian Age: A Survey from Coleridge to Gore* (2nd edn., London, 1995), pp. 237ff.
242 C. W. Goodwin, 'On the Mosaic Cosmogony', in F. Temple et al., *Essays and Reviews* (London, 1860), pp. 207–53.
243 *Ibid.*, p. 208.
244 *Ibid.*, p. 233.
245 B. Powell, 'On the Study of the Evidences of Christianity', in Temple, *op. cit.*, p. 142.
246 See Reardon, *op. cit.*, p. 241.
247 Powell, *op. cit.*, p. 139.
248 B. Jowett, 'On The Interpretation of Scripture, in Temple, *op. cit.*, p. 340.
249 *Ibid.*, p. 367.
250 Desmond, *op. cit.*, p. 278.
251 *Ibid.*, p. 298.
252 Reardon, *op. cit.*, p. 241.
253 See *The Reasoner* XXIV (1861), pp. 113–14.
254 *Ibid.*, p. 113.
255 *Ibid.*
256 *Ibid.*, p. 198.
257 Ward, *op. cit.*, p. 199.
258 *Ibid.*, p. 280.
259 *Ibid.*, p. 364.
260 *Ibid.*, p. 182.
261 *Ibid.*
262 *Ibid.*, p. 365.
263 S. Butler, *The Way of All Flesh* (London, 1903; Oxford, 1993), p. 411ff.
264 *Ibid.*, p. 414.
265 A significantly contrary perspective is expressed in B. Hilton, 'The Politics of Anatomy and an Anatomy of Politics, *c.*1825–1850', in S. Collini et al. (eds.), *History, Religion and Culture: British Intellectual History, 1750–1950* (Cambridge, 2000), p. 180 – Hilton claims that the most serious crisis of faith occurred in the 1840s and that Darwin's *Origin* was absorbed 'fairly painlessly'. However, this perspective rather flies in the face of much contemporary reportage.
266 *Corr.*, VIII, Leonard Jenyns to Charles Darwin, 4/1/1860 (p. 14).
267 *Ibid.*
268 Houghton, *op. cit.*, p. 50.
269 *Ibid.*, p. 59.
270 *Fraser's Magazine* n.s. IV (1871), p. 63.
271 J. T. Bonner and R. M. May, 'Introduction', in C. Darwin, *The Descent of Man, and Selection in Relation to Sex* (Princeton, N.J., 1981), p. xxiii.
272 Desmond, *op. cit.*, p. 399.
273 Darwin, *Descent, op. cit.*, p. 10.
274 *Ibid.*, p. 11.
275 *Ibid.*, p. 14.
276 *Ibid.*, pp. 20–21.
277 *Ibid.*, p. 26.
278 *Ibid.*, pp. 26–7.
279 *Ibid.*, p. 32.
280 *Ibid.*, p. 41.
281 *Ibid.*, p. 42.
282 *Ibid.*, p. 47.
283 *Ibid.*, p. 51.
284 *Ibid.*, p. 52.
285 *Ibid.*, pp. 74–5.
286 *Ibid.*, pp. 104–5.
287 *Ibid.*, pp. 108–9.
288 *Ibid.*, p. 154.
289 Desmond and Moore, *op. cit.*, p. 579.
290 *Ibid.*
291 *Ibid.*, p. 580.
292 *Ibid.*
293 *Ibid.*, p. 529.
294 *Westminster Review* XXXIX (1871), p. 551.
295 *Ibid.*
296 Desmond and Moore, *op. cit.*, pp. 579ff.
297 See the summary in *Westminster Review* XXXIX (1871), pp. 552ff.
298 See D. R. Oldroyd, *Darwinian Impacts: an introduction to the Darwinian Revolution* (2nd edn., Milton Keynes, 1983), pp. 250–51.
299 Desmond and Moore, *op. cit.*, p. 585.

300 *Ibid.*, p. 590.
301 *Westminster Review* xxxix (1871), p. 552.
302 J. A. V. Chapple, *Science and Literature in the Nineteenth Century* (London, 1986), p. 95.
303 See Beer, *op. cit.*, pp. 142ff – this was the theme of Richard Jefferies's *After London* (1885).
304 See Newsome, *op. cit.*, pp. 212–13.
305 S. Gill, 'Introduction' to C. Dickens, *Bleak House* (Oxford World's Classics edn., 1998), p. xiii.
306 *Ibid.*
307 Dickens, *Bleak House*, *op. cit.*, p. 14.
308 See chapter two.
309 Chapple, *op. cit.*, p. 87.
310 G. Eliot, *Middlemarch* (London, 1871–2; Penguin edn., Harmondsworth, 1994), pp. 25–6.
311 Beer, *op. cit.*, p. 238.
312 *Ibid.*
313 Chapple, *op. cit.*, p. 90.
314 T. Hardy, *A Pair of Blue Eyes* (London, 1872–3; Penguin Popular Classic edn., Harmondsworth, 1994), p. 241.
315 Beer, *op. cit.*, p. 248.
316 See T. Hardy, *The Return of the Native* (London, 1878; Macmillan edn., London, 1985), pp. 31–4.

Chapter Seven

1 W. H. Flower, *Essays on Museums and other Subjects Connected with Natural History* (London, 1898), p. 13.
2 'Science – its position and prospects', *Chambers's Journal* vi (July–Dec., 1856), p. 300.
3 See M. Foucault, 'Of other spaces', *Diacritics* xvi (1986), pp. 22–3.
4 See M. J. S. Rudwick, *Scenes from Deep Time: Early Pictorial Representations of the Prehistoric World* (Chicago, 1992), p. 1.
5 See p. 54.
6 See p. 211.
7 See T. Bennett, *The Birth of the Museum: History, Theory, Politics* (London, 1995), p. 178.
8 *Ibid.*
9 Useful discussion of this idea can be found in T. Mitchell, *Colonising Egypt* (Berkeley, 1988), pp. xiii–xiv, 10–13; the idea, in general, can be traced to Heidegger.
10 Sir T. Browne, quoted in H. W. Acland, *The Oxford Museum* (4th edn., Oxford, 1867), p. 13.
11 *Report of the Twenty-Eighth Meeting of the British Association for the Advancement of Science* (London, 1859), p. xcv.
12 *Ibid.*
13 N. A. Rupke, *Richard Owen: Victorian Naturalist* (New Haven and London, 1994), p. 81.
14 *Ibid.*, p. 83.
15 *Ibid.*, p. 84.
16 See p. 204.
17 *Report of the Twenty-Eighth Meeting of the British Association for the Advancement of Science*, *op. cit.*, p. xcv.
18 C. Yanni, *Nature's Museums: Victorian Science and the Architecture of Display* (London, 1999), p. 143.
19 see chapters three and four.
20 See p. 129.
21 *Illustrated London News* xxiv (1859), advertisement, 26 March.
22 See chapter one, under 'Artistic Visions'.
23 See chapter two.
24 See P. Thompson, *William Butterfield* (London, 1971), p. 145.
25 *Ibid.*
26 *Ibid.*, p. 35.
27 *Ibid.*, p. 145.
28 H. De la Beche, *Report on the Geology of Cornwall, Devon and Somerset* (London, 1834).
29 See 'Cornish Stone', *Household Words* x (1854), p. 96.
30 *Ibid.*
31 *Ibid.*
32 *Ibid.*, p. 95.
33 See p. 85.
34 H. Miller, *The Testimony of the Rocks, or Geology in its Bearing on the Two Theologies, Natural and Revealed* (Edinburgh, 1871), p. 216; see also the commentary in J. H. Brooke, 'Like Minds:

the God of Hugh Miller', in M. Shortland (ed.), *Hugh Miller and the Controversies of Victorian Science* (Oxford, 1996), pp. 179–80.

35   *Ibid.*, p. 219.

36   *Ibid.*, p. 220.

37   See C. Wainwright, 'The Early Victorian Interior', in S. M. Wright (ed.), *The Decorative Arts in the Victorian Period* (The Society of Antiquaries of London Occasional Paper, n.s. XII, 1989), pp. 10–11.

38   Quoted in T. Greenwood, *Museums and Art Galleries* (London, 1888), p. 5.

39   See Bodleian Library, Oxford, John Johnson Collection, Animals on Show 2, Menageries (3).

40   See the account in R. D. Altick, *The Shows of London* (Cambridge, Mass., 1978), chapter twenty-three.

41   Bodleian Library, Oxford, John Johnson Collection, London Play Places 5 (26, 27).

42   *Ibid.*, Entertainments 6.

43   A. Desmond and J. Moore, *Darwin* (London, 1991; Penguin edn., Harmondsworth, 1992), p. 65.

44   J. Barker, *The Brontës* (London, 1994; Phoenix edn., London, 1995), p. 640.

45   Desmond and Moore, *op. cit.*, p. 65.

46   *Ibid.*

47   *Illustrated London News* VI (Jan.–June, 1845), pp. 355-6.

48   *Ibid.*, p. 12.

49   *Punch* XV (July–Dec., 1848), p. 181.

50   Bodleian Library, Oxford, John Johnson Collection, London Play Places 9 (5, 8).

51   *Ibid.* London Play Places 3 (54, 55); see also Altick, *op. cit.*, pp. 28–32.

52   When Wombwell's Menagerie came to Oxford, there was 'Free admission to Protestants', according to one surviving handbill – Bodleian Library, Oxford, John Johnson Collection, Animals on Show 2, Menageries (2).

53   See Bennett, *op. cit.*, p. 6.

54   *Ibid.*

55   See note 35.

56   Quoted in Greenwood, *op. cit.*, p. 30.

57   See Yanni, *op. cit.*, p. 145.

58   See Altick, *op. cit.*, chapter eighteen.

59   See Greenwood, *op. cit.*, pp. 40–42.

60   *Ibid.*, p. 42.

61   *Ibid.*

62   cross-ref chapter four.

63   T. Greenwood, *op. cit.*, p. 42.

64   cross-ref. Chapter two.

65   See Yanni, *op. cit.*, p. 46.

66   *Ibid.*, pp. 47–8; see also Rupke, *op. cit.*, chapter one.

67   See Desmond and Moore, *op. cit.*, pp. 222–3.

68   See *Illustrated London News* VII (July–Dec, 1845), p. 210.

69   *Ibid.*

70   *Ibid.*

71   Yanni, *op. cit.*, p. 51; also S. J. Knell, *The Culture of English Geology, 1815–1851* (Aldershot, 2000), pp. 299–302.

72   See *Penny Magazine* XII (1843), pp. 319–20.

73   Yanni, *op. cit.*, p. 56.

74   A. Desmond, *Huxley: From Devil's Disciple to Evolution's High Priest* (London, 1997, Penguin edn., Harmondsworth, 1998), p. 199.

75   Desmond and Moore, *op. cit.*, p. 411.

76   *Chambers's Journal* VI (July–Dec., 1856), p. 9.

77   *Ibid.*, p. 12.

78   *Ibid.*

79   *Ibid.*, 'Science – its Position and Prospects', VI (July–Dec., 1856), p. 298.

80   *Ibid.*, pp. 297–300.

81   See A. V. Simcock, *The Ashmolean Museum and Oxford Science, 1683–1983* (Oxford, 1984), p. 14.

82   Yanni, *op. cit.*, p. 64.

83   See *Illustrated London News* XXVII (July–Dec., 1860), p. 320; also A. V. Simcock, *op. cit.*, p. 14.

84   See Yanni, *op. cit.*, p. 65.

85   See p. 181.

86   Yanni, *op. cit.*, p. 64.

87   Acland, *op. cit.*, p. 18.

88   See S. Muthesius, *The High Victorian Movement in Architecture, 1850–1870* (London, 1972), p. 165.

89   Acland, *op. cit.*, pp. 20–21.

90   Bennett, *op. cit.*, p. 70.

91   *Ibid.*, pp. 59ff.

92   Yanni, *op. cit.*, p. 112; also Rupke, *op. cit.*, chapter one.

93 Yanni, *op. cit.*, p. 112.
94 *Ibid.*, pp. 117ff.
95 See *The Builder* XXII (1864), p. 393.
96 *Ibid.*
97 Yanni, *op. cit.*, p. 123.
98 *Ibid.*, p. 124.
99 *Ibid.*, pp. 125ff.
100 *Ibid.*, pp. 130–31.
101 R. Dixon and S. Muthesius, *Victorian Architecture* (London, 1978), p. 174.
102 See Rupke, *op. cit.*, p. 13.
103 *Ibid.*
104 See 'Report of the Committee upon the Provincial Museums of the United Kingdom', *Report of the Fifty-Seventh Meeting of the British Association for the Advancement of Science, Manchester, 1887* (London, 1888), p. 97.
105 Yanni, *op. cit.*, p. 109.
106 See p. 19.
107 P. H. M. Cooper, *Fossils, Faith and Farming* (Sudbury, 1999), p. 91.
108 *Ibid.*, p. 94.
109 J. Phillips, writing about the new Oxford Museum in Acland, *op. cit.*, p. 36.
110 'A morning at the Museum of Practical Geology', *Chambers's Journal* VI (July–Dec, 1856), pp. 9–11.
111 *Ibid.*, p. 11.
112 See Bennett, *op. cit.*, p. 77.
113 Greenwood, *op. cit.*, pp. 180–81.
114 See Yanni, *op. cit.*, p. 23.
115 See F. O'Dwyer, *The Architecture of Deane & Woodward* (Cork, 1997), p. 220.
116 Acland, *op. cit.*, p. 21.
117 See *Illustrated London News* XXVII (July–Dec, 1860), p. 320.
118 *Ibid.*
119 Yanni, *op. cit.*, p. 81.
120 See Dixon and Muthesius, *op. cit.*, p. 160.
121 *Illustrated London News* XXVII (July–Dec., 1860), p. 320.
122 *Ibid.*
123 Yanni, *op. cit.*, p. 85.
124 *Ibid.*, p. 86, quoting Acland.
125 See O'Dwyer, *op. cit.*, p. 220.
126 Muthesius, *op. cit.*, p. 165.
127 Acland, *op. cit.*, p. 29.
128 See Thompson, *op. cit.*, p. 163.
129 *Ibid.*, p. 35.
130 O'Dwyer, *op. cit.*, p. 242.

131 *Ibid.*, p. 220.
132 See M. Girouard, *Alfred Waterhouse and the Natural History Museum* (New Haven and London, 1981), p. 36.
133 *Ibid.*, p. 56.
134 See Yanni, *op. cit.*, p. 132.
135 R. Owen, *On the Extent and Aims of a National Museum of Natural History* (2nd edn., London, 1862), p. 11.
136 C. Cunningham and P. Waterhouse, *Alfred Waterhouse, 1830–1905: Biography of a Practice* (Oxford, 1992), pp. 74–5.
137 Quoted in C. Yanni, *op. cit.*, pp. 141–2.
138 Cunningham and Waterhouse, *op. cit.*, p. 75.
139 See B. Haward, *Oxford University Museum: Its Architecture and Art* (Oxford, 1991), p. 16.
140 L. Carroll, *Alice's Adventures in Wonderland and Through the Looking Glass* (Children's Classic edn., London, 1948), p. 141.
141 *Ibid.*, p. 143.
142 See p. 252.
143 Girouard, *op. cit.*, pp. 36–8.
144 Quoted in Bennett, *op. cit.*, p. 101.
145 cross-ref earlier.
146 Cunningham and Waterhouse, *op. cit.*, p. 74 note 7.
147 'Report of the Committee upon the Provincial Museums of the United Kingdom', *op. cit.*, p. 114.
148 See, for example, L. Jenyns, *Natural History Museums* (Bath, 1865); H. Goodwin, *The Museum Question Briefly Considered* (London, 1869).
149 'Report of the Committee upon the Provincial Museums of the United Kingdom', *Report of the Fifty-Eighth Meeting of the British Association for the Advancement of Science, Bath, 1888* (London, 1889), p. 130.
150 *Ibid.*, p. 129.
151 *Ibid.*, p. 130.
152 P. H. Gosse, *The Romance of Natural History* (London, c. 1860), preface.
153 'Report of the Committee upon the Provincial Museums of the United Kingdom', *Report of the Fifty-Seventh Meeting of the BAAS*, p. 127.
154 *Ibid.*, p. 128.

155 See *London and Fashionable Resorts* (London, 1896), p. 35.
156 See Yanni, *op. cit.*, p. 108, figure 4.13.
157 See M. N. Cohen, *Lewis Carroll: A Biography* (London, 1995); also D. Thomas, *Lewis Carroll: A Portrait with Background* (London, 1996).
158 D. Hudson, *Lewis Carroll: An Illustrated Biography* (London, 1976), p. 129.
159 See M. Batey, *The Story of Alice: The Fascinating Background to Lewis Carroll's Famous Book* (London, 1991), pp. 34–5.
160 Hudson, *op. cit.*, p. 126.
161 Batey, *op. cit.*, p. 34.
162 *Ibid.*
163 See D. Rackin, *'Alice's Adventures in Wonderland' and 'Through the Looking Glass': Nonsense, Sense and Meaning* (New York, 1991), p. 35.
164 Thomas, *op. cit.*, p. 152.
165 Rackin, *op. cit.*, p. 36.
166 Thomas, *op. cit.*, p. 152.
167 *Ibid.*, pp. 4–6.
168 See M. Freeman, *Railways and the Victorian Imagination* (New Haven and London, 1999); S. Kern, *The Culture of Time and Space, 1880–1918* (Cambridge, Mass., 1983).
169 Carroll, *op. cit.*, p. 146.
170 Rackin, *op. cit.*, p. 37.
171 *Ibid.*, p. 54.
172 See pp. 54–5.
173 Rackin, *op. cit.*, p. 55.
174 *Ibid.*
175 Dodgson had in his library no less than nineteen volumes of works by Darwin and his critics – see Cohen, *op. cit.*, p. 350.
176 Rackin, *op. cit.*, pp. 51–2.
177 Thomas, *op. cit.*, p. 158.
178 *Ibid.*
179 Carroll, *op. cit.*, p. 59.
180 *Ibid.*, p. 57.
181 See Thomas, *op. cit.*, pp. 75–7; also Rupke, *The Great Chain of History: William Buckland and the English School of Geology, 1814–1849* (Oxford, 1983).
182 *Ibid.*, p. 77.
183 See pp. 181–3.
184 Carroll, *op. cit.*, p. 41.
185 *Ibid.*, p. 42.
186 *Ibid.*, p. 145.
187 See Rackin, *op. cit.*, p. 3.
188 See Batey, *op. cit.*, p. 30.
189 Rackin, *op. cit.*, p. 42.

Epilogue

1   T. Haines, *Walking with Dinosaurs: A Natural History* (London, 1999), p. 24.
2   *Ibid.*, p. 7.
3   *Ibid.*
4   See R. P. Turco, *Earth under Siege: From Air Pollution to Global Change* (Oxford, 2002), pp. 99–101.
5   Haines, *op. cit.*, p. 280.
6   *Ibid.*, p. 281.
7   *Ibid.*
8   *Ibid.*, p. 282.
9   See p. 173.
10   'Donati's Comet', *Chambers's Journal* XI (Jan.–June, 1859), p. 57; also *Illustrated London News* XXXIII (July–Dec., 1858), p. 309.
11   See M. Pointon, 'The Representation of Time in Painting: A Study of William Dyce's *Pegwell Bay*: A Recollection of October 5th, 1858', *Art History* I (1978), pp. 99–103.
12   See, for example, P. J. Currie and K. Padian, *Encyclopaedia of Dinosaurs* (London, 1997); also M. Benton, *The Penguin Historical Atlas of Dinosaurs* (Harmondsworth, 1996).
13   See J. C. Cargil, *The Fairy Tales of Science: A Book for Youth* (London, 1859), p. 3.
14   *Ibid.*, p. 13.
15   *Ibid.*
16   W. S. Symonds, *Old Bones or, Notes for Young Naturalists* (London, 1861), p. 1.
17   See Turco, *op. cit.*, p. 362.
18   See chapter seven.
19   See, for example, F. H. T. Rhodes and P. R. Shaffer, *Fossils: A Guide to Prehistoric Life* (London, 1962), p. 30.
20   See pp. 150–51.
21   See p. 70.
22   S. Kern, *The Culture of Time and Space* (Cambridge, Mass., 1983), p. 38.
23   *Ibid.*

Select Bibliography

The following list offers preliminary reading for those unfamiliar with the subject areas covered in the book. More detailed guides to reading are to be found embedded in the reference and footnote lists for each chapter.

Chapter 1 – Tracks to a Lost World

Fortey, R., *The Hidden Landscape: A Journey into the Geological Past* (London, 1993)

Insole, A., et al., *The Isle of Wight* – Geologist's Association Guide No. 60 (London, 1998)

Winchester, S., *The Map that Changed the World: The Tale of William Smith and the Birth of a Science* (London, 2001)

Chapter 2 – Time . . . That Unfathomable Abyss

Davies, G. L., *The Earth in Decay: A History of British Geomorphology, 1578–1878* (London, 1969)

Dean, D. R., *James Hutton and the History of Geology* (Ithaca, 1992)

Gould, S. J., *Time's Arrow, Time's Cycle: Myth and Metaphor in the Discovery of Geological Time* (Cambridge, Mass., 1987)

Millhauser, M., *Just Before Darwin: Robert Chambers and Vestiges* (Middletown, Conn., 1959)

Rossi, P., *The Dark Abyss of Time: The History of the Earth and the History of Nations from Hooke to Vico* (Chicago, 1984)

Rudwick, M. J. S., *Scenes from Deep Time: Early Pictorial Representations of the Prehistoric World* (Chicago, 1992)

Secord, J. A., *Victorian Sensation: The Extraordinary Publication, Reception and Secret Authorship of Vestiges of the Natural History of Creation* (Chicago, 2000)

Wilson, L. G., *Charles Lyell, The Years to 1841: The Revolution in Geology* (London, 1969)

Chapter 3 – The Testimony of the Rocks

Klonk, C,. *Science and the Perception of Nature: British Landscape Art in the Late Eighteenth and Early Nineteenth Centuries* (New Haven and London, 1996)

Knell, S. J., *The Culture of English Geology, 1815–1851: A Science Revealed Through its Collecting* (Aldershot, 2000)

Laudan, R., *From Mineralogy to Geology: The Foundations of a Science, 1650–1830* (Chicago, 1987)

Morell, J., and A. Thackray, *Gentlemen of Science: Early Years of the British Association for the Advancement of Science* (Oxford, 1981)

Oldroyd, D. R., *Thinking about the Earth: A History of Ideas in Geology* (London, 1996)

Porter, R., *The Making of Geology: Earth Science in Britain, 1660–1815* (Cambridge, 1997)

Chapter 4 – 'Let There be Dragons'

Bowler, P. J., *Fossils and Progress* (New York, 1976)

Cadbury, D., *The Dinosaur Hunters: A Story of Scientific Rivalry and the Discovery of the Prehistoric World* (London, 2000)

Dean, D. R., *Gideon Mantell and the Discovery of the Dinosaurs* (Cambridge, 1999)

Fortey, R., *Life; An Unauthorised Biography* (London, 1997)

Jardine, N., J. Secord and E. C. Spary (eds.), *Cultures of Natural History* (Cambridge, 1996)

McGowan, C., *The Dragon Seekers: The Discovery of the Dinosaurs during the Prelude to Darwin* (London, 2002)

Chapter 5 – 'Washing Away a World'

Cohn, N., *Noah's Flood: The Genesis Story in Western Thought* (New Haven and London, 1996)

Gage, J., *J. M. W. Turner: 'A Wonderful Range of Mind'* (New Haven and London, 1987)

Gillispie, G. L., *Genesis and Geology: A Study of the Relations of Scientific Thought, Natural Theology, and Social Opinion in Great Britain* (Cambridge, Mass., 1951)

Paley, M. D., *The Apocalyptic Sublime* (New Haven and London, 1986)

Rupke, N. A., *The Great Chain of History: William Buckland and the English School of Geology* (Oxford, 1983)

Chapter 6 – Competition, Competition . . .

Beer, G., *Darwin's Plots: Evolutionary Narrative in Darwin, George Eliot and Nineteenth-Century Fiction* (London, 1983)

Bowler, P. J., *Charles Darwin: The Man and his Influence* (Oxford, 1990)

Brooke, J. H., *Science and Religion: Some Historical Perspectives* (Cambridge, 1991)

Browne, J., *Charles Darwin: Voyaging* (London; 1995)

Desmond, A., *Huxley: From Devil's Disciple to Evolution's High Priest* (London, 1998)

Desmond A., and J. Moore, *Darwin* (London, 1991)

Oldroyd, D. R., *Darwinian Impacts: An Introduction to the Darwinian Revolution* (Milton Keynes, 1983)

Reardon, B. M. G., *Religious Thought in the Victorian Age: A Survey from Coleridge to Gore* (London, 1995)

Rupke, N. A., *Richard Owen: Victorian Naturalist* (New Haven and London, 1994)

Chapter 7 – The Prehistoric World as Exhibition

Bennett, T., *The Birth of the Museum: History, Theory, Politics* (London, 1995)

Dixon, R., and S. Muthesius, *Victorian Architecture* (London, 1978)

Girouard, M., *Alfred Waterhouse and the Natural History Museum* (New Haven and London, 1981)

Haward, B., *Oxford University Museum; Its Architecture and Art* (Oxford, 1991)

Rackin, D., *'Alice's Adventures in Wonderland' and 'Through the Looking Glass': Nonsense, Sense and Meaning* (New York, 1991)

Yanni, C., *Nature's Museums: Victorian Science and the Architecture of Display* (London, 1999)

Index